IEE CONTROL ENGINEERING SERIES 69

Series Editors: Professor D. Atherton
Professor S. Spurgeon

Advances in Unmanned Marine Vehicles

Other volumes in print in this series:

Volume 2	**Elevator traffic analysis, design and control** G. C. Barney and S. M. dos Santos
Volume 8	**A history of control engineering, 1800–1930** S. Bennett
Volume 14	**Optimal relay and saturating control system synthesis** E. P. Ryan
Volume 15	**Self-tuning and adaptive control: theory and application** C. J. Harris and S. A. Billings (Editors)
Volume 18	**Applied control theory** J. R. Leigh
Volume 20	**Design of modern control systems** D. J. Bell, P. A. Cook and N. Munro (Editors)
Volume 21	**Computer control of industrial processes** S. Bennett and D. A. Linkens (Editors)
Volume 23	**Robotic technology** A. Pugh (Editor)
Volume 27	**Process dynamics estimation and control** A. Johnson
Volume 28	**Robots and automated manufacture** J. Billingsley (Editor)
Volume 30	**Electromagnetic suspension – dynamics and control** P. K. Sinha
Volume 31	**Modelling and control of fermentation processes** J. R. Leigh (Editor)
Volume 32	**Multivariable control for industrial applications** J. O'Reilly (Editor)
Volume 34	**Singular perturbation methodology in control systems** D. S. Naidu
Volume 35	**Implementation of self-tuning controllers** K. Warwick (Editor)
Volume 37	**Industrial digital control systems (revised edition)** K. Warwick and D. Rees (Editors)
Volume 38	**Parallel processing in control** P. J. Fleming (Editor)
Volume 39	**Continuous time controller design** R. Balasubramanian
Volume 40	**Deterministic control of uncertain systems** A. S. I. Zinober (Editor)
Volume 41	**Computer control of real-time processes** S. Bennett and G. S. Virk (Editors)
Volume 42	**Digital signal processing: principles, devices and applications** N. B. Jones and J. D. McK. Watson (Editors)
Volume 43	**Trends in information technology** D. A. Linkens and R. I. Nicolson (Editors)
Volume 44	**Knowledge-based systems for industrial control** J. McGhee, M. J. Grimble and A. Mowforth (Editors)
Volume 47	**A history of control engineering, 1930–1956** S. Bennett
Volume 49	**Polynomial methods in optimal control and filtering** K. J. Hunt (Editor)
Volume 50	**Programming industrial control systems using IEC 1131-3** R. W. Lewis
Volume 51	**Advanced robotics and intelligent machines** J. O. Gray and D. G. Caldwell (Editors)
Volume 52	**Adaptive prediction and predictive control** P. P. Kanjilal
Volume 53	**Neural network applications in control** G. W. Irwin, K. Warwick and K. J. Hunt (Editors)
Volume 54	**Control engineering solutions: a practical approach** P. Albertos, R. Strietzel and N. Mort (Editors)
Volume 55	**Genetic algorithms in engineering systems** A. M. S. Zalzala and P. J. Fleming (Editors)
Volume 56	**Symbolic methods in control system analysis and design** N. Munro (Editor)
Volume 57	**Flight control systems** R. W. Pratt (Editor)
Volume 58	**Power-plant control and instrumentation** D. Lindsley
Volume 59	**Modelling control systems using IEC 61499** R. Lewis
Volume 60	**People in control: human factors in control room design** J. Noyes and M. Bransby (Editors)
Volume 61	**Nonlinear predictive control: theory and practice** B. Kouvaritakis and M. Cannon (Editors)
Volume 62	**Active sound and vibration control** M. O. Tokhi and S. M. Veres
Volume 63	**Stepping motors: a guide to theory and practice** P. P. Acarnley
Volume 64	**Control theory, 2nd edition** J. R. Leigh
Volume 65	**Modelling and parameter estimation of dynamic systems** J. R. Raol, G. Girija and J. Singh
Volume 66	**Variable structure systems: from principles to implementation** A. Sabanovic, L. Fridman and S. Spurgeon
Volume 67	**Motion vision: design of compact motion sensing solutions for autonomous systems navigation** J. Kolodko, L. Vlacic

Advances in Unmanned Marine Vehicles

Edited by
Geoff Roberts and Robert Sutton

The Institution of Electrical Engineers

Published by:

The Institution of Electrical Engineers,
Michael Faraday House,
Six Hills Way, Stevenage,
Herts. SG1 2AY, United Kingdom

© 2006: The Institution of Electrical Engineers

This publication is copyright under the Berne Convention and the Universal Copyright Convention. All rights reserved. Apart from any fair dealing for the purposes of research or private study, or criticism or review, as permitted under the Copyright, Designs and Patents Act, 1988, this publication may be reproduced, stored or transmitted, in any forms or by any means, only with the prior permission in writing of the publishers, or in the case of reprographic reproduction in accordance with the terms of licences issued by the Copyright Licensing Agency. Inquiries concerning reproduction outside those terms should be sent to the IEE at the address above.

While the authors and the publisher believe that the information and guidance given in this work are correct, all parties must rely upon their own skill and judgment when making use of them. Neither the authors nor the publishers assume any liability to anyone for any loss or damage caused by any error or omission in the work, whether such error or omission is the result of negligence or any other cause. Any and all such liability is disclaimed.

The moral right of the authors to be identified as authors of this work have been asserted by them in accordance with the Copyright, Designs and Patents Act 1988.

British Library Cataloguing in Publication Data

Geoff Roberts, Robert Sutton
   Advances in unmanned marine vehicles
     1. Remote submersibles 2. Vehicles, Remotely piloted 3. Research Vessels
     I. Roberts, Geoff II. Sutton, Robert, 1948- III. Institution of Electrical Engineers
     629'.046

ISBN 10: 0 86341 4508
ISBN 13: 978-086341-450-3

Typeset in India by Newgen Imaging Systems (P) Ltd., Chennai, India
Printed in the UK by MPG Books Limited, Bodmin, Cornwall

Contents

List of Authors   xv

1 Editorial: navigation, guidance and control of unmanned marine vehicles   1
G.N. Roberts and R. Sutton

    1.1 Introduction   1
    1.2 Contributions   4
    1.3 Concluding Remarks   11

2 Nonlinear modelling, identification and control of UUVs   13
T.I. Fossen and A. Ross

    2.1 Introduction   13
       2.1.1 Notation   13
    2.2 Modelling of UUVs   14
       2.2.1 Six DOF kinematic equations   14
       2.2.2 Kinetics   16
       2.2.3 Equations of motion   16
       2.2.4 Equations of motion including ocean currents   19
       2.2.5 Longitudinal and lateral models   20
    2.3 Identification of UUVs   24
       2.3.1 *A priori* estimates of rigid-body parameters   25
       2.3.2 *A priori* estimates of hydrodynamic added mass   25
       2.3.3 Identification of damping terms   25
    2.4 Nonlinear control of UUVs   31
       2.4.1 Speed, depth and pitch control   32
       2.4.2 Heading control   37
       2.4.3 Alternative methods of control   40
    2.5 Conclusions   40

3 Guidance laws, obstacle avoidance and artificial potential functions 43
A.J. Healey

- 3.1 Introduction 43
- 3.2 Vehicle guidance, track following 44
  - 3.2.1 Vehicle steering model 45
  - 3.2.2 Line of sight guidance 46
  - 3.2.3 Cross-track error 47
  - 3.2.4 Line of sight with cross-track error controller 49
  - 3.2.5 Sliding mode cross-track error guidance 50
  - 3.2.6 Large heading error mode 51
  - 3.2.7 Track path transitions 52
- 3.3 Obstacle avoidance 52
  - 3.3.1 Planned avoidance deviation in path 52
  - 3.3.2 Reactive avoidance 54
- 3.4 Artificial potential functions 59
  - 3.4.1 Potential function for obstacle avoidance 61
  - 3.4.2 Multiple obstacles 62
- 3.5 Conclusions 64
- 3.6 Acknowledgements 65

4 Behaviour control of UUVs 67
M. Carreras, P. Ridao, R. Garcia and J. Batlle

- 4.1 Introduction 67
- 4.2 Principles of behaviour-based control systems 69
  - 4.2.1 Coordination 71
  - 4.2.2 Adaptation 72
- 4.3 Control architecture 72
  - 4.3.1 Hybrid coordination of behaviours 73
  - 4.3.2 Reinforcement learning-based behaviours 75
- 4.4 Experimental set-up 76
  - 4.4.1 URIS *UUV* 76
  - 4.4.2 Set-up 78
  - 4.4.3 Software architecture 78
  - 4.4.4 Computer vision as a navigation tool 79
- 4.5 Results 80
  - 4.5.1 Target tracking task 80
  - 4.5.2 Exploration and mapping of unknown environments 82
- 4.6 Conclusions 83

5	**Thruster control allocation for over-actuated, open-frame underwater vehicles**		**87**
	E. Omerdic and G.N. Roberts		
	5.1	Introduction	87
	5.2	Problem formulation	88
	5.3	Nomenclature	90
		5.3.1 Constrained control subset Ω	90
		5.3.2 Attainable command set Φ	91
	5.4	Pseudoinverse	92
	5.5	Fixed-point iteration method	95
	5.6	Hybrid approach	96
	5.7	Application to thruster control allocation for over-actuated thruster-propelled UVs	98
	5.8	Conclusions	103
6	**Switching-based supervisory control of underwater vehicles**		**105**
	G. Ippoliti, L. Jetto and S. Longhi		
	6.1	Introduction	105
	6.2	Multiple models switching-based supervisory control	106
	6.3	The EBSC approach	109
		6.3.1 An implementation aspect of the EBSC	110
	6.4	The HSSC approach	111
		6.4.1 The switching policy	111
	6.5	Stability analysis	112
		6.5.1 Estimation-based supervisory control	112
		6.5.2 Hierarchically supervised switching control	113
	6.6	The ROV model	114
		6.6.1 The linearised model	116
	6.7	Numerical results	116
	6.8	Conclusions	121
7	**Navigation, guidance and control of the *Hammerhead* autonomous underwater vehicle**		**127**
	D. Loebis, W. Naeem, R. Sutton, J. Chudley and A. Tiano		
	7.1	Introduction	127
	7.2	The *Hammerhead* AUV navigation system	129
		7.2.1 Fuzzy Kalman filter	129
		7.2.2 Fuzzy logic observer	130
		7.2.3 Fuzzy membership functions optimisation	131
		7.2.4 Implementation results	131
		7.2.5 GPS/INS navigation	136
	7.3	System modelling	145
		7.3.1 Identification results	146

7.4	Guidance	147
7.5	*Hammerhead* autopilot design	148
	7.5.1 LQG/LTR controller design	149
	7.5.2 Model predictive control	150
7.6	Concluding remarks	155

8 Robust control of autonomous underwater vehicles and verification on a tethered flight vehicle 161
Z. Feng and R. Allen

8.1	Introduction	161
8.2	Design of robust autopilots for torpedo-shaped AUVs	162
	8.2.1 Dynamics of Subzero III (excluding tether)	163
	8.2.2 Plant models for control design	165
	8.2.3 Design of reduced-order autopilots	166
8.3	Tether compensation for Subzero III	169
	8.3.1 Composite control scheme	169
	8.3.2 Evaluation of tether effects	170
	8.3.3 Reduction of tether effects	177
	8.3.4 Verification of composite control by nonlinear simulations	179
8.4	Verification of robust autopilots via field tests	181
8.5	Conclusions	183

9 Low-cost high-precision motion control for ROVs 187
M. Caccia

9.1	Introduction	187
9.2	Related research	189
	9.2.1 Modelling and identification	189
	9.2.2 Guidance and control	189
	9.2.3 Sensing technologies	190
9.3	Romeo ROV mechanical design	192
9.4	Guidance and control	193
	9.4.1 Velocity control (dynamics)	194
	9.4.2 Guidance (task kinematics)	195
9.5	Vision-based motion estimation	196
	9.5.1 Vision system design	196
	9.5.2 Three-dimensional optical laser triangulation sensor	199
	9.5.3 Template detection and tracking	200
	9.5.4 Motion from tokens	201
	9.5.5 Pitch and roll disturbance rejection	201
9.6	Experimental results	202
9.7	Conclusions	208

10	**Autonomous manipulation for an intervention AUV**		**217**
	G. Marani, J. Yuh and S.K. Choi		
	10.1	Introduction	217
	10.2	Underwater manipulators	218
	10.3	Control system	218
		10.3.1 Kinematic control	218
		10.3.2 Kinematics, inverse kinematics and redundancy resolution	223
		10.3.3 Resolved motion rate control	223
		10.3.4 Measure of manipulability	224
		10.3.5 Singularity avoidance for a single task	225
		10.3.6 Extension to inverse kinematics with task priority	227
		10.3.7 Example	230
		10.3.8 Collision and joint limits avoidance	230
	10.4	Vehicle communication and user interface	232
	10.5	Application example	233
	10.6	Conclusions	236
11	**AUV 'r2D4', its operation, and road map for AUV development**		**239**
	T. Ura		
	11.1	Introduction	239
	11.2	AUV 'r2D4' and its no. 16 dive at Rota Underwater Volcano	240
		11.2.1 R-Two project	240
		11.2.2 AUV 'r2D4'	241
		11.2.3 Dive to Rota Underwater Volcano	244
	11.3	Future view of AUV research and development	248
		11.3.1 AUV diversity	250
		11.3.2 Road map of R&D of AUVs	252
	11.4	Acknowledgements	253
12	**Guidance and control of a biomimetic-autonomous underwater vehicle**		**255**
	J. Guo		
	12.1	Introduction	255
	12.2	Dynamic modelling	257
		12.2.1 Rigid body dynamics	258
		12.2.2 Hydrodynamics	263
	12.3	Guidance and control of the BAUV	265
		12.3.1 Guidance of the BAUV	266
		12.3.2 Controller design	267
		12.3.3 Experiments	270
	12.4	Conclusions	273

13 Seabed-relative navigation by hybrid structured lighting 277
F. Dalgleish, S. Tetlow and R.L. Allwood

- 13.1 Introduction 277
- 13.2 Description of sensor configuration 279
- 13.3 Theory 279
  - 13.3.1 Laser stripe for bathymetric and reflectivity seabed profiling 281
  - 13.3.2 Region-based tracker 283
- 13.4 Constrained motion testing 283
  - 13.4.1 Laser altimeter mode 283
  - 13.4.2 Dynamic performance of the laser altimeter process 285
  - 13.4.3 Dynamic performance of region-based tracker 286
  - 13.4.4 Dynamic imaging performance 288
- 13.5 Summary 291
- 13.6 Acknowledgements 291

14 Advances in real-time spatio-temporal 3D data visualisation for underwater robotic exploration 293
S.C. Martin, L.L. Whitcomb, R. Arsenault, M. Plumlee and C. Ware

- 14.1 Introduction 293
  - 14.1.1 The need for real-time spatio-temporal display of quantitative oceanographic sensor data 294
- 14.2 System design and implementation 295
  - 14.2.1 Navigation 295
  - 14.2.2 Real-time spatio-temporal data display with GeoZui3D 295
  - 14.2.3 Real-time fusion of navigation data and scientific sensor data 297
- 14.3 Replay of survey data from Mediterranean expedition 300
- 14.4 Comparison of real-time system implemented on the JHU ROV to a laser scan 301
  - 14.4.1 Real-time survey experimental set-up 301
  - 14.4.2 Laser scan experimental set-up 302
  - 14.4.3 Real-time system experimental results 303
  - 14.4.4 Laser scan experimental results 303
  - 14.4.5 Comparison of laser scan to real-time system 305
- 14.5 Preliminary field trial on the *Jason 2* ROV 305
- 14.6 Conclusions and future work 308

Contents   xi

15 Unmanned surface vehicles – game changing technology for naval operations   **311**
S.J. Corfield and J.M. Young

15.1	Introduction	311
15.2	Unmanned surface vehicle research and development	312
15.3	Summary of major USV subsystems	313
	15.3.1 The major system partitions	313
	15.3.2 Major USV subsystems	314
	15.3.3 Hulls	314
	15.3.4 Auxiliary structures	316
	15.3.5 Engines, propulsion subsystems and fuel systems	316
	15.3.6 USV autonomy, mission planning and navigation, guidance and control	317
15.4	USV payload systems	318
15.5	USV launch and recovery systems	319
15.6	USV development examples: MIMIR, SWIMS and FENRIR	319
	15.6.1 The MIMIR USV system	319
	15.6.2 The SWIMS USV system	321
	15.6.3 The FENRIR USV system and changing operational scenarios	325
15.7	The game changing potential of USVs	326

16 Modelling, simulation and control of an autonomous surface marine vehicle for surveying applications Measuring Dolphin MESSIN*   **329**
J. Majohr and T. Buch

16.1	Introduction and objectives	329
16.2	Hydromechanical conception of the MESSIN	330
16.3	Electrical developments of the MESSIN	332
16.4	Hierarchical steering system and overall steering structure	333
16.5	Positioning and navigation	336
16.6	Modelling and identification	337
	16.6.1 Second-order course model [16]	338
	16.6.2 Fourth-order track model [17]	338
16.7	Route planning, mission control and automatic control	342
16.8	Implementation and simulation	344
16.9	Test results and application	346

17 Vehicle and mission control of single and multiple autonomous marine robots   **353**
A. Pascoal, C. Silvestre and P. Oliveira

17.1	Introduction	353
17.2	Marine vehicles	354
	17.2.1 The *Infante* AUV	354

		17.2.2	The *Delfim* ASC	355
		17.2.3	The *Sirene* underwater shuttle	356
		17.2.4	The Caravela 2000 autonomous research vessel	357
	17.3	Vehicle control		358
		17.3.1	Control problems: motivation	359
		17.3.2	Control problems: design techniques	362
	17.4	Mission control and operations at sea		375
		17.4.1	The CORAL mission control system	376
		17.4.2	Missions at sea	379
	17.5	Conclusions		380

18 Wave-piercing autonomous vehicles — 387
H. Young, J. Ferguson, S. Phillips and D. Hook

	18.1	Introduction		387
		18.1.1	Abbreviations and definitions	387
		18.1.2	Concepts	388
		18.1.3	Historical development	388
	18.2	Wave-piercing autonomous underwater vehicles		390
		18.2.1	Robotic mine-hunting concept	391
		18.2.2	Early tests	393
		18.2.3	US Navy RMOP	393
		18.2.4	The Canadian 'Dorado' and development of the French 'SeaKeeper'	394
	18.3	Wave-piercing autonomous surface vehicles		396
		18.3.1	Development programme	398
		18.3.2	Command and control	400
		18.3.3	Launch and recovery	401
		18.3.4	Applications	402
	18.4	Daughter vehicles		403
		18.4.1	Applications	404
	18.5	Mobile buoys		405
		18.5.1	Applications	405
	18.6	Future development of unmanned wave-piercing vehicles		405

19 Dynamics, control and coordination of underwater gliders — 407
R. Bachmayer, N.E. Leonard, P. Bhatta, E. Fiorelli and J.G. Graver

	19.1	Introduction		407
	19.2	A mathematical model for underwater gliders		408
	19.3	Glider stability and control		412
		19.3.1	Linear analysis	412
		19.3.2	Phugoid-mode model	415
	19.4	*Slocum* glider model		417
		19.4.1	The *Slocum* glider	417
		19.4.2	Glider identification	419

19.5	Coordinated glider control and operations		424
	19.5.1	Coordinating gliders with virtual bodies and artificial potentials	425
	19.5.2	VBAP glider implementation issues	426
	19.5.3	AOSN II sea trials	426
19.6	Final remarks		429

Index   **433**

List of Authors

R. Allen
Institute of Sound and Vibration Research
University of Southampton
United Kingdom

R.L. Allwood
Offshore Technology Centre
Cranfield University
United Kingdom

R. Arsenault
Data Visualization Research Laboratory
University of New Hampshire
USA

R. Bachmayer
Institute for Ocean Technology
National Research Council
St. John's
Canada

J. Batlle
Computer Vision and Robotics Group
University of Girona
Spain

P. Bhatta
Department of Mechanical and Aerospace Engineering
Princeton University
USA

T. Buch
Institute of Telecommunications and Information Engineering
University of Rostock
Germany

M. Caccia
Consiglio Nazionale della Ricerche (CNR)
Genova
Italy

M. Carreras
Computer Vision and Robotics Group
University of Girona
Spain

S.K. Choi
MASE Inc.
Honolulu
USA

J. Chudley
Marine and Industrial Dynamic Analysis Research Group
The University of Plymouth
United Kingdom

S.J. Corfield
Underwater Platforms and Signature Group
QinetiQ Ltd. United Kingdom

List of Authors

F. Dalgleish
Offshore Technology Centre
Cranfield University
United Kingdom

Z. Feng
Institute of Sound and Vibration
 Research
University of Southampton
United Kingdom

J. Ferguson
International Submarine Engineering
 Ltd. British Columbia
Canada

E. Fiorelli
Department of Mechanical and
 Aerospace Engineering
Princeton University
USA

T.I. Fossen
Department of Engineering Cybernetics
Norwegian University of Science and
 Technology
Norway

R. Garcia
Computer Vision and Robotics Group
University of Girona
Spain

J. Guo
Department of Engineering Science and
 Ocean Engineering
National Taiwan University
Republic of China

J.G. Graver
Department of Mechanical and
 Aerospace Engineering
Princeton University
USA

A.J. Healey
Department of Mechanical and
 Aeronautical Engineering
Naval Post Graduate School
Monterey
USA

D. Hook
Autonomous Surface Vehicles Ltd
Chichester
United Kingdom

G. Ippoliti
Dipartimento di Ingegneria Informatica
Gestionale e dell'Automazione
Università Politecnica delle Marche
Ancona
Italy

L. Jetto
Dipartimento di Ingegneria Informatica
Gestionale e dell'Automazione
Università Politecnica delle Marche
Ancona
Italy

N.E. Leonard
Department of Mechanical and
 Aerospace Engineering
Princeton University
USA

D. Loebis
Marine and Industrial Dynamic
 Analysis Research Group
The University of Plymouth
United Kingdom

S. Longhi
Dipartimento di Ingegneria Informatica
Gestionale e dell'Automazione
Università Politecnica delle Marche
Ancona
Italy

J. Majohr
Institute of Automation
College for Informatics and Electrical
    Engineering
University of Rostock
Germany

G. Marani
Autonomous Systems Laboratory
University of Hawaii
USA

S.C. Martin
Dynamic Systems and Control
    Laboratory
Johns Hopkins University
USA

W. Naeem
Marine and Industrial Dynamic
    Analysis Research Group
The University of Plymouth
United Kingdom

E. Omerdic
Electronics & Computer Engineering
    Department
University of Limerick
Republic of Ireland

P. Oliveira
Institute for Systems and Robotics
Instituto Superior Técnico (IST)
Lisbon
Portugal

A. Pascoal
Institute for Systems and Robotics
Instituto Superior Técnico (IST)
Lisbon
Portugal

S. Phillips
Autonomous Surface Vehicles Ltd
Chichester
United Kingdom

M. Plumlee
Data Visualization Research Laboratory
University of New Hampshire
USA

P. Ridao
Computer Vision and Robotics Group
University of Girona
Spain

G.N. Roberts
Control Theory and Applications Centre
    (CTAC)
Coventry University, United Kingdom

A. Ross
Department of Engineering Cybernetics
Norwegian University of Science and
Technology. Norway

C. Silvestre
Institute for Systems and Robotics
Instituto Superior Técnico (IST)
Lisbon
Portugal

R. Sutton
Marine and Industrial Dynamic
    Analysis Research Group
The University of Plymouth
United Kingdom

S. Tetlow
Offshore Technology Centre
Cranfield University
United Kingdom

A. Tiano
Department of Information and Systems
University of Pavia
Italy

T. Ura
Underwater Technology Research
    Centre
The University of Tokyo
Japan

C. Ware
Data Visualization Research Laboratory
University of New Hampshire
USA

L.L. Whitcomb
Dynamic Systems and Control
    Laboratory
Johns Hopkins University
USA

H. Young
Autonomous Surface Vehicles Ltd
Chichester
United Kingdom

J.M. Young
Underwater Platforms and Signature
    Group
QinetiQ Ltd. United Kingdom

J. Yuh
Division of Information and Intelligent
    Systems
National Science Foundation
Arlington
USA

Chapter 1

Editorial: navigation, guidance and control of unmanned marine vehicles

G.N. Roberts and R. Sutton

1.1 Introduction

Although Bourne can be credited with producing the first conceptual design for a submarine in 1578, the first one built was constructed in 1620 by Van Drebbel. Nevertheless, it was not until 1776 that a submarine was specifically launched to take part in naval operations. Bushnell's submarine, the *Turtle*, was designed to destroy the Royal Navy men-of-war which were participating in naval blockages during the American War of Independence. Fortunately for the British fleet, the attacks by the human powered *Turtle* were unsuccessful. The *Turtle's* single crew member blamed the ineffectiveness of the assaults on the inability to lay 150 lb charges against the hulls of the ships owing to their reputed copper sheathing. In actual fact, the British warships were not sheathed. A more probable explanation has been postulated by Coverdale and Cassidy [1], who propose it was due to the crew member being physically exhausted and affected by the build up of unacceptable carbon dioxide levels in the vessel by the time it reached an intended target. Reader *et al.* [2] light-heartedly suggest that this may have been the initial impetus for the search for unmanned underwater vehicles (UUVs)! Clearly, since those pioneering days, manned submarine technology has advanced dramatically. However, the common potential weakness throughout their evolution has been the reliance on humans to perform operational tasks.

More realistically, the ancestry of the modern UUV may be traced back to the self-propelled torpedo perfected in 1868 by Whitehead, whilst some of the earliest more significant developments in this type of technology can be attributed to the cable controlled underwater recovery vehicle design and construction programme instigated by the US Navy in 1958. In 1963, such a craft was used in the search for the ill-fated USS *Thresher* which tragically sank off the New England coast in 1400 fathoms of water. Later, in 1966, another was used to help recover the US Navy hydrogen bomb lost off the coast of Palomares, Spain. Notwithstanding those successes and the accompanying publicity, the commercial potential of UUVs was not recognized until the discovery of the offshore oil and gas supplies in the North Sea. Specifically, remotely operated vehicles (ROVs) began and continue to be used extensively throughout the offshore industry, whereas in both the naval and commercial sectors, autonomous underwater vehicle (AUV) usage was limited. Nonetheless, interest in the possible use of both types has heightened, prompted by the needs of the offshore industry to operate and explore at extreme depths in a hostile environment, and the operational requirements of navies to have low-cost assets capable of undertaking covert surveillance missions, and performing mine laying and disposal tasks.

The reader should note that the phrase 'unmanned underwater vehicle' as used here is a generic expression to describe both an AUV and an ROV. An AUV is a marine craft which fulfils a mission without being constantly monitored and supervised by a human operator, whilst an ROV is a marine vessel that requires instructions from an operator via a tethered cable or acoustic link. Also, in some more recent literature the term 'unmanned' is being replaced by 'uninhabited'.

It should be borne in mind that this book is also devoted to the design and development of unmanned surface vehicles (USVs) and unmanned semi-submersible craft (USSC). From a historical perspective as highlighted herein by Corfield and Young, the use of a USV provided a solution to the problem of a naval blockade in 1718. They also mention briefly the fire ships employed by Howard and Drake. In this particular instance during the summer of 1588, Howard and Drake arranged for eight fire ships (USVs) to be sent against the Armada which caused the majority of the Spanish ships to break formation and leave their safe anchorage in Calais for the open sea. Soon after this incident the Battle of Gravelines ensued. Of course, the process of automatically steering ships has its origins going back to the time when early fishermen would bind the tiller or rudder of their boats in a fixed position to produce an optimal course in order to release extra manpower to assist with the launching and recovery of nets. Most likely the criteria for selecting the optimum course would be to minimise induced motions and to maintain a course which would help with the deployment or recovery of the nets. However, by way of a contradiction, Slocum [3] recounts an incident where a highly experienced shipmaster of that era served as an expert witness in a famous murder trial in Boston, USA, in which he stated that a particular sail ship could not hold her course long enough for the steersman to leave the helm to cut the throat of its captain! For some it will also be of interest to know that Slocum was the first solo global circumnavigator and to honour this achievement a class of unmanned underwater glider has been named after him.[1]

An essential part of any USV or USSC is an onboard automatic control system for steering a set course or maintaining a given track. It was not until the industrial revolution that methods for automatically steering ships were first contemplated, whilst the first ship autopilots did not come into use until the first part of the twentieth century. Even so, before the development of autopilots could progress it was necessary to

---

[1] Chapter 19 describes the control and coordination of *Slocum* gliders.

devise a powered rudder or steering engine. The motivation for a steering engine came primarily from the naval requirement for warships to undertake high-speed manoeuvres. Bennett [4], in his chapter on the development of servo mechanisms, reports that in the late 1860s the British Admiralty equipped several of their sailing ships with various types of steering engines, many of which were based on a steam-hydraulic system.

Once powered rudders or steering machines became commonplace, attention turned to providing inputs to these devices from suitable heading-seeking equipment. Fossen [5] describes how the development of the electrically driven gyrocompass was pivotal in the evolution of ship autopilots. Electrically driven gyrocompasses overcame the problems associated with magnet compasses which had their readings corrupted by local magnetic fields and the electrical systems in ships, torpedoes and submarines. The first electrically driven gyroscope was demonstrated in 1890, in 1908 Anschutz patented the first North-seeking gyrocompass and in 1911 Sperry patented his ballistic compass, which included vertical damping. The invention of the electrically driven gyrocompass is arguably the most important breakthrough in ship control systems design.

During the first quarter of the twentieth century a number of automatic steering systems for ships were devised. The idea of 'check helm' or derivative action was introduced by Sir James Henderson who was granted a patent in 1913 for his 'automatic steering device' which used both heading error and heading error rate in the feedback loop. However, the major contributions to the development of practical ship steering systems were made by the Sperry Gyroscope Company. Allensworth [6] describes how Elmer Sperry constructed the first automatic ship steering mechanism in 1911 which he called the automatic pilot or gyropilot [7]. Sperry's gyropilot was known as *Metal Mike* because in its operation it appeared to replicate the actions of an experienced helmsman. Sperry had observed that a helmsman 'eases' the rudder as the ship responds and tended to apply rudder proportional to heading error. He also observed that an experienced helmsman would reduce the rudder angle as the heading error reduced and would apply counter rudder to negate the possibility of overshoot. To account for these human actions he included an 'anticipator' that automatically adjusted the backlash of the controller according to the heading angle, in effect resulting in a simple adaptive autopilot.

The work of Minorski [8] is also regarded as making key contributions to the development of automatic ship steering systems. Minorski's main contribution was the theoretical analysis of automatic steering and the specification of the three-term or proportional–integral–derivative (PID) controller for automatic ship steering. As in the case of Sperry's work, his PID controller designs were predicated on visual observation of the way the experienced helmsman would steer a ship. He acknowledged that helmsmen would anticipate ship motions before applying rudder corrections but more importantly he postulated that they possessed the ability to judge angular velocity, that is, yaw rate, thereby effecting derivative control. Indeed it is worth noting that the ability of a human operator to estimate innately velocity is now widely accepted in human factors research.

Since the early pioneering days which have been briefly mentioned above, worldwide interest is gathering a pace into the design and development of unmanned marine vehicles (UMVs) as they are now considered by many as being able to provide cost-effective solutions to a number of commercial, naval and scientific problems. The attention now being given to such vehicles is also coupled with the current and ongoing advances being made in control systems engineering, artificial intelligence and sensor technology, the availability of low-cost computer systems which have high performance specifications, and the advent of communication networks such as the global positioning system (GPS). Verification of the aforementioned statement will be forthcoming in the following 18 chapters of this book.

1.2 Contributions

Until very recently, the main emphasis in research and development of UMVs concerned ROVs and AUVs and this level of activity is reflected in the following 11 chapters of the book. This is followed by two chapters which specifically address underwater vision and its role in UMV navigation. USVs are discussed in Chapters 15, 16, 17 and 18 and the final chapter is concerned with underwater gliders. As each chapter is a self-contained exposition of the work undertaken, some introductory observations on the contributions of each chapter is provided here.

In Chapter 2, the equations of motion are developed for UUVs, including the aspects of ocean currents. These generalised equations are decoupled into longitudinal and lateral models, and the characteristics of each are then explained. Methods for identification of unknown parameters are presented, consisting of least-square solutions, and nonlinear mapping techniques. These techniques are applied in a case-study simulation of the *Infante* UUV. The chapter concludes with a control system methodology, where a backstepping algorithm is developed for application in developing autopilots in speed, depth and pitch. Finally, a pole placement algorithm is presented for the development of a heading autopilot.

Chapter 3 suggests that in the context of autonomy for underwater vehicles, it is assumed that a usual suite of feedback controllers are present in the form of autopilot functions that provide for the regulation of vehicle speed, heading and depth or altitude. The notion of a guidance system is based around the idea that the heading command is taken from the guidance system. Guidance is driven from the perceptory inputs modified by sonar signal processing, and algorithms for obstacle detection, obstacle tracking, location and mapping. It allows for track following, cross track error control and for obstacle avoidance, and as such requires knowledge of vehicle position from a navigation system, and knowledge of the mission so that track plans can be made and modified. In the discussion of obstacle avoidance guidance, the methods of artificial potential fields, curved path deviation planning and reactive avoidance are included. The chapter considers the topic of guidance laws, obstacle avoidance and the use of artificial potential functions and deals with the computations required to plan and develop paths and commands, which are used by these autopilots.

Chapter 4 considers guidance and control using a behavioural approach. The authors argue that tremendous progress has been made recently in the field of behaviour-based control architectures. This chapter presents an overview of architectures for controlling UUVs and focuses on the use of behaviour-based control techniques. In a behaviour-based controller, a set of independent behaviours compete for commanding an autonomous robot. Each behaviour represents a basic conduct such as 'avoid obstacles', 'go to a destination position', 'follow a target', etc. The coordination of a properly selected set of behaviours has shown to be suitable to accomplish autonomously a robotic task. In this chapter, the basic principles of these kinds of controllers are presented. The authors also present the behaviour-based control system designed to control the *URIS* AUV. The main features of the robot, its vision-based navigation system, the control architecture and the development methodology are detailed. Finally, some results show a practical case in which a behaviour-based controller is applied to a real robot and the performance of the navigation system is shown.

Chapter 5 presents a hybrid approach for thruster control allocation of over-actuated, open-framed ROVs. It is shown that a standard pseudoinverse method for the solution of the thruster control allocation problem is able to find a feasible solution only on a subset of the attainable command set. This subset is called the feasible region for pseudoinverse. Some other methods, like direct control allocation or fixed-point iteration method, are able to find the feasible solution on the entire attainable command set. A novel, hybrid approach for control allocation, proposed in this chapter, is based on integration of the pseudoinverse and the fixed-point iteration method and is implemented as a two-step process. The pseudoinverse solution is found in the first step. Then the feasibility of the solution is examined analysing its individual components. If violation of actuator constraint(s) is detected, the fixed-point iteration method is activated in the second step. In this way, the hybrid approach is able to allocate the exact solution, optimal in the l_2 sense, inside the entire attainable command set. This solution minimises a control energy cost function, which is the most suitable criterion for underwater applications where the optimal use of energy has important ramifications for mission duration.

Chapter 6 considers the position control problem of an underwater vehicle used in the exploitation of combustible gas deposits at great water depths. The vehicle is subjected to different load configurations, which from time to time introduce considerable variations of its mass and inertial parameters. Depending on the experimental situation, the different possible vehicle configurations may or may not be known in advance. However, in general, it is not *a priori* known either when the operating conditions are changed or which is the new vehicle configuration after the change. The control of this kind of mode-switch process cannot be adequately faced with traditional adaptive control techniques because of the long time needed for adaptation. In this chapter, two supervised, switched, control strategies are proposed. The main advantage of switching control is that one builds up a bank of alternative candidate controllers and switches among them based on on-line collected measurements.

Navigation, guidance and control of the *Hammerhead* AUV is considered in Chapter 7. The navigation system is based on an integrated use of the global positioning system (GPS) and several inertial navigation system (INS) sensors. A simple Kalman filter (SKF) and an extended Kalman filter (EKF) are proposed to be used subsequently to fuse the data from the INS sensors and to integrate them with the GPS data. The chapter highlights the use of fuzzy logic techniques optimised by a multi-objective genetic algorithm (GA) to the adaptation of the initial statistical assumption of both the SKF and EKF caused by possible changes in sensor noise characteristics. For controller design, the *Hammerhead* models are extracted using system identification techniques on actual vehicle data obtained from full-scale experiments. Two guidance laws are proposed which are designed for cable/pipeline inspection task for cruising type vehicles. The control systems developed for *Hammerhead* are a combination of optimal control and artificial intelligence strategies. A discrete time linear quadratic Gaussian controller with loop transfer recovery is formulated. In addition, two forms of model predictive controllers (MPCs) blended with a GA and fuzzy logic are designed and tested. Simulation as well as real-time results are presented.

Chapter 8 describes the development and testing of a robust control approach, H-infinity, for the design of autopilots for a flight vehicle, *Subzero III*. The authors suggest that automatic control of an AUV, either a flight vehicle or a hovering vehicle, presents several difficulties due to the fundamentally nonlinear dynamics of the vehicle, the model uncertainty resulting from, for example, inaccurate hydrodynamic coefficients and the external disturbances such as underwater currents. Depending on the degree of parameter uncertainty and disturbances, different levels of machine intelligence, from robust control to adaptive control are required. In this chapter, the vehicle dynamics are modelled and simulated in order to develop and test the autopilots, and water trials are reported to demonstrate the effectiveness of robust control in practice. *Subzero III* has a thin communication cable attached which has an effect on vehicle control. A model-based approach is developed to overcome the tether effects, and this could have application to ROVs where the effects of the umbilical are much more pronounced.

Chapter 9 addresses the problem of developing a low-cost station-keeping system for ROVs in the proximity of the seabed through vision-based motion estimation. Owing to the different nature of disturbances affecting the vehicle dynamics (external actions of sea currents and tether, uncertainty in system dynamics related to the poor knowledge of hydrodynamic derivatives, noisy and low sampling rate sensor measurements, propeller–propeller and propeller–hull interactions), a practical approach, consisting of the integration of a set of suitable sub-systems in a harmonious blend of vehicle mechanical design, system modelling and identification, model-based motion estimation, guidance and control, and computer vision, is proposed and discussed with respect to its application to the *Romeo* ROV. Thus, the need of revising the conventional ROV model to include thruster installation coefficients, taking into account propeller–hull and propeller–propeller interactions, and of developing on-board sensor-based identification techniques, defining a set of suitable manoeuvres for observing drag and inertia parameters, is discussed before the adopted dual-loop guidance and control architecture, which is integrated with a specially developed laser-triangulation optical-correlation sensor, guarantees satisfactory performance when operating the *Romeo* ROV in the proximity of the seabed. Experimental results are shown and discussed to demonstrate the effectiveness of the proposed approach.

In Chapter 10, intervention tasks and the evolution of the hardware of underwater manipulators is described. Currently many intervention tasks in the underwater world are performed by manned submersibles or ROVs in the tele-operation mode. AUVs with no physical link and no human occupant permit operating in areas where humans cannot go, such as under ice, in classified areas or on missions to retrieve hazardous objects. However, the low bandwidth and significant time delay inherent in acoustic, sub-sea communication are a significant obstacle to operate remotely a manipulation system, making it impossible for remote controllers to react to problems in a timely manner. Autonomous manipulation is the key element in underwater intervention performed with autonomous vehicles. Today there are a few AUVs equipped with manipulators, for example, *SAUVIM* (Semi Autonomous Underwater Vehicle for Intervention Mission, University of Hawaii) and *ALIVE* (AUV for light interventions on deepwater sub-sea fields, Cybernétix, France). The chapter introduces the electromechanical arm of *SAUVIM* and discusses some theoretical issues with the arm control system, addressing the required robustness in different situations that the manipulator may face during intervention missions. An advanced user interface, which assists in handling the communication limits and providing a remote programming environment where the interaction with the manipulator is limited only to a very high level, will then be briefly discussed. An example application utilising *SAUVIM* is presented.

Chapter 11 describes the AUV r2D4, which was built in 2003 and successfully observed NW Rota 1 Underwater Volcano in May 2004. r2D4 is a middle-sized cruising type AUV, approximately 1.5 tonnes in weight and 4.4 m in length, and has a large payload bay for scientists. The chapter introduces the robot systems, especially focusing at the datasets for operation and the emergency countermeasures by software, and also details the practical considerations for planning and scheduling r2D4's dives at the Rota Underwater Volcano. Drawing on these experiences at sea and a long history of research and development of AUVs at the Institute of Industrial Science of the University of Tokyo, the author discusses 'AUV Diversity', which is one of the important keywords for designing ideal AUVs for practical use. Three groups of AUVs being developed are Cruising, Bottom Reference and Advance Autonomy, by taking account of the level of interaction with the sea floor and underwater behaviours. Finally, by considering AUV applications for sea floor observation, a road map for future development of AUVs is presented.

A test-bed biometric AUV is described in Chapter 12. Recently, research and development on biomimetic AUVs (BAUVs) has progressed rapidly to solve the problems of AUV propulsion by propellers, that is, AUVs with rigid hulls and powered by rotary propellers suffer problems such as inefficient propulsion, positioning difficulties, limited turning agility and imprecise hovering. This chapter describes a biomimetic submarine design with efficient swimming capability and high manoeuvrability. A planar model of the BAUV has a slender body divided into three segments with waves passing from its nose to its tail. Reaction forces due to momentum change, friction and cross-flow drag acting on each segment are considered. Equations of motion described by the body-fixed coordinate are obtained by summing the longitudinal force, lateral force and yaw moment impacting each segment of the BAUV. A simple mathematical model based on a blade element synthesis model that evaluates the hydrodynamic forces acting on pectoral fin motions is proposed. The pectoral fin is considered as a number of moving blade elements; the lift, cross-flow drag and added inertia acting on each blade element are examined as a two-dimensional oscillating thin foil. Parameters suitable for control of the surge speed and turning rate are then identified for both the body/caudal fin and for a pair of pectoral fins. A coordinated controller that integrates the tail fin and pectoral fins is then proposed. This controller combines two mechanisms for high-speed swimming and low-speed hovering. The control design employs features that include an inner local control loop and an outer global control loop. The local control law generates forward and yawing velocities using the body/caudal fin and a pair of pectoral fins in local coordinates. Five parameters are identified that coordinate joint angles of body/caudal fin and pectoral fins for forward and turning motions. A global control law is developed for solving the waypoint tracking problem. Finally, successful waypoint tracking, turning, hovering and braking behaviours of the BAUV are demonstrated with water tank experiments.

Seabed navigation using structured light is outlined in Chapter 13. Here, as an alternative to mosaic-based navigation and map building, a laser stripe imaging (LSI) system has been installed on the *Hammerhead* AUV. The imaging system works by projecting a stripe of laser light onto the seabed to produce a high-contrast, optical waterfall image. Simultaneously, the stripe position is used to measure accurately the altitude of the vehicle above the seabed. This then allows a region-based tracker working between successive images to determine the motion of the vehicle. This information can then be fed back to the image generation process to correct the image for motions of the vehicle. The system has been tested during constrained motion trials in a seawater test facility where the imaging sensor was used up to a range of 18 m. The dynamic performance of the optical altimeter, region-based tracker and modified image building process in realistic conditions are described and results are presented.

Chapter 14 outlines the development of a real-time, human–computer interface (HCI) system that enables an operator to more effectively utilise the large volume of quantitative data (i.e., navigation, scientific and vehicle status) generated in real-time by the sensor suites of underwater robotic vehicles. The system provides an interactive, three-dimensional graphical interface that displays, under user control, the quantitative spatial and temporal sensor data presently available to pilots and users only as two-dimensional plots and numerical displays. The system can display real-time bathymetric renderings of the sea floor based upon vehicle navigation and sonar sensor data; vehicle trajectory data; a variety of scalar values sensor data; and geo-referenced targets and way-points. Accuracy of real-time navigation and bathymetric sonar processing is evaluated by comparing real-time sonar bathymetry in a test tank with a high-resolution laser scan of the same tank floor. The real-time sonar bathymetry is shown to compare favourably to the laser scan data. Results of preliminary engineering trials of this system in operation on the Woods Hole Oceanographic Institution's *Jason 2* ROV are also reported.

The next four chapters concern USVs. The first of these, Chapter 15, considers that over the past 5 years the world's navies have started to examine the potential utility of USVs in a wide variety of roles. It is suggested that USVs offer users force multiplication with significantly reduced through life costs in comparision with more standard options and opportunities for reducing the risk of deployments in hazardous areas. A number of low-cost USV systems have been developed based on existing small hulls and this trend appears set to continue in the near term. However, future USVs may use new types of advanced hull configuration to gain advantage of the reduced constraints due to the removal of human crew. USV systems can take advantage of modern satellite positioning systems, wide area communications technologies and advanced control system development tools within their command and control systems, allowing human-in-the-loop situation analysis and decision making and leading to reduced requirements for on-board intelligence. This means that fieldable systems can be produced now from commercial off-the-shelf elements at relatively low-cost and in very short timescales. A prime example of this has been the operational Royal Navy *SWIMS* mine countermeasures system which was used in the recent Iraq conflict. A prototype was designed and built by QinetiQ within 3 weeks of contract placement and operator training completed within another 3 weeks. Fourteen conversion kits and 11 operational systems were all produced during this timescale. The systems were transported into the operational area and system acceptance into service and handover was achieved within 2 weeks. The cost advantages and reduced time to in-service timescales demonstrated by USV developments to date confirm that they represent a game changing technology which will become ever more prominent and integrated into the navies of the twenty-first century.

Chapter 16 describes the modelling and control of a USV, the Catamaran–Measuring Dolphin (*MESSIN*), a highly efficient autonomous surface craft designed to carry out multiple measuring and surveying tasks in the field of ocean research and water monitoring especially in shallow waters, that is, water depths of less than 5 m. The chapter gives a comprehensive account of the following aspects: hydromechanical design; electric and electronic sub-systems; hierarchical control for positioning, navigation route planning and mission control. Results of the manoeuvring tests and the actual applications of *MESSIN* are represented and it is argued that the systems developed for this application could be used in other manned and unmanned marine craft.

Chapter 17 addresses the topics of marine vehicle and mission control for single and multiple autonomous UMVs, from both a theoretical and a practical point of view. The mission scenarios envisioned call for the control of single or multiple USVs and AUVs acting in cooperation to execute challenging tasks without the close supervision of human operators. The material presented is rooted in practical developments and experiments carried out with the *Delfim* USV and the *Infante* AUV. Examples of mission scenarios with the above vehicles working alone or in cooperation set the stage for the main contents of the chapter. The missions described motivate the four basic categories of theoretical control problems addressed in the text: vertical and horizontal plane control, pose control, trajectory tracking and path following, and coordinated motion control. Challenging topics in these areas and current research trends are discussed. For a selected number of representative problems, the linear and nonlinear control design techniques used to solve them are briefly summarised. Linear control design borrows from recent advances in gain-scheduling control theory and from the theory of linear matrix inequalities (LMIs). Nonlinear control design builds on Lyapunov-based techniques and backstepping. Design examples and results of experimental tests at sea with the controllers developed are given. After covering the development of 'time-driven' systems for vehicle control, the chapter then provides a brief overview of the 'event-driven' systems that must be in place in order to perform mission programming and mission execution reliably, that is, mission control. The mission control systems developed build on Petri Net theory and allow for programming single and multiple vehicle missions using graphical interfaces. The hardware and software tools used for distributed system implementation are described. Results of real missions with the *Delfim* USV and the *Infante* AUV illustrate the performance of the systems developed.

Chapter 18 introduces a new range of unmanned vehicles, which have been developed comparatively recently. These are the unmanned wave-piercing vehicles (UWVs), which lie midway between the USVs and UUVs. They combine the increased stability and stealth associated with UUVs with the speed, endurance, broad bandwidth communications and GPS positioning obtainable with USVs, although, as a compromise design, they lack the full characteristics of either. Two approaches have been adopted in the development of UWVs. The first has been to drive a submarine hull to just below the surface and to pierce the waves with a permanent schnorkel resulting in an unmanned wave-piercing underwater vehicle (UWUV). A later approach has been to take a surface vehicle and give it a very small water-plane area by fitting a fared spar of small cross-section to penetrate the waves, giving an unmanned wave-piercing surface vehicle (UWSV). Each of these approaches leads to its own characteristics and applications, which are described in the chapter.

Finally, Chapter 19 covers the field of autonomous underwater gliders from the perspective of dynamics, control and multi-vehicle operations. The chapter begins with modelling glider dynamics from first principles and concludes with presentation and discussion of results from sea trials with a coordinated fleet of underwater gliders. Underwater gliders are endurance vehicles and are thus well suited to ocean sensing applications. The mathematical model for underwater glider dynamics restricted to the longitudinal plane is presented. Use of this model enables the calculation of equilibrium motions of the underwater glider. These correspond to upward or downward steady glide paths at constant speed, glide angle and pitch. Linear and nonlinear analysis tools are used to study the stability of steady glides. With reference to linear controllability and observability, a set of control laws for stabilising glider motion to a desired steady glide are derived. The glider model is adapted to more closely represent the commercially available *Slocum* glider. The basic hydrodynamic coefficients of this adapted model, such as lift and drag coefficients, are identified using an interesting approach that blends experimental data collected during sea trials with aerodynamic reference data, computational fluid dynamics (CFD) model results and wind tunnel tests. The chapter concludes with a discussion on the issues relating to multi-glider operations and summarises a methodology for coordinating gliders into coherent arrays that can translate, rotate, expand and contract. An overview of multi-vehicle sea trials, conducted in Monterey Bay during the summer of 2003, is presented, which demonstrate this cooperative control methodology in uncertain and sometimes adverse sea conditions. In this field experiment, part of the Autonomous Ocean Sampling Network (AOSN) project, the glider array served as a mobile sensor network, and it is shown how the glider array was able to estimate gradients in measured fields such as temperature and salinity and how this can be used to track features in the ocean.

1.3 Concluding Remarks

This book covers aspects of navigation guidance and control of UMVs. A wide range of vehicle types and applications are presented, together with a wide range of innovative approaches and techniques for enhancing the performance of UMVs. There are many challenges for the future and whilst it is difficult to predict in which particular areas, and when, advance will be made, it is clear that improvements in a number of areas are imminent. In particular advances in areas such as underwater communications, battery technology, fuel cells, propulsion systems, autonomous underwater docking, sensor fusion and swarms (cooperative UMVs) will impact on the future use and deployment of UMVs. Guidance and control of UMVs and associated technologies is therefore a very active and productive research area, and it is clear that significant advances will be made.

However, before UMVs become commonplace throughout the marine industry as a whole, there is a need to settle certain legal and practical issues. In particular, the legal responsibilities and liabilities concerning the deployment of such vehicles at sea need to be firmly resolved. To their credit, the Society for Underwater Technology (SUT) have confounded this matter regarding AUVs and have published guidelines in a series consisting of three volumes [9–11]. Also from a practical viewpoint, consideration should be given to standardisation to achieve compatibility between vehicles in order to provide maximum operational flexibility by permitting interchangeability of equipment, and the design and development of plug and play systems.

Finally, the Editors would like to thank all the contributors for their friendly assistance in the preparation of this book. Also, the authors are to be congratulated on producing chapters of such excellent quality. It is sincerely hoped that readers from all disciplines will find this book interesting, stimulating and useful in whatever aspect of UMV design they are involved and that it will also provide the genesis for the importation of ideas into other related fields of study.

References

1 Coverdale, A. and S. Cassidy (1987). The use of scrubbers in submarines. *Journal of Naval Engineering,* pp. 528–544.

2 Reader, G.T., G. Walker, and J.G. Hawley (1989). Non-nuclear power plants for AUVs. In Proceedings of the 6th International Symposium on Untethered Submersible Technology, University of New Hampshire, pp. 88–99.
3 Slocum, J. (1900). *Sailing Alone Around the World*. Hazell, Watson and Viney Ltd.
4 Bennett, S. (1997). *A History of Control Engineering 1800–1930*. Peter Peregrinus Ltd.
5 Fossen, T.I. (2000). A survey of nonlinear ship control: from theory to practice. In Proceedings of the 5th IFAC Conference on Manoeuvring and Control of Marine Craft, pp. 1–16.
6 Allensworth, T. (1999). A short history of Sperry Marine. http://www.sperry-marine.com/pages/history.html.
7 Sperry, E.A. (1922). Automatic steering. *Transactions: Society of Naval Architects and Marine Engineers*, pp. 53–61.
8 Minorski, N. (1922). Directional stability of automatic steered bodies. *Journal of the American Society of Naval Engineers*, 34(2), pp. 280–309.
9 SUT Publication (2000). *The Operation of Autonomous Underwater Vehicles*. Volume One: Recommended Code of Practice. SUT Publications, London.
10 SUT Publication (2000). *The Operation of Autonomous Underwater Vehicles*. Volume Two: Report on the Law. SUT Publications, London.
11 SUT Publication (2001). *The Operation of Autonomous Underwater Vehicles*. Volume Three: The Law Governing AUV Operations – Questions and Answers. SUT Publications, London.

Chapter 2

Nonlinear modelling, identification and control of UUVs

T.I. Fossen and A. Ross

2.1 Introduction

This chapter presents the unmanned underwater vehicle (UUV) equations of motion using the results of Fossen [1,2]. The nonlinear model presented in this chapter is mainly intended for control systems design in combination with system identification and parameter estimation. The resulting model is decoupled into longitudinal and lateral motions such that autopilots for speed, depth/diving and heading control can be designed.

The motivation for using nonlinear theory is that one model can cover the whole flight envelope of the UUV, instead of linearising the model about many working points and using gain scheduling between these. In addition, physical model properties in terms of first principles are not destroyed which is a primary problem associated with linearising a plant. Finally, nonlinear optimisation methods are used as a tool for system identification, while proportional–integral–derivative (PID) control and backstepping demonstrate how nonlinear autopilots can be designed.

2.1.1 Notation

In this chapter, vectors of dimension n in the Euclidean space are denoted as $\mathbf{x} \in \mathbb{R}^n$, matrices are denoted as $\mathbf{A} \in \mathbb{R}^{m \times n}$ and scalars are written as $a \in \mathbb{R}$.

A matrix $\mathbf{S} \in SS(n)$, that is the set of skew-symmetric matrices of order n, is said to be skew-symmetrical if:

$$\mathbf{S} = -\mathbf{S}^\mathrm{T}$$

This implies that the off-diagonal elements of $\mathbf{S}$ satisfy $s_{ij} = -s_{ji}$ for $i \neq j$ while the diagonal elements are zero.

The vector cross product $\times$ is defined by

$$\lambda \times \mathbf{a} := \mathbf{S}(\lambda)\mathbf{a} \tag{2.1}$$

where $\mathbf{S} \in SS(3)$ is defined as

$$\mathbf{S}(\lambda) = -\mathbf{S}^\mathrm{T}(\lambda) = \begin{bmatrix} 0 & -\lambda_3 & \lambda_2 \\ \lambda_3 & 0 & -\lambda_1 \\ -\lambda_2 & \lambda_1 & 0 \end{bmatrix}, \quad \lambda = \begin{bmatrix} \lambda_1 \\ \lambda_2 \\ \lambda_3 \end{bmatrix} \tag{2.2}$$

For marine vehicles moving in six degrees of freedom (DOF), six independent coordinates are necessary to determine the position and orientation. The first three coordinates, and their time derivatives, correspond to the position and translational motion along the x-, y-, and z-axes, respectively, while the last three coordinates, and their time derivatives, are used to describe orientation and rotational motions. The six motion components are conveniently defined as surge, sway, heave, roll, pitch and yaw [3].

The general motion of an UUV in 6 DOF is modelled by using the notation of Fossen [4]. Consider the following vectors:

$$\eta = [n, e, d, \phi, \theta, \psi]^\mathrm{T}, \quad \nu = [u, v, w, p, q, r]^\mathrm{T} \tag{2.3}$$

where n, e, d denote the North–East–Down (NED) positions in Earth-fixed coordinates (n-frame), ϕ, θ, ψ are the Euler angles (orientation) and u, v, w, p, q, r are the six DOF generalised velocities in body-fixed coordinates (b-frame).

2.2 Modelling of UUVs

The UUV equations of motion involve the study of two fields of mechanics: statics and dynamics. Statics deals with the forces that produce equilibrium, such as buoyancy or gravity, while dynamics focuses on the analysis of forces primarily in the context of the motions produced in bodies.

Statics is the oldest of the engineering sciences, dating back to Archimedes (287–212 BC), who derived the basic law of hydrostatic buoyancy. The study of dynamics started much later, since accurate measurements of time are necessary to perform dynamic experiments. The scientific basis of dynamics was provided by Sir Isaac Newton's *Principia*, published in 1687.

The study of dynamics can be divided into two parts: kinematics (Section 2.2.1), which treats only geometric aspects of motion, without consideration of mass or forces, while kinetics (Section 2.2.2) is the analysis of the forces causing these motions.

2.2.1 Six DOF kinematic equations

In Fossen [1,2] it was shown that the six DOF kinematic equations can be expressed in vector form as:

$$\dot{\eta} = \mathbf{J}(\eta)\nu \tag{2.4}$$

where

$$\mathbf{J}(\eta) = \begin{bmatrix} \mathbf{R}_b^n(\Theta) & \mathbf{0}_{3\times 3} \\ \mathbf{0}_{3\times 3} & \mathbf{T}_\Theta(\Theta) \end{bmatrix} \quad (2.5)$$

with $\eta \in \mathbb{R}^3 \times \mathcal{S}^3$ (torus of dimension 3 implying that there are three angles defined on the interval $[0, 2\pi]$) and $\nu \in \mathbb{R}^6$. The Euler angle rotation matrix $\mathbf{R}_b^n(\Theta) \in \mathbb{R}^{3\times 3}$ is defined in terms of the principal rotations:

$$\mathbf{R}_{x,\phi} = \begin{bmatrix} 1 & 0 & 0 \\ 0 & c\phi & -s\phi \\ 0 & s\phi & c\phi \end{bmatrix}, \quad \mathbf{R}_{y,\theta} = \begin{bmatrix} c\theta & 0 & s\theta \\ 0 & 1 & 0 \\ -s\theta & 0 & c\theta \end{bmatrix},$$

$$\mathbf{R}_{z,\psi} = \begin{bmatrix} c\psi & -s\psi & 0 \\ s\psi & c\psi & 0 \\ 0 & 0 & 1 \end{bmatrix} \quad (2.6)$$

where $s\cdot = \sin(\cdot)$ and $c\cdot = \cos(\cdot)$ using the *zyx*-convention:

$$\mathbf{R}_b^n(\Theta) := \mathbf{R}_{z,\psi}\mathbf{R}_{y,\theta}\mathbf{R}_{x,\phi} \quad (2.7)$$

or

$$\mathbf{R}_b^n(\Theta) = \begin{bmatrix} c\psi c\theta & -s\psi c\phi + c\psi s\theta s\phi & s\psi s\phi + c\psi c\phi s\theta \\ s\psi c\theta & c\psi c\phi + s\phi s\theta s\psi & -c\psi s\phi + s\theta s\psi c\phi \\ -s\theta & c\theta s\phi & c\theta c\phi \end{bmatrix} \quad (2.8)$$

The inverse transformation satisfies:

$$\mathbf{R}_b^n(\Theta)^{-1} = \mathbf{R}_n^b(\Theta) = \mathbf{R}_{x,\phi}^T \mathbf{R}_{y,\theta}^T \mathbf{R}_{z,\psi}^T \quad (2.9)$$

The Euler angle attitude transformation matrix is:

$$\mathbf{T}_\Theta(\Theta) = \begin{bmatrix} 1 & s\phi t\theta & c\phi t\theta \\ 0 & c\phi & -s\phi \\ 0 & s\phi/c\theta & c\phi/c\theta \end{bmatrix} \implies \mathbf{T}_\Theta^{-1}(\Theta) = \begin{bmatrix} 1 & 0 & -s\theta \\ 0 & c\phi & c\theta s\phi \\ 0 & -s\phi & c\theta c\phi \end{bmatrix},$$

$$\theta \neq \pm 90° \quad (2.10)$$

Notice that $\mathbf{T}_\Theta(\Theta)$ is undefined for a pitch angle of $\theta = \pm 90°$ and that $\mathbf{T}_\Theta^{-1}(\Theta) \neq \mathbf{T}_\Theta^T(\Theta)$. For UUVs operating close to this singularity two Euler angle representations with different singularities can be used to avoid the singular point by simply switching between these. Another possibility is to use a quaternion representation [1,2], which approaches this problem using a four parameter method.

The transformation matrix $\mathbf{J}$ and its elements $\mathbf{J}_1 = \mathbf{R}_b^n(\Theta)$ and $\mathbf{J}_2 = \mathbf{T}_\Theta(\Theta)$ can be computed in Matlab using the MSS toolbox [5]:

```
[J,J1,J2] = eulerang(phi,theta,psi)
```

Alternatively, we can write the equation (2.4) in component form as:

$$\dot{n} = u\cos\psi\cos\theta + v(\cos\psi\sin\theta\sin\phi - \sin\psi\cos\phi)$$
$$+ w(\sin\psi\sin\phi + \cos\psi\cos\phi\sin\theta) \quad (2.11)$$
$$\dot{e} = u\sin\psi\cos\theta + v(\cos\psi\cos\phi + \sin\phi\sin\theta\sin\psi)$$
$$+ w(\sin\theta\sin\psi\cos\phi - \cos\psi\sin\phi) \quad (2.12)$$
$$\dot{d} = -u\sin\theta + v\cos\theta\sin\phi + w\cos\theta\cos\phi \quad (2.13)$$
$$\dot{\phi} = p + q\sin\phi\tan\theta + r\cos\phi\tan\theta \quad (2.14)$$
$$\dot{\theta} = q\cos\phi - r\sin\phi \quad (2.15)$$
$$\dot{\psi} = q\frac{\sin\phi}{\cos\theta} + r\frac{\cos\phi}{\cos\theta}, \quad \theta \neq \pm 90° \quad (2.16)$$

2.2.2 Kinetics

For UUVs it is desirable to derive the equations of motion for an arbitrary origin in the *b*-frame to take advantage of the vessel's geometric properties. Since the hydrodynamic forces and moments are given in the *b*-frame, Newton's laws will be formulated in the *b*-frame as well. The *b*-frame coordinate system rotates with respect to the *n*-frames, and the velocities in the *b*- and *n*-frames are related through (2.4).

2.2.3 Equations of motion

In Fossen [1,2] it is shown that Newton's second law can be expressed in an arbitrary body-fixed coordinate frame as:

$$\mathbf{M}_{RB}\dot{v} + \mathbf{C}_{RB}(v)v = \tau_H + \tau \quad (2.17)$$

where $\tau_H = [X_H, Y_H, Z_H, K_H, M_H, N_H]^T$ is a vector of the hydrostatic and hydrodynamic forces and moments and τ is a vector of control inputs. The matrices are:

$$\mathbf{M}_{RB} = \mathbf{M}_{RB}^T = \begin{bmatrix} m\mathbf{I}_{3\times 3} & -m\mathbf{S}(\mathbf{r}_g^b) \\ m\mathbf{S}(\mathbf{r}_g^b) & \mathbf{I}_o \end{bmatrix}$$

$$= \begin{bmatrix} m & 0 & 0 & 0 & mz_g & -my_g \\ 0 & m & 0 & -mz_g & 0 & mx_g \\ 0 & 0 & m & my_g & -mx_g & 0 \\ 0 & -mz_g & my_g & I_x & -I_{xy} & -I_{xz} \\ mz_g & 0 & -mx_g & -I_{yx} & I_y & -I_{yz} \\ -my_g & mx_g & 0 & -I_{zx} & -I_{zy} & I_z \end{bmatrix} \quad (2.18)$$

$$\mathbf{C}_{RB}(v) = \begin{bmatrix} \mathbf{0}_{3\times 3} & -m\mathbf{S}(v_1) - m\mathbf{S}(v_2)\mathbf{S}(\mathbf{r}_g^b) \\ -m\mathbf{S}(v_1) + m\mathbf{S}(\mathbf{r}_g^b)\mathbf{S}(v_2) & -\mathbf{S}(\mathbf{I}_o v_2) \end{bmatrix}, \quad (2.19)$$

where $v_1 = [u, v, w]^T$, $v_2 = [p, q, r]^T$, $\mathbf{r}_g^b = [x_g, y_g, z_g]$ is the centre of gravity (CG) with respect to the *b*-frame origin, and

$$\mathbf{I}_o := \begin{bmatrix} I_x & -I_{xy} & -I_{xz} \\ -I_{yx} & I_y & -I_{yz} \\ -I_{zx} & -I_{zy} & I_z \end{bmatrix}, \quad \mathbf{I}_o = \mathbf{I}_o^T > 0 \qquad (2.20)$$

is the *inertia matrix* about the *b*-frame origin. The rigid-body dynamics can be written in component form according to:

$$\begin{aligned}
m[\dot{u} - vr + wq - x_g(q^2 + r^2) + y_g(pq - \dot{r}) + z_g(pr + \dot{q})] &= X_H + X \\
m[\dot{v} - wp + ur - y_g(r^2 + p^2) + z_g(qr - \dot{p}) + x_g(qp + \dot{r})] &= Y_H + Y \\
m[\dot{w} - uq + vp - z_g(p^2 + q^2) + x_g(rp - \dot{q}) + y_g(rq + \dot{p})] &= Z_H + Z \\
I_x \dot{p} + (I_z - I_y)qr - (\dot{r} + pq)I_{xz} + (r^2 - q^2)I_{yz} + (pr - \dot{q})I_{xy} & \\
+ m[y_g(\dot{w} - uq + vp) - z_g(\dot{v} - wp + ur)] &= K_H + K \\
I_y \dot{q} + (I_x - I_z)rp - (\dot{p} + qr)I_{xy} + (p^2 - r^2)I_{zx} + (qp - \dot{r})I_{yz} & \\
+ m[z_g(\dot{u} - vr + wq) - x_g(\dot{w} - uq + vp)] &= M_H + M \\
I_z \dot{r} + (I_y - I_x)pq - (\dot{q} + rp)I_{yz} + (q^2 - p^2)I_{xy} + (rq - \dot{p})I_{zx} & \\
+ m[x_g(\dot{v} - wp + ur) - y_g(\dot{u} - vr + wq)] &= N_H + N
\end{aligned}$$
(2.21)

For a deeply submerged vehicle the hydrodynamic forces and moments will be due to added mass and damping, while the hydrostatic forces and moments are due to weight and buoyancy. This suggests that

$$\underbrace{\mathbf{M}_{RB}\dot{v} + \mathbf{C}_{RB}(v)v}_{\text{rigid-body terms}} + \underbrace{\mathbf{M}_A\dot{v} + \mathbf{C}_A(v)v + \mathbf{D}(v)v}_{\text{hydrodynamic terms}} + \underbrace{\mathbf{g}(\eta)}_{\text{hydrostatic terms}} = \tau \qquad (2.22)$$

such that

$$\mathbf{M}\dot{v} + \mathbf{C}(v)v + \mathbf{D}(v)v + \mathbf{g}(\eta) = \tau \qquad (2.23)$$

where $\mathbf{M} = \mathbf{M}_{RB} + \mathbf{M}_A$, $\mathbf{C}(v) = \mathbf{C}_{RB}(v) + \mathbf{C}_A(v)$ and

M – system inertia matrix (including added mass)
C(v) – Coriolis-centripetal matrix (including added mass)
D(v) – damping matrix
g(η) – vector of gravitational/buoyancy forces and moments.

For UUVs the constant zero frequency added mass terms are used. The matrices are:

$$\mathbf{M}_A = \mathbf{M}_A^T = -\begin{bmatrix} X_{\dot{u}} & X_{\dot{v}} & X_{\dot{w}} & X_{\dot{p}} & X_{\dot{q}} & X_{\dot{r}} \\ Y_{\dot{u}} & Y_{\dot{v}} & Y_{\dot{w}} & Y_{\dot{p}} & Y_{\dot{q}} & Y_{\dot{r}} \\ Z_{\dot{u}} & Z_{\dot{v}} & Z_{\dot{w}} & Z_{\dot{p}} & Z_{\dot{q}} & Z_{\dot{r}} \\ K_{\dot{u}} & K_{\dot{v}} & K_{\dot{w}} & K_{\dot{p}} & K_{\dot{q}} & K_{\dot{r}} \\ M_{\dot{u}} & M_{\dot{v}} & M_{\dot{w}} & M_{\dot{p}} & M_{\dot{q}} & M_{\dot{r}} \\ N_{\dot{u}} & N_{\dot{v}} & N_{\dot{w}} & N_{\dot{p}} & N_{\dot{q}} & N_{\dot{r}} \end{bmatrix} \quad (2.24)$$

$$\mathbf{C}_A(\nu) = -\mathbf{C}_A(\nu)^T = \begin{bmatrix} 0 & 0 & 0 & 0 & -a_3 & a_2 \\ 0 & 0 & 0 & a_3 & 0 & -a_1 \\ 0 & 0 & 0 & -a_2 & a_1 & 0 \\ 0 & -a_3 & a_2 & 0 & -b_3 & b_2 \\ a_3 & 0 & -a_1 & b_3 & 0 & -b_1 \\ -a_2 & a_1 & 0 & -b_2 & b_1 & 0 \end{bmatrix} \quad (2.25)$$

where

$$\begin{aligned} a_1 &= X_{\dot{u}}u + X_{\dot{v}}v + X_{\dot{w}}w + X_{\dot{p}}p + X_{\dot{q}}q + X_{\dot{r}}r \\ a_2 &= Y_{\dot{u}}u + Y_{\dot{v}}v + Y_{\dot{w}}w + Y_{\dot{p}}p + Y_{\dot{q}}q + Y_{\dot{r}}r \\ a_3 &= Z_{\dot{u}}u + Z_{\dot{v}}v + Z_{\dot{w}}w + Z_{\dot{p}}p + Z_{\dot{q}}q + Z_{\dot{r}}r \\ b_1 &= K_{\dot{u}}u + K_{\dot{v}}v + K_{\dot{w}}w + K_{\dot{p}}p + K_{\dot{q}}q + K_{\dot{r}}r \\ b_2 &= M_{\dot{u}}u + M_{\dot{v}}v + M_{\dot{w}}w + M_{\dot{p}}p + M_{\dot{q}}q + M_{\dot{r}}r \\ b_3 &= N_{\dot{u}}u + N_{\dot{v}}v + N_{\dot{w}}w + N_{\dot{p}}p + N_{\dot{q}}q + N_{\dot{r}}r \end{aligned} \quad (2.26)$$

The gravitational force will act through the CG while the buoyancy force will act through the centre of buoyancy (CB) defined by $\mathbf{r}_b^b = [x_b, y_b, z_b]^T$. For a UUV, the submerged weight of the body W and buoyancy force B are given by:

$$W = mg, \quad B = \rho g \nabla \quad (2.27)$$

where ∇ is the volume of fluid displaced by the vehicle, g is the acceleration of gravity (positive downwards) and ρ is the water density. It can be shown that [2]

$$\mathbf{g}(\eta) = \begin{bmatrix} (W - B)\sin\theta \\ -(W - B)\cos\theta \sin\phi \\ -(W - B)\cos\theta \cos\phi \\ -(y_g W - y_b B)\cos\theta \cos\phi + (z_g W - z_b B)\cos\theta \sin\phi \\ (z_g W - z_b B)\sin\theta + (x_g W - x_b B)\cos\theta \cos\phi \\ -(x_g W - x_b B)\cos\theta \sin\phi - (y_g W - y_b B)\sin\theta \end{bmatrix} \quad (2.28)$$

The damping forces for UUVs can be written as the sum of a linear damping term $\mathbf{D}$ and a nonlinear damping term $\mathbf{D}_n(\nu)$ such that

$$\mathbf{D}(\nu) = \mathbf{D} + \mathbf{D}_n(\nu) \quad (2.29)$$

where

$$\mathbf{D} = \mathbf{D}^{\mathrm{T}} = -\begin{bmatrix} X_u & X_v & X_w & X_p & X_q & X_r \\ Y_u & Y_v & Y_w & Y_p & Y_q & Y_r \\ Z_u & Z_v & Z_w & Z_p & Z_q & Z_r \\ K_u & K_v & K_w & K_p & K_q & K_r \\ M_u & M_v & M_w & M_p & M_q & M_r \\ N_u & N_v & N_w & N_p & N_q & N_r \end{bmatrix} \qquad (2.30)$$

The nonlinear manoeuvring coefficients $\mathbf{D}_n(v)$ are usually modelled by using a third-order Taylor series expansion or modulus functions (quadratic drag). If the xz-plane is a plane of symmetry (starboard/port symmetry) an odd Taylor series expansion containing first- and third-order terms in velocity will be sufficient to describe most manoeuvres.

2.2.4 Equations of motion including ocean currents

When simulating UUVs it is necessary to include environmental disturbances. For deeply submerged vessels wave-induced disturbances can usually be neglected. Hence, the only environmental load will be ocean currents. Ocean currents are horizontal and vertical circulating systems of ocean waters produced by gravity, wind friction and water density variation in different parts of the ocean.

The effect of ocean currents can be incorporated in the six DOF equations of motion (2.23) by using the concept of relative motion under the assumption of an irrotational flow. Let:

$$v_r = [u - u_c^b, v - v_c^b, w - w_c^b, p, q, r]^{\mathrm{T}} \qquad (2.31)$$

where $u_c^b, v_c^b,$ and w_c^b are the body-fixed current velocities. The rigid-body dynamics and generalised hydrodynamic forces are written:

$$\underbrace{\mathbf{M}_{\mathrm{RB}} \dot{v} + \mathbf{C}_{\mathrm{RB}}(v) v}_{\text{rigid-body terms}} + \underbrace{\mathbf{M}_A \dot{v}_r + \mathbf{C}_A(v_r) v_r + \mathbf{D}(v_r) v_r}_{\text{hydrodynamic terms}} + \underbrace{\mathbf{g}(\eta)}_{\text{hydrostatic terms}} = \tau$$

$$(2.32)$$

It is common to assume that the current velocity vector is slowly varying, that is, $\dot{v}_c \approx \mathbf{0}$ such that $\dot{v}_r \approx \mathbf{0}$. Hence, the equation of motion with currents becomes

$$\mathbf{M} \dot{v} + \mathbf{C}_{\mathrm{RB}}(v) v + \mathbf{C}_A(v_r) v_r + \mathbf{D}(v_r) v_r + \mathbf{g}(\eta) = \tau \qquad (2.33)$$

The current speed V_c is usually defined in the Earth-fixed reference frame using flow axes, that is $[V_c, 0, 0]^{\mathrm{T}}$ such that the current speed is directed in the x-direction. The transformation from flow axes to NED velocities can be performed by defining α_c as the flow angle of attack and β_c as the flow sideslip angle. Hence, the three-dimensional (3D) current velocities are found by performing two principal rotations [2]:

$$\begin{bmatrix} u_c^n \\ v_c^n \\ w_c^n \end{bmatrix} = \mathbf{R}_{y,\alpha_c}^T \mathbf{R}_{z,-\beta_c}^T \begin{bmatrix} V_c \\ 0 \\ 0 \end{bmatrix} \quad (2.34)$$

where the rotation matrices $\mathbf{R}_{y,\alpha_c}$ and $\mathbf{R}_{z,-\beta_c}$ are defined as:

$$\mathbf{R}_{y,\alpha_c} = \begin{bmatrix} \cos\alpha_c & 0 & \sin\alpha_c \\ 0 & 1 & 0 \\ -\sin\alpha_c & 0 & \cos\alpha_c \end{bmatrix}, \quad \mathbf{R}_{z,-\beta_c} = \mathbf{R}_{z,\beta_c}^T = \begin{bmatrix} \cos\beta_c & \sin\beta_c & 0 \\ -\sin\beta_c & \cos\beta_c & 0 \\ 0 & 0 & 1 \end{bmatrix} \quad (2.35)$$

Expanding (2.34) using (2.35), yields:

$$u_c^n = V_c \cos\alpha_c \cos\beta_c \quad (2.36)$$
$$v_c^n = V_c \sin\beta_c \quad (2.37)$$
$$w_c^n = V_c \sin\alpha_c \cos\beta_c \quad (2.38)$$

The NED current velocities are then transformed to body-fixed velocities using the Euler angle rotation matrix. Consequently:

$$\begin{bmatrix} u_c^b \\ v_c^b \\ w_c^b \end{bmatrix} = \mathbf{R}_b^n(\mathbf{\Theta})^T \begin{bmatrix} u_c^n \\ v_c^n \\ w_c^n \end{bmatrix} \quad (2.39)$$

For the 2D case, the 3D equations (2.36)–(2.38) with $\alpha_c = 0$ reduce to:

$$u_c^n = V_c \cos\beta_c \quad (2.40)$$
$$v_c^n = V_c \sin\beta_c \quad (2.41)$$

since the component w_c^n is not used in the horizontal plane. Hence (2.39) reduces to:

$$u_c^b = V_c \cos(\beta_c - \psi) \quad (2.42)$$
$$v_c^b = V_c \sin(\beta_c - \psi) \quad (2.43)$$

2.2.5 Longitudinal and lateral models

For UUVs, the six DOF equations of motion can in many cases be divided into two non-interacting (or lightly interacting) subsystems:

- Longitudinal subsystem: states (u, w, q) and (n, d, θ)
- Lateral subsystem: states (v, p, r) and (e, ϕ, ψ)

This decomposition is good for slender bodies (large length/width ratio) as shown in Figure 2.1. Consequently, it is a common assumption for a submarine [6–8]. Other examples are flying vehicles like the NPS AUV II [9] shown in Figure 2.2 and the *MARIUS* vehicle [10].

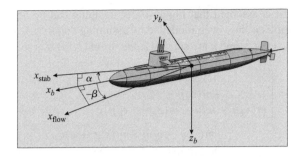

Figure 2.1 Submarine

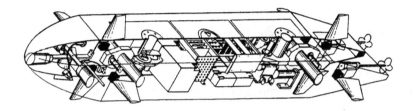

Figure 2.2 The NPS AUV II [9]

The longitudinal and lateral decompositions are found by considering a slender body with starboard–port symmetry, implying the system inertia matrix takes the following form:

$$\mathbf{M} = \begin{bmatrix} m_{11} & 0 & m_{13} & 0 & m_{15} & 0 \\ 0 & m_{22} & 0 & m_{24} & 0 & m_{26} \\ m_{31} & 0 & m_{33} & 0 & m_{35} & 0 \\ 0 & m_{42} & 0 & m_{44} & 0 & m_{46} \\ m_{51} & 0 & m_{53} & 0 & m_{55} & 0 \\ 0 & m_{62} & 0 & m_{64} & 0 & m_{66} \end{bmatrix} \quad (2.44)$$

This gives the two subsystems:

$$\mathbf{M}_{\text{long}} = \begin{bmatrix} m_{11} & m_{13} & m_{15} \\ m_{31} & m_{33} & m_{35} \\ m_{51} & m_{53} & m_{55} \end{bmatrix}, \quad \mathbf{M}_{\text{lat}} = \begin{bmatrix} m_{22} & m_{24} & m_{26} \\ m_{42} & m_{44} & m_{46} \\ m_{62} & m_{64} & m_{66} \end{bmatrix} \quad (2.45)$$

which are evidently decoupled.

2.2.5.1 Longitudinal subsystem

Assuming that the states v, p, r, ϕ are small, the longitudinal kinematics for surge, heave and pitch are, see (2.8) and (2.10):

$$\begin{bmatrix} \dot{d} \\ \dot{\theta} \end{bmatrix} = \begin{bmatrix} \cos\theta & 0 \\ 0 & 1 \end{bmatrix} \begin{bmatrix} w \\ q \end{bmatrix} + \begin{bmatrix} -\sin\theta \\ 0 \end{bmatrix} u \quad (2.46)$$

For simplicity, it is assumed that higher-order damping can be neglected, that is, $\mathbf{D}_n(v) = \mathbf{0}$. Coriolis is, however, modelled by assuming that $u \gg 0$ and that second-order terms in v, w, p, q and r are small. Hence, from (2.19) it is seen that:

$$\mathbf{C}_{\text{RB}}(v)v = \begin{bmatrix} m(y_g q + z_g r)p \\ -m(z_g p - v)p \\ m(x_g q - w)u - m(z_g r + x_g p)v \end{bmatrix}$$

$$\begin{matrix} -m(x_g q - w)q - m(x_g r + v)r \\ -m(z_g q + u)q + m(x_g p + y_g q)r \\ +m(z_g q + u)w + (I_{yz}q + I_{xz}p - I_z r)p + (-I_{xz}r - I_{xy}q + I_x p)r \end{matrix}$$

such that

$$\mathbf{C}_{\text{RB}}(v)v \approx \begin{bmatrix} 0 & 0 & 0 \\ 0 & 0 & -mu \\ 0 & 0 & mx_g u \end{bmatrix} \begin{bmatrix} u \\ w \\ q \end{bmatrix} \qquad (2.47)$$

Notice that $\mathbf{C}_{\text{RB}}(v) \neq -\mathbf{C}_{\text{RB}}^T(v)$ for the decoupled model. For simplicity assume that $\mathbf{M}_A$ is diagonal. Hence, the corresponding added mass terms are:

$$\mathbf{C}_A(v)v = \begin{bmatrix} -Z_{\dot{w}} wq + Y_{\dot{v}} vr \\ -Y_{\dot{v}} vp + X_{\dot{u}} uq \\ (Z_{\dot{w}} - X_{\dot{u}})uw + (N_{\dot{r}} - K_{\dot{p}})pr \end{bmatrix}$$

$$\approx \begin{bmatrix} 0 & 0 & 0 \\ 0 & 0 & X_{\dot{u}} u \\ 0 & (Z_{\dot{w}} - X_{\dot{u}})u & 0 \end{bmatrix} \begin{bmatrix} u \\ w \\ q \end{bmatrix} \qquad (2.48)$$

For a UUV with weight equal to buoyancy ($W = B$) and coincident centres of gravity and buoyancy ($x_g = x_b$), the dynamics then become:

$$\begin{bmatrix} m - X_{\dot{u}} & -X_{\dot{w}} & mz_g - X_{\dot{q}} \\ -X_{\dot{w}} & m - Z_{\dot{w}} & -mx_g - Z_{\dot{q}} \\ mz_g - X_{\dot{q}} & -mx_g - Z_{\dot{q}} & I_y - M_{\dot{q}} \end{bmatrix} \begin{bmatrix} \dot{u} \\ \dot{w} \\ \dot{q} \end{bmatrix}$$

$$+ \begin{bmatrix} -X_u & -X_w & -X_q \\ -Z_u & -Z_w & -Z_q \\ -M_u & -M_w & -M_q \end{bmatrix} \begin{bmatrix} u \\ w \\ q \end{bmatrix}$$

$$+ \begin{bmatrix} 0 & 0 & 0 \\ 0 & 0 & -(m - X_{\dot{u}})u \\ 0 & (Z_{\dot{w}} - X_{\dot{u}})u & mx_g u \end{bmatrix} \begin{bmatrix} u \\ w \\ q \end{bmatrix} + \begin{bmatrix} 0 \\ 0 \\ WBG_z \sin\theta \end{bmatrix} = \begin{bmatrix} \tau_X \\ \tau_Z \\ \tau_M \end{bmatrix}$$
$$(2.49)$$

where $BG_z = z_g - z_b$. This model is the basis for forward speed control (state u) and depth/diving autopilot design (states w, q, θ).

If the forward speed is stabilised by a forward speed controller such that

$$u = u_o = \text{constant} \qquad (2.50)$$

forward speed can be eliminated from the longitudinal equations of motion such that

$$\begin{bmatrix} m - Z_{\dot{w}} & -mx_g - Z_{\dot{q}} \\ -mx_g - Z_{\dot{q}} & I_y - M_{\dot{q}} \end{bmatrix} \begin{bmatrix} \dot{w} \\ \dot{q} \end{bmatrix} + \begin{bmatrix} -Z_w & -Z_q \\ -M_w & -M_q \end{bmatrix} \begin{bmatrix} w \\ q \end{bmatrix}$$
$$+ \begin{bmatrix} 0 & -(m - X_{\dot{u}})u_o \\ (Z_{\dot{w}} - X_{\dot{u}})u_o & mx_g u_o \end{bmatrix} \begin{bmatrix} w \\ q \end{bmatrix} + \begin{bmatrix} 0 \\ BG_z W \sin\theta \end{bmatrix} = \begin{bmatrix} \tau_Z \\ \tau_M \end{bmatrix} \quad (2.51)$$

Moreover, if $\dot{w} = w = 0$ (constant depth) and θ is small such that $\sin\theta \approx \theta$, the linear pitch dynamics become:

$$(I_y - M_{\dot{q}})\ddot{\theta} - M_q\dot{\theta} + BG_z W\theta = \tau_M \quad (2.52)$$

where the natural frequency is

$$\omega_\theta = \sqrt{\frac{BG_z W}{I_y - M_{\dot{q}}}} \quad (2.53)$$

2.2.5.2 Lateral subsystem

Under the assumption that the states u, w, p, r, ϕ and θ are small, the lateral kinematics, see (2.8) and (2.10), reduce to:

$$\dot{\phi} = p \quad (2.54)$$
$$\dot{\psi} = r \quad (2.55)$$

Again it is assumed that higher-order velocity terms can be neglected so that $\mathbf{D}_n(v) = 0$, and that the Coriolis terms in $u = u_o$ are the most important. This gives:

$$\mathbf{C}_{RB}(v)v = \begin{bmatrix} -m(y_g p + w)p + m(z_g r + x_g p)q - m(y_g r - u)r \\ -m(y_g q + z_g r)u + m(y_g p + w)v + m(z_g p - v)w \\ m(x_g r + v)u + m(y_g r - u)v - m(x_g p + y_g q)w \end{bmatrix}$$

$$\begin{matrix} +(-I_{yz}q - I_{xz}p + I_z r)q + (I_{yz}r + I_{xy}p - I_y q)r \\ +(-I_{yz}r - I_{xy}p + I_y q)p + (I_{xz}r + I_{xy}q - I_x p)q \end{matrix}$$

Hence

$$\mathbf{C}_{RB}(v)v \approx \begin{bmatrix} 0 & 0 & mu_o \\ 0 & 0 & 0 \\ 0 & 0 & mx_g u_o \end{bmatrix} \begin{bmatrix} v \\ p \\ r \end{bmatrix} \quad (2.56)$$

Under the assumption of a diagonal $\mathbf{M}_A$, the corresponding added mass terms are:

$$\mathbf{C}_A(v)v = \begin{bmatrix} Z_{\dot{w}}wp - X_{\dot{u}}ur \\ (Y_{\dot{v}}-Z_{\dot{w}})vw + (M_{\dot{q}}-N_{\dot{r}})qr \\ (X_{\dot{u}}-Y_{\dot{v}})uv + (K_{\dot{p}}-M_{\dot{q}})pq \end{bmatrix}$$

$$\approx \begin{bmatrix} 0 & 0 & -X_{\dot{u}}u \\ 0 & 0 & 0 \\ (X_{\dot{u}}-Y_{\dot{v}})u & 0 & 0 \end{bmatrix} \begin{bmatrix} v \\ p \\ r \end{bmatrix} \quad (2.57)$$

Next, assume that $W = B$, $x_g = x_b$, and $y_g = y_b$. Then the UUV dynamics reduce to

$$\begin{bmatrix} m - Y_{\dot{v}} & -mz_g - Y_{\dot{p}} & mx_g - Y_{\dot{r}} \\ -mz_g - Y_{\dot{p}} & I_x - K_{\dot{p}} & -I_{zx} - K_{\dot{r}} \\ mx_g - Y_{\dot{r}} & -I_{zx} - K_{\dot{r}} & I_z - N_{\dot{r}} \end{bmatrix} \begin{bmatrix} \dot{v} \\ \dot{p} \\ \dot{r} \end{bmatrix}$$

$$+ \begin{bmatrix} -Y_v & -Y_p & -Y_r \\ -M_v & -M_p & -M_r \\ -N_v & -N_p & -N_r \end{bmatrix} \begin{bmatrix} v \\ p \\ r \end{bmatrix}$$

$$+ \begin{bmatrix} 0 & 0 & (m - X_{\dot{u}})u \\ 0 & 0 & 0 \\ (X_{\dot{u}} - Y_{\dot{v}})u & 0 & mx_g u \end{bmatrix} \begin{bmatrix} v \\ p \\ r \end{bmatrix} + \begin{bmatrix} 0 \\ WBG_z \sin\phi \\ 0 \end{bmatrix} = \begin{bmatrix} \tau_Y \\ \tau_K \\ \tau_N \end{bmatrix}$$
$$(2.58)$$

For vehicles where $\dot{p}$ and p are small (small roll motions) and the speed is $u = u_o$, this reduces to:

$$\begin{bmatrix} m - Y_{\dot{v}} & mx_g - Y_{\dot{r}} \\ mx_g - Y_{\dot{r}} & I_z - N_{\dot{r}} \end{bmatrix} \begin{bmatrix} \dot{v} \\ \dot{r} \end{bmatrix} + \begin{bmatrix} -Y_v & -Y_r \\ -N_v & -N_r \end{bmatrix} \begin{bmatrix} v \\ r \end{bmatrix}$$

$$+ \begin{bmatrix} 0 & (m - X_{\dot{u}})u_o \\ (X_{\dot{u}} - Y_{\dot{v}})u_o & mx_g u_o \end{bmatrix} \begin{bmatrix} v \\ r \end{bmatrix} = \begin{bmatrix} \tau_Y \\ \tau_N \end{bmatrix} \quad (2.59)$$

which is the sway–yaw manoeuvring model. The decoupled linear roll equation under the assumption of a small ϕ is

$$(I_x - K_{\dot{p}})\ddot{\phi} - K_p \dot{\phi} + WBG_z \phi = \tau_K \quad (2.60)$$

for which the natural frequency is

$$\omega_\phi = \sqrt{\frac{BG_z W}{I_x - K_{\dot{p}}}} \quad (2.61)$$

2.3 Identification of UUVs

The longitudinal and lateral models presented in Section 2.2.5 involve a large number of parameters. When estimating parameters using system identification (SI) or nonlinear optimisation methods, the main problem is overparametrisation. Moreover, if all the parameters are to be estimated from time-series, it is likely that several of the estimates will depend on each other due to this overparametrisation. The practical solution to this problem is to exploit *a priori* information on the nature of the model, and estimate a model of partially known parameters. We will briefly illustrate such an approach in this chapter.

2.3.1 A priori *estimates of rigid-body parameters*

For a UUV, the mass parameters are easily obtained since these are computed when constructing the vehicle. Therefore we will assume that the following quantities are known:

$$m, I_x, I_y, I_z, I_{xy}, I_{xz}, I_{yz}, x_g, y_g, z_g \qquad (2.62)$$

The moment and product of inertia are usually computed by assuming that the equipment on the vehicle consists of point masses or known geometrical shapes located at known coordinates. A useful tool for this is the parallel axes theorem. Similarly, the displacement and centre of buoyancy must be computed, implying that we know:

$$\nabla, x_b, y_b, z_b \qquad (2.63)$$

For a neutrally buoyant vehicle $m = \rho \nabla$.

This means that the elements of

$$\mathbf{M}_{RB}, \mathbf{C}_{RB}(v), \mathbf{g}(\eta) \qquad (2.64)$$

are also available *a priori*.

2.3.2 A priori *estimates of hydrodynamic added mass*

For a UUV the hydrodynamic mass terms can be computed by using a CAD/CAM drawing of the vehicle as input to a hydrodynamic software program for instance by using AutoCad (www.autocad.com). This data format can usually be transformed to other formats if necessary.

There are several commercially available software programs for computation of hydrodynamic added mass, for example, the 3D-potential theory program of *WAMIT* (www.wamit.com), which assumes that the forward speed is zero (or at least small). This program is quite popular with oil companies since it can be applied to floating structures like rigs and production ships.

For flying vehicles moving at higher forward speeds, there are programs that can calculate the elements of $\mathbf{M}_A$ and $\mathbf{C}_A(v)$ using 2D-strip theory programs, such as the MARINTEK program *ShipX* (www.marintek.sintef.no) with the add-in for vessel repsonse calculations (*VERES*), which uses the strip theory of Salvesen *et al.* [11], or SeaWay by Amacom (www.amacom.com).

2.3.3 *Identification of damping terms*

The main problem when designing UUV control systems is to obtain an estimate of the damping terms. One first guess is to use simple formulae for linear damping, and identification methods applied to estimate the remaining terms. For control system design, the formulae for linear damping will be sufficient in many cases.

2.3.3.1 Formulae for linear damping

For a neutrally buoyant vehicle, the formulae for linear damping in surge, sway, heave and yaw can be derived by assuming that these modes are second-order linear decoupled mass–damper systems:

$$m\ddot{x} + d\dot{x} = \tau \tag{2.65}$$

The linear damping coefficient can be found by specifying the time constant $T > 0$ in

$$d = \frac{m}{T} \tag{2.66}$$

suggesting that

$$-X_u \approx \frac{m - X_{\dot{u}}}{T_{\text{surge}}} \tag{2.67}$$

$$-Y_v \approx \frac{m - X_{\dot{u}}}{T_{\text{sway}}} \tag{2.68}$$

$$-Z_z \approx \frac{m - Z_{\dot{w}}}{T_{\text{heave}}} \tag{2.69}$$

$$-N_r \approx \frac{m - N_{\dot{r}}}{T_{\text{yaw}}} \tag{2.70}$$

The time constants can be found by performing step responses in surge, sway, heave and yaw with the UUV.

The remaining modes, roll and pitch, are assumed to be second-order mass–damper–spring systems

$$m\ddot{x} + d\dot{x} + kx = \tau \tag{2.71}$$

where

$$\frac{d}{m} = 2\zeta\omega_n$$

$$= 2\zeta\sqrt{\frac{k}{m}} \tag{2.72}$$

or

$$d = 2\zeta\sqrt{km} \tag{2.73}$$

Here, the relative damping ratio ζ can be used to specify the linear damping coefficient d since $\omega_n = \sqrt{k/m}$ is known *a priori* for known m and k. This gives the following formulae:

$$-K_p \approx 2\zeta_{\text{roll}}\sqrt{BG_zW(I_x - K_{\dot{p}})}$$
$$-M_q \approx 2\zeta_{\text{pitch}}\sqrt{BG_zW(I_y - M_{\dot{q}})}$$

The relative damping ratios ζ_{roll} and ζ_{pitch} for marine vehicles are quite small, typically between 0.1 and 0.2.

Thus an initial estimate of the damping forces and moments will be

$$\mathbf{D} = -\text{diag}\{X_u, Y_v, Z_w, K_p, M_q, N_r\} \qquad (2.74)$$

2.3.3.2 Identification of damping terms using a partially known model

The unknown damping terms, both linear and nonlinear, can be estimated using SI or nonlinear optimisation methods. We will illustrate this by computing both sets of damping terms in Matlab.

2.3.3.2.1 Infante AUV model

The case study which follows is carried out on a simulation model of the *Infante* AUV also see Chapter 17 for further details, depicted in Figure 2.3. This craft was designed, developed and modelled at the Instituto Superior Tecnico in Lisbon, Portugal. It is a flying vehicle, making it ideal for the identification of quadratic damping here. For thorough information on the vehicle's development, testing and verification, see Silvestre and Pascoal [12,13].

Figure 2.3   The Infante *AUV (Picture courtesy of Silvestre and Pascoal [12,13])*

Consider the sway–yaw system with nonlinear quadratic damping:

$$\begin{bmatrix} m - Y_{\dot{v}} & mx_g - Y_{\dot{r}} \\ mx_g - Y_{\dot{r}} & I_z - N_{\dot{r}} \end{bmatrix} \begin{bmatrix} \dot{v} \\ \dot{r} \end{bmatrix} + \begin{bmatrix} -Y_v & -Y_r \\ -N_v & -N_r \end{bmatrix} \begin{bmatrix} v \\ r \end{bmatrix}$$
$$+ \begin{bmatrix} -Y_{|v|v}|v| & -Y_{|r|r}|v| \\ -N_{|v|v}|v| & -N_{|r|r}|v| \end{bmatrix} \begin{bmatrix} v \\ r \end{bmatrix}$$
$$+ \begin{bmatrix} 0 & (m - X_{\dot{u}})u_0 \\ (X_{\dot{u}} - Y_{\dot{v}})u_0 & (mx_g - Y_{\dot{r}})u_0 \end{bmatrix} \begin{bmatrix} v \\ r \end{bmatrix} = \begin{bmatrix} \tau_Y \\ \tau_N \end{bmatrix} \quad (2.75)$$

Assume that the only unknowns are the eight damping terms:

$$-Y_v, \ -Y_r, \ -N_v, \ -N_r, \ -Y_{|v|v}, \ -Y_{|r|r}, \ -N_{|v|v}, \ -N_{|r|r}$$

2.3.3.2.2 Least squares optimisation

It is elementary to parameterise the system according to

$$\mathbf{z} = \phi^T \theta^* \quad (2.76)$$

where the terms are defined as follows:

$$\mathbf{y} = [\mathbf{M}\dot{v} + \mathbf{C}(v)v - \tau] \in \mathbb{R}^2 \quad (2.77)$$

$$\mathbf{z} = \frac{\mathbf{y}}{\lambda(s)} \quad (2.78)$$

$$\phi = \begin{bmatrix} -\dfrac{v^T}{\lambda(s)} & 0 & -\dfrac{|v|v^T}{\lambda(s)} & 0 \\ 0 & -\dfrac{v^T}{\lambda(s)} & 0 & -\dfrac{|v|v^T}{\lambda(s)} \end{bmatrix}^T \in \mathbb{R}^{8 \times 2} \quad (2.79)$$

$$\theta^* = \begin{bmatrix} Y_v & Y_r & N_v & N_r & Y_{|v|v} & Y_{|r|r} & N_{|v|v} & N_{|r|r} \end{bmatrix}^T \in \mathbb{R}^8 \quad (2.80)$$

Essentially, the system is now in a linear parameterised form, with all the unknowns contained within the parameter vector θ^*. A stable filter $\lambda(s)$ is used to ensure that all signals are available, and therefore that the algorithm is implementable. There are many methods available, both online and offline, to find the best solution to this problem. To take an example of offline estimation define the cost function to be minimised as

$$\min_{\theta} J = \frac{1}{2} \int_0^T (\mathbf{z} - \phi^T \theta)^T (\mathbf{z} - \phi^T \theta) \, d\tau \quad (2.81)$$

where θ is the calculated estimate of the true parameter vector θ^*. Performing a minimisation with a least squares solution gives convergence to the correct parameters with a residual of the order of $1e^{-3}$. Furthermore, if we take the solution as $T \to \infty$, this result holds in the presence of any zero mean noise. This is an offline methodology: it is rudimentary to take this, and implement a continuous time recursive least squares algorithm, for example, the methods described in Reference 14.

2.3.3.2.3 Model based identification

One fairly modern approach to the problem of identification is to use a Lyapunov method, first proposed by Lyshevski and Abel in Reference 15, and further developed in Reference 16. This algorithm provides a method of identification for systems of the form:

$$\dot{\mathbf{x}}(t) = \mathbf{A} f[\mathbf{x}(t), \mathbf{u}(t)], \quad t \geq 0, \qquad \mathbf{x}(t_0) = \mathbf{x}_0 \tag{2.82}$$

$$\dot{\mathbf{x}}_M(t) = \mathbf{A}_M f[\mathbf{x}_M(t), \mathbf{u}(t)], \quad t \geq 0, \qquad \mathbf{x}_M(t_0) = \mathbf{x}_{M0} \tag{2.83}$$

where (2.82) is a parameterisation sectioning the knowns into the function $f(\cdot, \cdot)$ and the unknowns, or only partial knowns, into the matrix A, with A_M as an estimate of this. Equation (2.83) is essentially a mirror of (2.82), except that the state vector is that of the identification model, and not the true states. If we have $x_M = x$ and $A_M = A$, then (2.83) and (2.82) are identical and the problem of estimating the unknowns of A has been solved. The terms are of the following form:

$\mathbf{x}, \mathbf{x}_M \in \mathbb{R}^c, \quad u \in \mathbb{R}^m$

$\mathbf{f}(\cdot, \cdot) \in \mathbb{R}^c \times \mathbb{R}^m \to \mathbb{R}^n \cup C^0$ and is a known analytic vector field

$\mathbf{A}, \mathbf{A}_M \in \mathbb{R}^{c \times n}$

There are three error terms to be considered

$\Delta \mathbf{x}(t) = \mathbf{x}(t) - \mathbf{x}_M(t)$

$\Delta \mathbf{A} = \mathbf{A}(t) - \mathbf{A}_M(t)$

$\Delta \mathbf{f}[\mathbf{x}(t), \mathbf{x}_M(t), \mathbf{u}(t)] = \mathbf{f}[\mathbf{x}(t), \mathbf{u}(t)] - f[\mathbf{x}_M(t), \mathbf{u}(t)]$

It is easy to see that $\Delta \mathbf{x}(t)$ is the state error vector: the difference between the actual system states and the states of the identifier, while $\Delta \mathbf{A}(t)$ is the error in the parameter estimation matrix: the difference between the actual parameters and the estimates, and finally $\Delta \mathbf{f}[\mathbf{x}(t), \mathbf{x}_M(t), \mathbf{u}(t)]$ is the error present in the vector field. Clearly if $\Delta \mathbf{A} = 0$, we have the solution $\mathbf{A} = \mathbf{A}_M$. It has been shown in Reference 16 that a method for evaluating $\mathbf{A}_M$ can be realised using nonlinear error mappings defined such that

$$\Delta \dot{\mathbf{A}}_M(t) = -\sum_{i=0}^{\eta} \{\Delta \dot{\mathbf{x}}(t) - \mathbf{A}_M(t) \Delta \mathbf{f}[\mathbf{x}(t), \mathbf{x}_M(t), \mathbf{u}(t)]\}^{(2i+1)/(2\beta+1)}$$

$$\times \mathbf{f}[\mathbf{x}(t), \mathbf{u}(t)]^T \mathbf{K}(t) \tag{2.84}$$

$\Delta \mathbf{A}_M(0) = \Delta \mathbf{A}_{M0}$

where $\mathbf{K}(t) \in \mathbb{R}_+ \to \mathbb{R}^{n \times n}$, $\eta, \beta \in \mathbb{N}$ are chosen to guarantee convergence properties. For the sway–yaw subsystem of (2.75), the parameterisation (2.82) takes the form:

$$\mathbf{A} = \begin{bmatrix} \mathbf{I}_{2\times 2} & \mathbf{I}_{2\times 2} & \mathbf{M}^{-1}\mathbf{N}_L & \mathbf{M}^{-1}\mathbf{N}_Q \end{bmatrix} \quad (2.85)$$

$$\mathbf{f}[\mathbf{x}(t), \mathbf{u}(t)] = \begin{bmatrix} (\mathbf{M}^{-1}\boldsymbol{\tau})^T & -(\mathbf{M}^{-1}\mathbf{C}\nu)^T & -\nu^T & -|\nu|\nu^T \end{bmatrix}^T \quad (2.86)$$

where N_L is the linear damping matrix from (2.75) and N_Q is the matrix of quadratic coefficients from the same equation, by taking $|\nu|$ out as a common factor. A post-multiplication of the matrix estimates removes the mass terms from (2.85). After selecting η, $\beta = 2$ and

$$\mathbf{K} = \begin{bmatrix} \mathbf{0}_{4\times 4} & \mathbf{0}_{4\times 8} \\ \mathbf{0}_{8\times 4} & 10^5 \times \mathrm{diag}\{0.5 \quad 1 \quad 0.5 \quad 1 \quad 10 \quad 50 \quad 10 \quad 50\} \end{bmatrix}$$

simulations were performed for the *Infante* proceeding forwards at 2 m/s or at just under 4 knots, with sinusoidal inputs on both the bow thruster and rudder, and the identification algorithm used to estimate the unknowns in the system. The results of these simulations are presented in Figure 2.4 and 2.5, and show good convergence results to the true parameters.

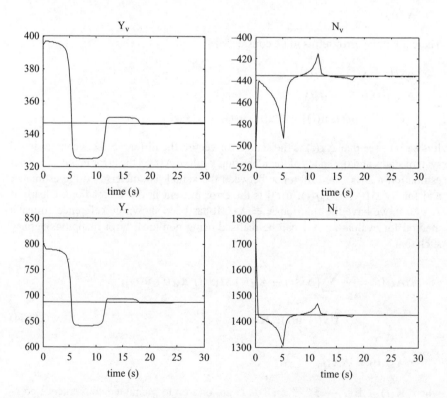

Figure 2.4 Convergence of linear damping coefficient estimates

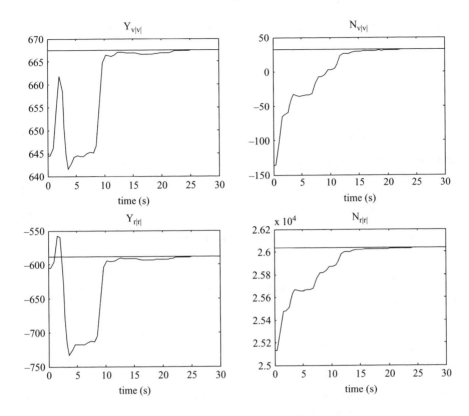

Figure 2.5 Convergence of quadratic damping coefficient estimates

2.4 Nonlinear control of UUVs

In this section it is shown how to design autopilots for

- speed, depth and pitch control
- heading control

using the longitudinal and lateral equations of motion.

When deriving the multivariable nonlinear backstepping controller the following theorem will be used.

Theorem 1 (MIMO nonlinear backstepping controller) *Consider the nonlinear plant*

$$\dot{\eta} = \mathbf{J}(\eta)v \tag{2.87}$$

$$\mathbf{M}\dot{v} + \mathbf{C}(v)v + \mathbf{D}(v)v + \mathbf{g}(\eta) = \tau \tag{2.88}$$

$$\tau = \mathbf{B}u \tag{2.89}$$

and assume that $\mathbf{BB}^T > 0$. Then the nonlinear backstepping controller

$$\tau = \mathbf{M}\dot{\nu}_r + \mathbf{C}(\nu)\nu_r + \mathbf{D}(\nu)\nu_r + \mathbf{g}(\eta) - \mathbf{J}^T(\eta)\mathbf{K}_p\tilde{\eta} - \mathbf{J}^T(\eta)\mathbf{K}_d\mathbf{s} \quad (2.90)$$

$$\mathbf{u} = \mathbf{B}^\dagger \tau \quad (2.91)$$

with $\mathbf{K}_p = \mathbf{K}_p^T > 0, \mathbf{K}_d > 0, \tilde{\eta} = \eta - \eta_d$ is the tracking error, and

$$\mathbf{s} = \dot{\eta} - \dot{\eta}_r = \dot{\tilde{\eta}} + \mathbf{\Lambda}\tilde{\eta} \quad (2.92)$$

and the reference trajectory (η_d, ν_d) satisfies $\dot{\eta}_d = \mathbf{J}(\eta)\nu_d$ such that:

$$\dot{\eta}_r = \dot{\eta}_d - \mathbf{\Lambda}\tilde{\eta}, \quad \mathbf{\Lambda} > 0 \quad (2.93)$$

$$\nu_r = \mathbf{J}^{-1}(\eta)\dot{\eta}_r, \quad \theta \neq \pm 90° \quad (2.94)$$

results in an exponentially stable equilibrium point $(\tilde{\eta}, \mathbf{s}) = (\mathbf{0}, \mathbf{0})$. Consequently, $\tilde{\eta} \to \mathbf{0}$ and $\tilde{\nu} \to \mathbf{0}$ as $t \to \infty$.

Proof The proof is found in Chapter 7 of Fossen [2], where the control Lyapunov function (CLF):

$$V_1 = \tfrac{1}{2}\tilde{\eta}^T\mathbf{K}_p\tilde{\eta}, \quad \mathbf{K}_p = \mathbf{K}_p^T > 0 \quad (2.95)$$

$$V_2 = \tfrac{1}{2}\mathbf{s}^T\mathbf{M}^*(\eta)\mathbf{s} + V_1, \quad \mathbf{M}^* = (\mathbf{M}^*)^T > 0 \quad (2.96)$$

with $\mathbf{M}^*(\eta) = \mathbf{J}^{-T}(\eta)\mathbf{M}\mathbf{J}^{-1}(\eta)$ is applied. It is straightforward to show that vectorial backstepping in 2 steps gives:

$$\dot{V}_2 = -\mathbf{s}^T(\mathbf{D}^*(\nu,\eta) + \mathbf{K}_d)\mathbf{s} - \tilde{\eta}^T\mathbf{K}_p\mathbf{\Lambda}\tilde{\eta}$$

where $\mathbf{D}^*(\nu,\eta) = \mathbf{J}^{-T}(\eta)\mathbf{D}(\nu)\mathbf{J}^{-1}(\eta)$. This proves that the equilibrium point $(\tilde{\eta}, \mathbf{s}) = (\mathbf{0}, \mathbf{0})$ is exponentially stable.

2.4.1 Speed, depth and pitch control

For UUVs, the longitudinal model (2.49) can be used to control speed, depth and pitch using the three controls τ_1, τ_3 and τ_5, or just speed and depth using τ_1 and τ_5 as control input. In this section, we will discuss both approaches using feedback linearisation and backstepping [17].

- Simultaneous speed, depth and pitch control using three controls.
- Speed and depth control using two controls (underactuated UUV).

The solution for the underactuated UUV will be based on the pure pitching equation (2.52) instead of (2.49). Speed will be controlled independently of these modes. This is based on the assumption that the UUV is stable in heave. The analysis for the underactuated UUV can be extended to include both the surge, heave and pitch dynamics by following the approach of Pattersen and Egeland [18].

2.4.1.1 Speed, depth and pitch control using three controls

Consider the longitudinal UUV dynamics (2.52) which can be written as:

$$\dot{\eta} = \mathbf{J}(\eta)v \tag{2.97}$$

$$\mathbf{M}\dot{v} + \mathbf{N}(v)v + \mathbf{g}(\eta) = \tau \tag{2.98}$$

where the Coriolis and drag terms are both contained within $\mathbf{N}(v)$ and the matrices and vectors are defined as follows:

$$\mathbf{J}(\eta) = \begin{bmatrix} c\psi c\theta & c\psi s\theta & 0 \\ -s\theta & c\theta & 0 \\ 0 & 0 & 1 \end{bmatrix} \tag{2.99}$$

$$\mathbf{M} = \begin{bmatrix} m - X_{\dot{u}} & -X_{\dot{w}} & mz_g - X_{\dot{q}} \\ -X_{\dot{w}} & m - Z_{\dot{w}} & -mx_g - Z_{\dot{q}} \\ mz_g - X_{\dot{q}} & -mx_g - Z_{\dot{q}} & I_y - M_{\dot{q}} \end{bmatrix} \tag{2.100}$$

$$\mathbf{N}(v) = \begin{bmatrix} -X_u & -X_w & -X_q \\ -Z_u & -Z_w & -(m - X_{\dot{u}})u - Z_q \\ -M_u & (Z_{\dot{w}} - X_{\dot{u}})u - M_w & mx_g u - M_q \end{bmatrix} \tag{2.101}$$

$$\mathbf{g}(\eta) = \begin{bmatrix} 0 \\ 0 \\ WBG_z \sin\theta \end{bmatrix} \tag{2.102}$$

When applying Theorem 1 we will not implement position feedback in surge since only forward speed is of interest here. Hence the controller must be implemented using velocity control in surge and position control in heave and pitch. The speed command is u_{ref} while the depth and pitch commands are denoted as d_{ref} and θ_{ref}, respectively. These commands are fed into a reference model exploiting the property $\dot{\eta}_d = \mathbf{J}(\eta)v_d$ where we delete the surge position (only velocity feedback in surge). Consequently,

$$\begin{bmatrix} \dot{d}_d \\ \dot{\theta}_d \end{bmatrix} = \begin{bmatrix} -s\theta & c\theta & 0 \\ 0 & 0 & 1 \end{bmatrix} \begin{bmatrix} u_d \\ w_d \\ q_d \end{bmatrix} \tag{2.103}$$

where the desired velocities $v_d = [u_d, w_d, q_d]^T$ are computed using the linear filters:

$$\dot{u}_d + \frac{1}{T_u}u_d = \frac{1}{T_u}u_{\text{ref}} \tag{2.104}$$

$$\ddot{w}_d + 2\zeta_w \omega_w \dot{w}_d + \omega_w^2 d_d = \omega_w^2 d_{\text{ref}} \tag{2.105}$$

$$\ddot{q}_d + 2\zeta_q \omega_q \dot{q}_d + \omega_q^2 \theta_d = \omega_q^2 \theta_{\text{ref}} \tag{2.106}$$

where $T_u, \zeta_w, \omega_w, \zeta_q, \omega_q$ are design parameters used to obtain the desired UUV response. Notice that the reference model must preserve the property $\dot{\eta}_d = \mathbf{J}(\eta)v_d$ in order to obtain exponential stability.

The controller is chosen according to Theorem 1 with $\mathbf{K}_d = \mathbf{J}^{-T}(\eta)\mathbf{K}_d^*\mathbf{J}^{-1}(\eta)$ such that:

$$\tau = \mathbf{M}\dot{v}_r + \mathbf{N}(v)v_r + \mathbf{g}(\eta) - \mathbf{J}^T(\eta)\mathbf{K}_p\tilde{\eta} - \mathbf{K}_d^*\mathbf{J}^{-1}(\eta)\mathbf{s}$$

$$\mathbf{u} = \mathbf{B}^\dagger \tau \tag{2.107}$$

If the controls are a propeller producing the thrust T, and two stern rudders with deflections δ_{S1} and δ_{S2}, respectively, the input model becomes

$$\begin{bmatrix} \tau_1 \\ \tau_3 \\ \tau_5 \end{bmatrix} = \underbrace{\begin{bmatrix} b_{11} & 0 & 0 \\ 0 & b_{22} & b_{23} \\ 0 & b_{32} & b_{33} \end{bmatrix}}_{\mathbf{B}} \begin{bmatrix} T \\ \delta_{S1} \\ \delta_{S2} \end{bmatrix} \tag{2.108}$$

which is invertible for physical values of the parameters b_{ij}, that is, $\mathbf{B}^\dagger = \mathbf{B}^{-1}$ exists. The controller gains are modified to obtain speed control in surge. Moreover

$$\mathbf{K}_p = \text{diag}\{0, K_p^{\text{heave}}, K_p^{\text{pitch}}\} \tag{2.109}$$

$$\mathbf{K}_d^* = \text{diag}\{K_d^{\text{surge}}, K_d^{\text{heave}}, K_d^{\text{pitch}}\} \tag{2.110}$$

$$\mathbf{\Lambda} = \text{diag}\{0, \lambda^{\text{heave}}, \lambda^{\text{pitch}}\} \tag{2.111}$$

such that

$$\begin{aligned} v_r &= \mathbf{J}^{-1}(\eta)\dot{\eta}_r \\ &= \mathbf{J}^{-1}(\eta)(\dot{\eta}_d - \mathbf{\Lambda}\tilde{\eta}) \\ &= v_d - \mathbf{J}^{-1}(\eta)\mathbf{\Lambda}\tilde{\eta} \end{aligned} \tag{2.112}$$

Expanding this expression gives

$$\begin{bmatrix} u_r \\ w_r \\ q_r \end{bmatrix} = \begin{bmatrix} u_d \\ w_d \\ q_d \end{bmatrix} + \begin{bmatrix} c\psi c\theta & c\psi s\theta & 0 \\ -s\theta & c\theta & 0 \\ 0 & 0 & 1 \end{bmatrix}^{-1} \begin{bmatrix} 0 \\ \lambda^{\text{heave}}(d - d_d) \\ \lambda^{\text{pitch}}(\theta - \theta_d) \end{bmatrix} \tag{2.113}$$

The feedback term $\mathbf{K}_d^*\mathbf{J}^{-1}(\eta)\mathbf{s}$ is rewritten as

$$\begin{aligned} \mathbf{K}_d^*\mathbf{J}^{-1}(\eta)\mathbf{s} &= \mathbf{K}_d^*\mathbf{J}^{-1}(\eta)(\dot{\eta} - \dot{\eta}_r) \\ &= \mathbf{K}_d^*(v - v_r) \\ &= \begin{bmatrix} K_d^{\text{surge}} & 0 & 0 \\ 0 & K_d^{\text{heave}} & 0 \\ 0 & 0 & K_d^{\text{pitch}} \end{bmatrix} \begin{bmatrix} u - u_r \\ w - w_r \\ q - q_r \end{bmatrix} \end{aligned} \tag{2.114}$$

Then it follows from Theorem 1 that $u \to u_d, d \to d_d$ and $\theta \to \theta_d$ exponentially fast.

2.4.1.2 Decoupled speed and depth control using two controls

In many cases only two controls τ_1 and τ_5 are available, that is, $\tau_3 = 0$. This problem can be solved by designing two decoupled controllers for speed and depth control at the price of undesirable interactions between the subsystems.

2.4.1.2.1 Forward speed control

The forward speed controller is usually designed using a decoupled surge model

$$(m - X_{\dot{u}})\dot{u} + R(u) = T \tag{2.115}$$

where $(m - X_{\dot{u}})$ is the mass of the vehicle included added mass, $R(u)$ is the added resistance, which can be modelled as quadratic drag

$$R(u) = -X_{|u|u}|u|u \tag{2.116}$$

for a 'flying vehicle', and T is control force. Hence, the forward speed controller can be chosen using feedback linearisation

$$T = R(u) - (m - X_{\dot{u}})\left[K_p(u - u_d) + K_i \int_0^t (u - u_d)\,d\tau\right] \tag{2.117}$$

implying that u converges to u_d = constant for positive controller gains $K_p > 0$ and $K_i > 0$. The resulting error dynamics consist of a second-order linear system:

$$\ddot{z} + K_p\dot{z} + K_i z = 0, \quad z = \int_0^t (u - u_d)\,d\tau \tag{2.118}$$

suggesting that $K_p = 2\zeta\omega_o$ and $K_i = \omega_o^2$ where ζ and ω_o can be used as tuning parameters.

2.4.1.2.2 Depth control

The depth controller will be designed under the assumption of perfect speed control, such that

$$u = u_o = \text{constant} \tag{2.119}$$

Consider (2.46) and (2.52) where the surge velocity u is treated as a known signal

$$\begin{bmatrix} \dot{d} \\ \dot{\theta} \end{bmatrix} = \begin{bmatrix} \cos\theta & 0 \\ 0 & 1 \end{bmatrix}\begin{bmatrix} w \\ q \end{bmatrix} + \begin{bmatrix} -\sin\theta \\ 0 \end{bmatrix} u \tag{2.120}$$

$$\begin{bmatrix} m - Z_{\dot{w}} & -mx_g - Z_{\dot{q}} \\ -mx_g - Z_{\dot{q}} & I_y - M_{\dot{q}} \end{bmatrix}\begin{bmatrix} \dot{w} \\ \dot{q} \end{bmatrix} + \begin{bmatrix} -Z_w & -Z_q \\ -M_w & -M_q \end{bmatrix}\begin{bmatrix} w \\ q \end{bmatrix}$$
$$+ \begin{bmatrix} 0 & -(m - X_{\dot{u}})u_o \\ (Z_{\dot{w}} - X_{\dot{u}})u_o & mx_g u_o \end{bmatrix}\begin{bmatrix} w \\ q \end{bmatrix} + \begin{bmatrix} 0 \\ BG_z W \sin\theta \end{bmatrix} = \begin{bmatrix} \tau_Z \\ \tau_M \end{bmatrix}$$

If a single stern rudder is used for depth control, the input model becomes

$$\begin{bmatrix} \tau_Z \\ \tau_M \end{bmatrix} = \begin{bmatrix} b_1 \\ b_2 \end{bmatrix}\delta_S \tag{2.121}$$

where δ_S is the stern rudder angle and b_1 and b_2 are two parameters. This problem can be solved by using the sliding mode controller of Healey and Lienard [9] or backstepping design for underactuated vehicles [19], for instance.

An even simpler approach is to replace the coupled heave and pitch equation with the pure pitching motion such that the governing model becomes

$$\dot{d} = w\cos\theta - u\sin\theta, \quad \theta \neq \pm 90° \tag{2.122}$$

$$\dot{\theta} = q \tag{2.123}$$

$$(I_y - M_{\dot{q}})\dot{q} - M_q q + BG_z W \sin\theta = b_2 \delta_S \tag{2.124}$$

This is of course based on the assumption that the heave dynamics are stable even for the inputs $\tau_3 = b_1 \delta_S$ such that we can use $\tau_5 = b_2 \delta_S$ for depth control. Backstepping will now be applied to produce the control law for δ_S.

Step 1: Let d_d be the desired depth. The overall control objective is $d = d_d$ but this is not straightforward, since backstepping with $\dot{d} = w\cos\theta - u\sin\theta$ suggests that $u\sin\theta$ should be used as virtual control in a three-step procedure. This problem can be circumvented by redefining the output to be controlled as a *sliding surface*:

$$z_1 = (d - d_d) + \lambda\theta, \quad \lambda > 0 \tag{2.125}$$

It is seen that $z_1 = 0$ only if $\theta = 0$ and $d = d_d$ since $\theta \neq 0$ changes the depth and consequently d moves away from d_d.

Consider the control Lyapunov function (CLF)

$$V_1 = \tfrac{1}{2} z_1^2 \tag{2.126}$$

Consequently,

$$\dot{V}_1 = z_1 \dot{z}_1$$
$$= z_1(w\cos\theta - u\sin\theta - w_d + \lambda q), \quad w_d = \dot{d}_d \tag{2.127}$$

Choosing λq as virtual controller in Step 1, that is

$$\lambda q := \alpha_1 + z_2 \tag{2.128}$$

gives

$$\dot{V}_1 = z_1(\alpha_1 + z_2 + w\cos\theta - u\sin\theta - w_d) \tag{2.129}$$

The stabilising function α_1 is chosen as:

$$\alpha_1 = u\sin\theta - w\cos\theta + w_d - k_1 z_1 \tag{2.130}$$

where $k_1 > 0$ is the feedback gain while the new state z_2 is left undefined until the next step. This finally gives

$$\dot{V}_1 = -k_1 z_1^2 + z_1 z_2 \tag{2.131}$$

Step 2: In Step 2, we start with the z_2 dynamics which can be obtained from (2.128). Moreover,

$$\dot{z}_2 = \lambda \dot{q} - \dot{\alpha}_1 \tag{2.132}$$

The next CLF is chosen as

$$V_2 = V_1 + \tfrac{1}{2} z_2^2 \tag{2.133}$$

such that

$$\begin{aligned}\dot{V}_2 &= \dot{V}_1 + z_2 \dot{z}_2 \\ &= -k_1 z_1^2 + z_1 z_2 + z_2(\lambda \dot{q} - \dot{\alpha}_1)\end{aligned} \tag{2.134}$$

Substituting the dynamics (2.122) into the expression for $\dot{q}$ gives

$$\dot{V}_2 = -k_1 z_1^2 + z_2 \left(z_1 + \frac{\lambda}{I_y - M_{\dot{q}}} (b_2 \delta_S - M_q q + B G_z W \sin\theta) - \dot{\alpha}_1 \right) \tag{2.135}$$

Then, the control law

$$\delta_S = \frac{1}{b_2} \left[M_q q - B G_z W \sin\theta - \frac{I_y - M_{\dot{q}}}{\lambda} (\dot{\alpha}_1 - z_1 - k_2 z_2^2) \right] \tag{2.136}$$

where $k_2 > 0$, finally gives

$$\dot{V}_2 = -k_1 z_1^2 - k_1 z_2^2 \tag{2.137}$$

Hence, the equilibrium point $(z_1, z_2) = (0, 0)$ is exponentially stable and $d \to d_d$ and $\theta \to 0$. When implementing the control law, the following approximation can be used:

$$\begin{aligned}\dot{\alpha}_1 &= \dot{u} \sin\theta + uq \cos\theta + wq \sin\theta - \dot{w} \cos\theta + \dot{w}_d - k_1((\dot{d} - \dot{d}_d) + \lambda \dot{\theta}), \\ &\stackrel{\dot{u}=\dot{w}=0}{\approx} uq \cos\theta + wq \sin\theta + \dot{w}_d - k_1(w \cos\theta - u \sin\theta - w_d + \lambda q)\end{aligned} \tag{2.138}$$

since $\dot{u}$ and $\dot{w}$ are small. Constant depth is obtained by choosing $d_d = $ constant and $\dot{w}_d = w_d = 0$.

2.4.2 Heading control

The heading control problem is usually solved using a Nomoto approximation [20] where the roll mode is left uncontrolled. From (2.59), we recognise the manoeuvring model of Davidson and Schiff [21]

$$\begin{bmatrix} m - Y_{\dot{v}} & m x_g - Y_{\dot{r}} \\ m x_g - Y_{\dot{r}} & I_z - N_{\dot{r}} \end{bmatrix} \begin{bmatrix} \dot{v} \\ \dot{r} \end{bmatrix} + \begin{bmatrix} -Y_v & -Y_r \\ -N_v & -N_r \end{bmatrix} \begin{bmatrix} v \\ r \end{bmatrix} + \begin{bmatrix} 0 & (m - X_{\dot{u}})u_o \\ (X_{\dot{u}} - Y_{\dot{v}})u_o & m x_g u_o \end{bmatrix} \begin{bmatrix} v \\ r \end{bmatrix} = \begin{bmatrix} \tau_Y \\ \tau_N \end{bmatrix} \tag{2.139}$$

Assume that the heading is controlled by using a rudder with deflection δ_R such that

$$\begin{bmatrix} \tau_Y \\ \tau_N \end{bmatrix} = \begin{bmatrix} b_1 \\ b_2 \end{bmatrix} \delta_R \tag{2.140}$$

where b_1 and b_2 are two parameters. Then, the rudder angle can be computed as

$$\delta_R = \frac{1}{b_2} \tau_N \tag{2.141}$$

where τ_N is computed by the control law. The second-order Nomoto models are obtained as

$$\frac{v}{\delta_R}(s) = \frac{K_v(1 + T_v s)}{(1 + T_1 s)(1 + T_2 s)} \tag{2.142}$$

$$\frac{r}{\delta_R}(s) = \frac{K(1 + T_3 s)}{(1 + T_1 s)(1 + T_2 s)} \tag{2.143}$$

where T_i ($i = 1, \ldots, 3$) and T_v are time constants, and K and K_v are the gain constants.

A first-order approximation to (2.142) is obtained by defining the effective time constant

$$T = T_1 + T_2 - T_3 \tag{2.144}$$

such that

$$\frac{r}{\delta_R}(s) = \frac{K}{(1 + T s)} \tag{2.145}$$

where T and K are known as the Nomoto time constant and gain constant, respectively. Including the yaw kinematics

$$\dot{\psi} = r \tag{2.146}$$

finally yields

$$\frac{\psi}{\delta_R}(s) = \frac{K(1 + T_3 s)}{s(1 + T_1 s)(1 + T_2 s)}$$

$$\approx \frac{K}{s(1 + T s)} \tag{2.147}$$

This model is the most popular model for ship autopilot design due to its simplicity and accuracy, and applies well to UUVs as well.

The gain and time constants in Nomoto's first- and second-order models can be made invariant with respect to length L and speed $U = \sqrt{u^2 + v^2 + w^2}$ by defining

$$K' = (L/U) K, \qquad T' = (U/L) T \tag{2.148}$$

This suggests that the first-order ship dynamics can be expressed as

$$\dot{\psi} = r \tag{2.149}$$

$$\dot{r} = -\frac{1}{T'}\left(\frac{U}{L}\right)r + \frac{K'}{T'}\left(\frac{U}{L}\right)^2 \delta_R \tag{2.150}$$

A robust control law for heading is

$$\delta_R = \frac{T'}{K'}\left(\frac{L}{U}\right)^2 \left(\tau_{FF} \underbrace{- K_p e - K_d \dot{e} - K_i \int_0^t e(\tau)d\tau}_{PID}\right), \quad e = \psi - \psi_d \tag{2.151}$$

$$\tau_{FF} = \frac{1}{T'}\left(\frac{U}{L}\right)\dot{r}_d + r_d \tag{2.152}$$

where τ_{FF} is reference feedforward from $r_d = \dot{\psi}_d$.

The controller gains are easily found, for instance by using pole placement. For the case $K_i = 0$ the error dynamics becomes

$$\ddot{e} + \left(\frac{1}{T'}\left(\frac{U}{L}\right) + K_d\right)\dot{e} + K_p e = 0 \tag{2.153}$$

This suggests that

$$2\omega_n \zeta = \left(\frac{U}{L}\right)\frac{1}{T'} + K_d \tag{2.154}$$

$$\omega_n^2 = K_p \tag{2.155}$$

where ω_n and ζ are pole placement design parameters. The time constant for the integrator ($K_i \neq 0$) is chosen approximately as $\omega_n/10$. This gives the following pole placement algorithm; see Reference 2

> *Pole placement algorithm for heading autopilot*
> 1. Specify the bandwidth $\omega_b > 0$ and the relative damping ratio $\zeta > 0$
> 2. Compute the natural frequency:
>
> $$\omega_n = \frac{1}{\sqrt{1-2\zeta^2+\sqrt{4\zeta^4-4\zeta^2+2}}}\omega_b$$
>
> 3. Compute the P-gain:
>
> $$K_p = \omega_n^2$$
>
> 4. Compute the D-gain:
>
> $$K_d = 2\zeta\omega_n - \left(\frac{U}{L}\right)\frac{1}{T'}$$
>
> 5. Compute the I-gain:
>
> $$K_i = \frac{\omega_s n}{10} K_p$$

2.4.3 Alternative methods of control

There are various other methods available for control of UUVs, which cannot, for reasons of space, be presented here. For comparisons of H_∞ control, μ-synthesis and sliding mode control, one may see Logan's work in Reference 22. Other comparisons of the various control techniques can be found in References 23 and 24. For additional information on sliding mode control see Reference 25, and for experimental work, Reference 26, which presents the results of using sliding mode control on the MUST UUV. The use of adaptive control can be seen, for instance, in References 27–29.

2.5 Conclusions

The six DOF equations of motion for UUVs have been presented as longitudinal and lateral subsystems for speed, diving, depth and heading control. Nonlinear optimisation has been used as a tool for determination of hydrodynamic coefficients in conjunction with hydrodynamic software packages. Nonlinear control theory has been applied to combined diving and depth control while heading control is solved using a conventional PID controller with reference feedforward. The presented solutions are intended for UUVs moving at forward speed (flying vehicles).

References

1 Fossen, T.I. (1994). *Guidance and Control of Ocean Vehicles*. John Wiley and Sons Ltd. ISBN 0-471-94113-1.

2 Fossen, T.I. (2002). *Marine Control Systems. Guidance, Navigation and Control of Ships, Rigs and Underwater Vehicles*. Marine Cybernetics AS, Trondheim, ISBN 82-92356-00-2.
3 The Society of Naval Architects and Marine Engineers (SNAME) (1950). Nomenclature for treating the motion of a submerged body through a fluid. *Technical and Research Bulletin No. 1-5*.
4 Fossen, T.I. (1991). Nonlinear modeling and control of underwater vehicles. Ph.D. thesis, Department of Engineering Cybernetics, Norwegian University of Science and Technology, Trondheim, Norway.
5 Fossen, T.I., T. Perez, Ø. Smogeli, and A.J. Sørensen (2004). Marine Systems Simulator, Norwegian University of Science and Technology, Trondheim <www.cesos.ntnu.no/mss>
6 Tinker, S.J. (1982). Identification of submarine dynamics from free-model test. In Proceedings of the DRG Seminar, The Netherlands.
7 Gertler, M. and G.R. Hagen (1967). Standard equations of motion for submarine simulation. Technical Report DTMB-2510. Naval Ship Research and Development Center, Washington, DC.
8 Feldman, J. (1979). DTMSRDC revised standard submarine equations of motion. Technical Report DTNSRDC-SPD-0393-09. Naval Ship Research and Development Center. Washington, DC.
9 Healey, A.J. and D. Lienard (1993). Multivariable sliding mode control for autonomous diving and steering of unmanned underwater vehicles. *IEEE Journal of Ocean Engineering*, JOE-18(3), pp. 327–339.
10 Pascoal, A., P. Oliveira, C. Silvestre, A. Bjerrum, A. Ishoy, J.-P. Pignon, G. Ayela, and C. Petzelt (1997). MARIUS: an autonomous underwater vehicle for coastal oceanography. *IEEE Robotics and Automation Magazine*, 4(4), pp. 46–59.
11 Salvesen, N., E.O. Tuck, and O.M. Faltinsen (1970). Ship motions and sea loads. *Transactions of SNAME*, 78, pp. 250–287.
12 Silvestre, C. and A. Pascoal (2001). The Infante AUV Dynamic model. Technical Report, Dynamical Systems and Ocean Robotics Laboratory, Institute for Systems and Robotics, Lisbon, Portugal, October.
13 Silvestre, C. and A. Pascoal (2004). Control of the INFANTE AUV using gain scheduled static output feedback. *Control Engineering Practice*, 12(12), pp. 1501–1509.
14 Ioannou, P. and J. Sun (1996). *Robust Adaptive Control*. Prentice-Hall, Inc., Englewood Cliffs, NJ. ISBN 0-13-439100-4.
15 Lyshevski, S. and L. Abel (1994). Nonlinear system identification using the Lyapunov method. In Proceedings of IFAC Symposium in System Identification, Copenhagen, pp. 307–312.
16 Lyshevski, S. (1998). State-space model identification of deterministic nonlinear systems: nonlinear mapping technology and application of the Lyapunov theory. *Automatica*, 34(5), pp. 659–664.
17 Krstic, M., I. Kanellakopoulos, and P.V. Kokotovic (1995). *Nonlinear and Adaptive Control Design*. John Wiley and Sons Ltd., New York.

18 Pettersen, K.Y. and O. Egeland (1999). Time-varying exponential stabilization of the position and attitude of an underactuated autonomous underwater vehicle. *IEEE Transactions on Automatic Control*, 44(1), pp. 112–115.
19 Fossen, T.I., M. Breivik, and R. Skjetne (2003). Line-of-sight path following of underactuated marine craft. In Proceedings of the 6th IFAC MCMC, Girona, Spain, pp. 244–249.
20 Nomoto, K., T. Taguchi, K. Honda, and S. Hirano (1957). On the steering qualities of ships. Technical Report, International Shipbuilding Progress, vol. 4.
21 Davidson, K.S.M. and L.I. Schiff (1946). Turning and course keeping qualities. *Transactions of the SNAME*, pp. 152–200.
22 Logan, C.L. (1994). A comparison between H-infinity/mu-synthesis control and sliding mode control for robust control of a small autonomous underwater vehicle. In Proceedings of the IEEE Symposium on Autonomous Underwater Vehicle Technology, USA.
23 Campa, G. and M. Innocenti (1998). Robust control of underwater vehicles: sliding mode control vs. mu synthesis. In Proceedings of OCEANS, France.
24 Innocenti, M. and G. Campa (1999). Robust control of underwater vehicles: sliding mode vs. LMI synthesis. In Proceedings of the American Control Conference, California.
25 Rodrigues, L., P. Tavares, and M.G. de Sousa Prado (1996). Sliding mode control of an AUV in the diving and steering planes. In Proceedings of OCEANS, Fort Lauderdale, USA.
26 Dougherty, F. and G. Woolweaver (1990). At-sea testing of an unmanned underwater vehicle flight control system. In Proceedings of the Symposium on Autonomous Underwater Vehicle Technology, Washington, DC.
27 Goheen, K.R. and E.R. Jefferys (1990). Multivariable self-tuning autopilots for autonomous and remotely operated underwater vehicles. *IEEE Journal of Oceanic Engineering*, 15(3), pp. 144–151.
28 Antonelli, G., F. Caccavalle, S. Chiaverini, and G. Fusco (2001). A novel adaptive control law for autonomous underwater vehicles. In Proceedings of the IEEE International Conference on Robotics and Automation, Seoul, Korea.
29 Mrad, F.T. and A.S. Majdalani (2003). Composite adaptive control of astable UUVs. *IEEE Journal of Oceanic Engineering*, 28(2), pp. 303–307.

Chapter 3

Guidance laws, obstacle avoidance and artificial potential functions

A.J. Healey

3.1 Introduction

In the context of autonomy for underwater vehicles, we assume that a usual suite of feedback controllers are present in the form of autopilot functions that provide for the regulation of vehicle speed, heading and depth or altitude. In this chapter, we consider the topic of guidance laws, obstacle avoidance and the use of artificial potential functions (APFs). This topic deals with the computations required to plan and develop paths and commands, which are used by these autopilots. Simple guidance laws such as 'proportional guidance' have been used for many years in missiles to provide interception with targets. Lateral accelerations are commanded proportional to the rate of change of line of sight. So long as the chaser vehicle has a speed advantage over the non-manoeuvring target, simply reducing the angle of line of sight (LOS) to zero will result in an interception.

For applications with unmanned underwater vehicles, guidance laws allow vehicles to follow paths constructed in conjunction with mission objectives. For instance, in a mine-hunting mission, we often design paths that 'mow the lawn'. An objective area is defined; tracks are developed, with track spacing defined according to the swath width of the side-scanning sonar and a required overlap, so that complete coverage of the area is obtained. Increasing overlap leads to increased probability of detection at the expense of overall search rate. These tracks are defined by starting and ending points (way-points) and the vehicle is guided along these tracks by a track following guidance law that drives the cross-track position error to zero. In this mode, the vehicle closely follows the defined path in position only, but no attempt is made to follow a specific trajectory in both time and position. Trajectory tracking [1], which may be necessary if formation control is required, implies that not only cross-track but also the corresponding along track errors are controlled. Clearly, this

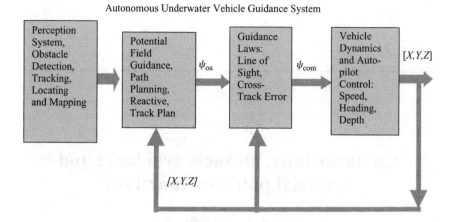

Figure 3.1   Overview of guidance functions

added requirement means that some position along the track determined by some moving point on the trajectory is to be maintained. In essence, trajectory following is more complex and requires speed as well as heading control. Another related topic involves the design of optimal paths to follow, for example, in Reference 2. In this sense, path length is minimised subject to curvature and other constraints rather than allowing the vehicle to respond to way-points, with no guaranteed coverage in tight spaces between closely positioned points. In open ocean conditions it is rare that such precision in an actual path is important, but it could be important in obstacle avoidance manoeuvring where minimum distances between vehicles and obstacles should be observed.

By reference to Figure 3.1, the autopilots for speed, heading and depth control are assumed to be already present in the vehicle dynamics. The notion of a guidance system is based around the idea that the heading command is taken from the guidance system. It allows for track following, cross-track error control and for obstacle avoidance, and as such requires knowledge of vehicle position from a navigation system, and knowledge of the mission so that track plans can be made and modified. In the discussion of obstacle avoidance guidance, we include the methods of artificial potential fields, curved path deviation planning and reactive avoidance. The system is driven from the perceptory inputs modified by sonar signal processing, and algorithms for obstacle detection, obstacle tracking, location and mapping.

3.2 Vehicle guidance, track following

To follow a set of straight line tracks, forms the basis of many simple guidance requirements. In this section, a simple LOS guidance law is described. It is supplemented with a simple cross-track error term for better performance in cross-currents, and a sliding mode controller is presented that has been experimentally validated under a

Guidance laws, obstacle avoidance and artificial potential functions

wide variety of conditions. Other works have studied similar problems for land robots (e.g., smooth path planning in Reference 3) and usually develop a stable guidance law based on cross-track error, which returns the vehicle to the desired path.

In all simulations shown later in this chapter, the same vehicle dynamics model is used, taken from the *REMUS* autonomous underwater vehicle (AUV) as explained in earlier work [4] and the steering autopilot is designed according to a sliding mode methodology [5]. The hydrodynamic coefficients are given in Reference 4.

3.2.1 Vehicle steering model

Using the coefficients given below, and with standard nomenclature for underwater vehicles [6],

$$\begin{bmatrix} m - Y_{\dot{v}_r} & -Y_{\dot{r}} & 0 \\ -N_{\dot{v}_r} & I_{zz} - N_{\dot{r}} & 0 \\ 0 & 0 & 1 \end{bmatrix} \begin{bmatrix} \dot{v}_r \\ \dot{r} \\ \dot{\psi} \end{bmatrix} = \begin{bmatrix} Y_{v_r} & Y_r - mU_0 & 0 \\ N_{v_r} & N_r & 0 \\ 0 & 1 & 0 \end{bmatrix} \begin{bmatrix} v_r \\ r \\ \psi \end{bmatrix}$$

$$+ \begin{bmatrix} Y_\delta \\ N_\delta \\ 0 \end{bmatrix} \delta_r(t) \begin{bmatrix} \dot{v}_r \\ \dot{r} \\ \dot{\psi} \end{bmatrix} = \begin{bmatrix} m - Y_{\dot{v}_r} & -Y_{\dot{r}} & 0 \\ -N_{\dot{v}_r} & I_{zz} - N_{\dot{r}} & 0 \\ 0 & 0 & 1 \end{bmatrix}^{-1} \begin{bmatrix} Y_{v_r} & Y_r - mU_0 & 0 \\ N_{v_r} & N_r & 0 \\ 0 & 1 & 0 \end{bmatrix}$$

$$\times \begin{bmatrix} v_r \\ r \\ \psi \end{bmatrix} + \begin{bmatrix} m - Y_{\dot{v}_r} & -Y_{\dot{r}} & 0 \\ -N_{\dot{v}_r} & I_{zz} - N_{\dot{r}} & 0 \\ 0 & 0 & 1 \end{bmatrix}^{-1} \begin{bmatrix} Y_\delta \\ N_\delta \\ 0 \end{bmatrix} \delta_r(t) \quad (3.1)$$

Table of data for the REMUS model use in simulations

$Y_{\dot{v}_r}$	-3.55 exp 01 kg
$Y_{\dot{r}}$	1.93 kg m/rad
Y_{v_r}	-6.66 exp 01 kg/s (same as Z_w)
Y_r	2.2 kg m/s (same as Z_q)
$N_{\dot{v}_r}$	1.93 kg m
$N_{\dot{r}}$	-4.88 kg m^2/rad
N_{v_r}	-4.47 kg m/s
N_r	-6.87 kg m^2/s (same as M_q)
N_δ	-3.46 exp 01/3.5 kg m/s^2
Y_δ	5.06 exp 01/3.5 kg m/s^2

$m = 30.48$ kg; $I_{zz} = 3.45$ kg m^2/rad.
Table of hydrodynamic and inertial parameters for *REMUS* steering dynamics with the modifications to the values taken initially from Reference 7.

and in more compact form

$$\dot{x}(t) = Ax(t) + bu(t) \quad (3.2)$$

with the path response being taken from

$$\dot{X} = U_0 \cos \psi - v_r \sin \psi + U_{cx}$$
$$\dot{Y} = U_0 \sin \psi + v_r \cos \psi + U_{cy} \quad (3.3)$$

and

$$A = \begin{bmatrix} m - Y_{\dot{v}_r} & -Y_{\dot{r}} & 0 \\ -N_{\dot{v}_r} & I_{zz} - N_{\dot{r}} & 0 \\ 0 & 0 & 1 \end{bmatrix}^{-1} \begin{bmatrix} Y_{v_r} & Y_r - mU_0 & 0 \\ N_{v_r} & N_r & 0 \\ 0 & 0 & 1 \end{bmatrix};$$

$$B = \begin{bmatrix} m - Y_{\dot{v}_r} & -Y_{\dot{r}} & 0 \\ -N_{\dot{v}_r} & I_{zz} - N_{\dot{r}} & 0 \\ 0 & 0 & 1 \end{bmatrix}^{-1} \begin{bmatrix} Y_\delta \\ N_\delta \\ 0 \end{bmatrix}$$

where

$$x(t) = [v_r \quad r \quad \psi]'; \quad u(t) = \delta_r(t) \quad (3.4)$$

In the above, U_{cx} and U_{cy} are northerly and easterly water currents, respectively. With this in mind, any steering control law may be used, but for the sake of further discussion let us use the sliding mode control,

$$\delta_r(t) = K_1 v(t) + K_2 r(t) + \eta \tanh(\sigma(t)/\phi)$$
$$\sigma(t) = s_1(\psi_{\text{com}(LOS(i))} - \psi(t)) + s_2(r_{\text{com}}(t) - r(t)) - s_3 v(t) \quad (3.5)$$

where, for the *REMUS* vehicle model, the constants, based on pole placement are taken to be:

$$s_1 = 0.8647; \quad s_2 = 0.5004; \quad s_3 = 0.0000; \quad \eta = 0.5000;$$
$$\phi = 0.1000; \quad K_1 = 0.0000; \quad K_2 = 0.6000 \quad (3.6)$$

The coefficients of the side slip velocity have been zeroed since it is not practical to include that particular sensor in the control because of noise levels, and it has been found that the stability of the heading controller is not impaired.

3.2.2 Line of sight guidance

The basic guidance law given in terms of the line of sight law for use with the autopilot in Eq. (3.6) is

$$\psi(t)_{\text{com}(LOS(i))} = \tan^{-1}(\tilde{Y}(t)_{\text{wpt}(i)} / \tilde{X}(t)_{\text{wpt}(i)})$$
$$\tilde{Y}(t)_{\text{wpt}(i)} = (Y_{\text{wpt}(i)} - Y(t))$$
$$\tilde{X}(t)_{\text{wpt}(i)} = (X_{\text{wpt}(i)} - X(t)) \quad (3.7)$$

The position definitions and nomenclature are as given by reference to Figure 3.2.

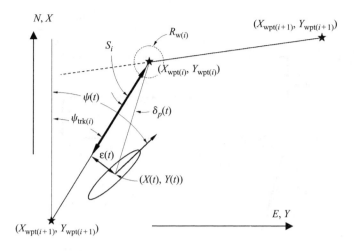

Figure 3.2 Cross-track position error $\varepsilon(t)$ definitions

A typical response for the vehicle track following in the presence of northerly and easterly currents of one-half a knot is illustrated in Figure 3.3. In this figure we see that the vehicle heads along the track but is seriously in error with the presence of the cross-currents.

3.2.3 Cross-track error

The variable of interest to minimise is the cross-track error, $\varepsilon(t)$, and is defined as the perpendicular distance between the centre of the vehicle (located at $(X(t), Y(t))$) and the adjacent track line. The total track length between way-points i and $i-1$ is given by

$$L_i = \sqrt{(X_{\text{wpt}(i)} - X_{\text{wpt}(i-1)})^2 + (Y_{\text{wpt}(i)} - Y_{\text{wpt}(i-1)})^2} \tag{3.8}$$

where the ordered pairs $(X_{\text{wpt}(i)}, Y_{\text{wpt}(i)})$ and $(Y_{\text{wpt}(i-1)}, X_{\text{wpt}(i-1)})$ are the current and previous way-points, respectively. The track angle, $\psi_{\text{trk}(i)}$, is defined by

$$\psi_{\text{trk}(i)} = \tan 2^{-1}(Y_{\text{wpt}(i)} - Y_{\text{wpt}(i-1)}, X_{\text{wpt}(i)} - X_{\text{wpt}(i-1)}). \tag{3.9}$$

The cross-track heading error $\tilde{\psi}(t)_{\text{CTE}(i)}$ for the ith track segment is defined as

$$\tilde{\psi}(t)_{\text{CTE}(i)} = \psi(t) - \psi_{\text{trk}(i)} + \beta(t) \tag{3.10}$$

where $\tilde{\psi}(t)_{\text{CTE}(i)}$ must be normalised to lie between $+180°$ and $-180°$. $\beta(t)$ is the angle of side slip and is defined here as

$$\tan(\beta(t)) = v(t)/U$$

When $\tilde{\psi}(t)_{\text{CTE}(i)}$ is nulled, it may be seen that the instantaneous velocity vector then heads along the track with $\psi(t) = \psi_{\text{trk}(i)} - \beta(t)$.

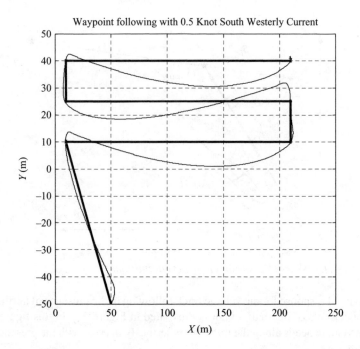

Figure 3.3 Line of sight guidance only in the presence of 0.5 knot cross-currents (design tracks in bold)

The difference between the current vehicle position and the next way-point is

$$\tilde{X}(t)_{\text{wpt}(i)} = X_{\text{wpt}(i)} - X(t)$$
$$\tilde{Y}(t)_{\text{wpt}(i)} = Y_{\text{wpt}(i)} - Y(t) \quad (3.11)$$

With the above definitions, the distance to the ith way-point projected to the track line $S(t)_i$, can be calculated as a percentage of track length using

$$S(t)_i = (\tilde{X}(t)_{\text{wpt}(i)}(X_{\text{wpt}(i)} - X_{\text{wpt}(i-1)}) + \tilde{Y}(t)_{\text{wpt}(i)}(Y_{\text{wpt}(i)} - Y_{\text{wpt}(i-1)}))/L_i \quad (3.12)$$

The cross-track error $\varepsilon(t)$ may now be defined in terms of the current vehicle position, the previous and next way-points as

$$\varepsilon(t) = S_i(t)\tan(\delta_p(t)) \quad (3.13)$$

where $\delta_p(t)$ is the angle between the line of sight to the next way-point and the current track line given by

$$\delta_p(t) = \tan 2^{-1}(Y_{\text{wpt}(i)} - Y_{\text{wpt}(i-1)}, X_{\text{wpt}(i)} - X_{\text{wpt}(i-1)})$$
$$- \tan 2^{-1}(\tilde{Y}(t)_{\text{wpt}(i)} - \tilde{X}(t)_{\text{wpt}(i)}) \quad (3.14)$$

and must be normalised to lie between $+180°$ and $-180°$.

3.2.4 Line of sight with cross-track error controller

Cross-track error is controlled by the addition of a term in the guidance law proportional to cross-track error, which provides additional heading to drive to the track path in spite of cross-currents. This topic is discussed in detail in Reference 8.

$$\psi(t)_{\text{com(LOS}(i))} = \tan^{-1}(\tilde{Y}(t)_{\text{wpt}(i)}/\tilde{X}(t)_{\text{wpt}(i)}) - \tan^{-1}(\varepsilon(t)/\rho) \quad (3.15)$$

In the cross-track error guidance law above, ρ is a parameter usually chosen depending on the vehicle turning radius and is commonly between 4 and 5 vehicle lengths. Too small a value leads to unstable steering response. It should be pointed out that in dealing with real-time control applications with unmanned underwater vehicles, the issues surrounding the wrapping of the heading are troublesome. For control and guidance work, we generally bound headings to the sector

$$-180 < \psi < 180$$

although for filtering work the heading state has to be continuous and wrapped. Figure 3.4 shows an improved tracking result from this guidance law, even in the presence of cross-currents. Steady state errors persist but can be eliminated by an integral of error term if so desired. Notice in Figure 3.5 that there is an over- and undershoot behaviour with integral control even though its magnitude is limited using anti-reset windup techniques.

One of the problems with this approach is that stability is not guaranteed as part of the design process and depends on the selection of the look-ahead distance ρ.

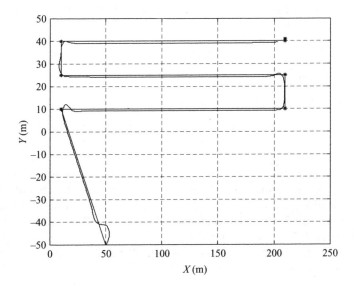

Figure 3.4   Line of sight with simple cross-track error term

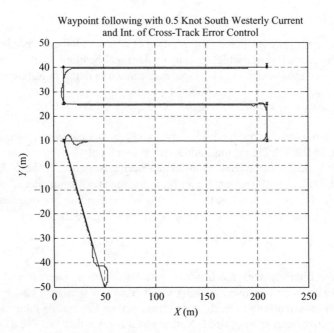

Figure 3.5   Line of sight with cross-track error term and integral of error in the presence of cross-currents

3.2.5   Sliding mode cross-track error guidance

With the cross-track error defined, a sliding surface can be cast in terms of derivatives of the errors such that

$$\varepsilon(t) = \varepsilon(t)$$
$$\dot{\varepsilon}(t) = U \sin(\tilde{\psi}(t)_{\text{CTE}(i)})$$
$$\ddot{\varepsilon}(t) = U(r(t) + \dot{\beta}(t)) \cos(\tilde{\psi}(t)_{\text{CTE}(i)})$$
$$\dddot{\varepsilon}(t) = U(\dot{r}(t) + \ddot{\beta}(t)) \cos(\tilde{\psi}(t)_{\text{CTE}(i)}) - U(r(t) + \dot{\beta}(t))^2 \sin(\tilde{\psi}(t)_{\text{CTE}(i)})$$
(3.16)

The question now arises as to how to include the effects of the side slip and its rate of change. We generally make the assumption that the time rate of change of side slip is small compared to the turn rate, $r(t)$, although this depends somewhat on the particular vehicle considered. For the NPS *ARIES* vehicle, it is typically less than 0.1 rad/s and may be neglected. Thus, we assume that $\dot{\beta}(t)$ and $\ddot{\beta}(t)$ in Eqs. (3.16) are negligible.

A sliding surface for the cross track error controller is selected to be a second-order polynomial of the form

$$\sigma(t) = \ddot{\varepsilon}(t) + \lambda_1 \dot{\varepsilon}(t) + \lambda_2 \varepsilon(t) \tag{3.17}$$

The reaching condition for reduction of error is

$$\dot{\sigma}(t) = \dddot{\varepsilon}(t) + \lambda_1 \ddot{\varepsilon}(t) + \lambda_2 \dot{\varepsilon}(t) = -\eta \tanh(\sigma/\phi) \tag{3.18}$$

and to recover the input for control, the heading dynamics Eq. (3.1) are simplified to neglect side slip dynamics, reducing to

$$\dot{r}(t) = ar(t) + b\delta_r(t); \quad \dot{\psi}(t) = r(t)$$

and substituting into (3.18) using (3.16) we get an expression for $\dot{\sigma}(t)$:

$$U(ar(t) + b\delta_r)\cos(\tilde{\psi}(t)_{CTE(i)}) - Ur(t)^2 \sin(\tilde{\psi}(t)_{CTE(i)}) \\ + \lambda_1 Ur(t) \cos(\tilde{\psi}(t)_{CTE(i)}) + \lambda_2 U \sin(\tilde{\psi}(t)_{CTE(i)}) \tag{3.19}$$

Rewriting Eq. (3.17), the sliding surface becomes

$$\sigma(t) = Ur(t) \cos(\tilde{\psi}(t)_{CTE(i)}) + \lambda_1 U \sin(\tilde{\psi}(t)_{CTE(i)}) + \lambda_2 \varepsilon(t) \tag{3.20}$$

The rudder input can then be expressed using Eqs. (3.19) and (3.18),

$$\delta_r(t) = \left(\frac{1}{Ub \cos(\tilde{\psi}(t)_{CTE(i)})} \right) \\ \times \begin{Bmatrix} -Uar(t) \cos(\tilde{\psi}(t)_{CTE(i)}) \\ +U(r(t))^2 \sin(\tilde{\psi}(t)_{CTE(i)}) - \lambda_1 Ur(t) \cos(\tilde{\psi}(t)_{CTE(i)}) \\ -\lambda_2 U \sin(\tilde{\psi}(t)_{CTE(i)}) - \eta \tanh(\sigma(t)/\phi) \end{Bmatrix} \tag{3.21}$$

where $\lambda_1 = 0.6$, $\lambda_2 = 0.1$, $\eta = 0.1$ and $\phi = 0.5$. To avoid division by zero, in the rare case where $\cos(\tilde{\psi}(t)_{CTE}) = 0.0$ (i.e., the vehicle heading is perpendicular to the track line) the rudder command is set to zero since this condition is transient in nature.

3.2.6 Large heading error mode

Heading errors larger than say 40° often occur in operations, especially when transitioning between tracks. It is not uncommon for a 90° heading change to be instituted suddenly. Under these conditions, the sliding mode guidance law breaks down, and a return to the basic LOS control law is required. In this situation, the heading command can be determined from

$$\psi(t)_{com(LOS(i))} = \tan 2^{-1}(\tilde{Y}(t)_{wpt(i)}, \tilde{X}(t)_{wpt(i)}) \tag{3.22}$$

and the LOS error from

$$\tilde{\psi}(t)_{LOS(i)} = \psi(t)_{com(LOS(i))} - \psi(t) \tag{3.23}$$

and the control laws used for heading control, Eqs. (3.5) and (3.6) may be used.

3.2.7 Track path transitions

Based on operational experience, we find it best to seek a combination of conditions in order for the way-point index to be incremented. The first and most usual case is if the vehicle has penetrated the way-point watch radius $R_{w(i)}$. This is likely if the vehicle is tracking well on its present track with small cross-track error. Second, most usually at start up, if a large amount of cross-track error is present, the next way-point will become active if the projected distance to the way-point $S(t)_i$ reached some minimum value $S_{\min(i)}$, or in simple cross-track error guidance, that the projected distance $S(t)_i$ becomes less than the cross-track error control distance, ρ, This is encoded with the transition conditional,

$$\text{if}\left(\sqrt{(\tilde{X}(t)_{\text{wpt}(i)})^2 + (\tilde{Y}(t)_{\text{wpt}(i)})^2} \leq R_{w(i)} \| S(t)_i < S_{\min(i)} \| S(t)_i < \rho \right)$$

then activate next way-point (3.24)

3.3 Obstacle avoidance

Three forms of obstacle avoidance guidance are considered. First, when an obstacle is detected using the forward looking sonar perception system, some metrology analysis will give size information and if it is determined to be an object to be avoided by steering, a planned path deviation can be used to override the current path requiring the vehicle to follow a deviated path [11,14]. Criteria are then checked to see if it is safe to return to the original path [13,15–17]. Avoidance paths can be evaluated using many possibilities but circular paths are convenient. Planning and replanning takes place continuously until no further need to avoid is reached when the original path is regained. In the deviation plan, an added heading command is prescribed for the vehicle to follow a circular path until a sufficient deviation from the obstacle is reached at which time the avoidance plan is ended, returning the vehicle to the original path-following behaviour.

3.3.1 Planned avoidance deviation in path

In this section, a proposed planned path is determined from a circular path that deviates the heading 50° off course over the planning horizon of 20 m. This is followed by a linear segment until a predetermined cross-track deviation is made. This planned approach has the advantage that a predetermined deviation from path is made before the vehicle is released to return to the original path. The amount of deviation from the original path is determined from the characteristics as well as the position of the obstacle as seen by the forward-looking sonar. The generation of the planned deviated path is triggered by the sonar detection of an obstacle coming within the planning horizon (in this case, 20 m) and at a bearing within a planning sector of 45°. The offset of the object from the path is sensed and used in the calculation of the avoidance distance needed (Figure 3.6).

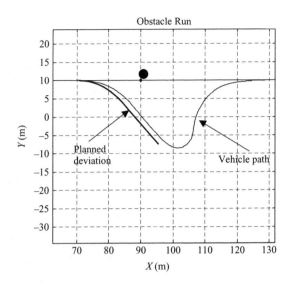

Figure 3.6   Response of vehicle to a planned deviation in track

A predetermined deviation is described using the following pseudocode

```
if (Range(i,1)<20 & abs(Bearing(i,1)) <  pi/4 &
    inplan==0),                inplan=1;end;

if (inplan ==1),
     count=count+1;
     offset=Range(i,1)*sin(Bearing(i,1));

% replace heading command with path planning command with
Radius R=20m, until 50 degrees deviation is met followed by a
straight line until offset criterion is met;

psi_errorLOS(i)= psi_comLOS(i) - psi_cont(i) + (count)*dt*U/R;
Xdev(i)=R*sin(count*dt*U/20);
Ydev(i)=R*(1-cos(count*dt*U/20));
end;

% limit the heading deviation

if (((count)*dt*U/R) > 50*pi/180) ,
psi_errorLOS(i)= psi_comLOS(i) - psi_cont(i) + 50*pi/180;
Xdev(i)=Xdev(i-1)+dt*U*sin(50*pi/180);
Ydev(i)=Ydev(i-1)++dt*U*sin(50*pi/180);
end;

if (abs(cte(i))>15),
inplan=0;count=0;r_com(i)=0;end;
% back to default plan
```

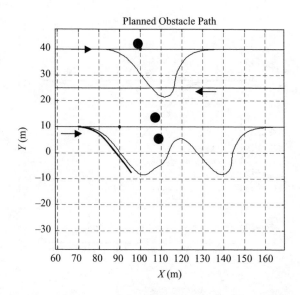

Figure 3.7 Multiple obstacles using the planned avoidance deviation method

Many different varieties of planned paths are possible. The above is only an example. The point is that using different planning methods, the paths selected are guaranteed to be free of obstacles. Planning for each segment of travel, and considering whether or not to deviate the path can be done continually in sequence every 20 m or so. If subsequent obstacles are found, deviational paths can be instituted as needed. Figure 3.7 shows the response of the vehicle when a second obstacle is detected while returning to the original track from an initial deviation.

3.3.2 Reactive avoidance

Reactive avoidance is based on the processing of the forward-look sonar image and detection of obstacles from which an avoidance or threat level can be assessed. Based on the threat level the strength of the avoidance action is determined to be either strong or weak. The response is immediate and hence is called reactive.

The model uses a two-dimensional forward-looking sonar with a 120° horizontal scan and a 110-m radial range. The probability of detection is based on a cookie-cutter approach in which the probability of detection is unity within the scan area and zero anywhere else. Bearing is resolved to the nearest degree and range is resolved every 20 cm.

The obstacle avoidance model developed in this work is based on the product of bearing and range weighting functions that form the gain factor for a dynamic obstacle avoidance behaviour. The basis for the weighting functions lies in a fuzzy logic methodology. The weighting functions are MATLAB membership functions from the fuzzy logic toolbox with the parameters selected to maximize obstacle avoidance behaviour. The membership function for bearing is a Gaussian curve function of

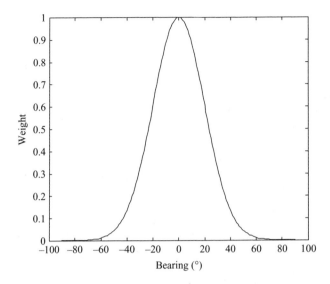

Figure 3.8   Bearing weighting function

the form:

$$w_1 = e^{(-(x-c)^2)/(2\sigma^2)} \tag{3.25}$$

where the parameters x, c and σ are position (or angular position in degrees for the purpose of this model), centre and shape, respectively. Shape defines the steepness of the Gaussian curve. The values selected for these parameters to provide sufficient tuning in this membership function were $-90{:}90$, 0, 20, respectively. The bearing weighting function can be seen in Figure 3.8.

The membership function for range is an asymmetrical polynomial spline-based curve called zmf and is of the form:

$$w_2 = \text{zmf}(x, [a\ b]) \tag{3.26}$$

where a and b are parameters that locate the extremes of the sloped portions of the curve. These parameters are called breakpoints and define where the curve changes concavity. In order to maximise obstacle avoidance behaviour, these values were tuned to be (*sonrange-99*) and (*sonrange-90*). With this selection, the range weight is approximately unity for anything closer than 20 m and zero for anything farther than 40 m from *REMUS* as seen in Figure 3.9.

A final weight based on both bearing and range is calculated from the product of w_1 and w_2. This weight becomes the gain coefficient that is applied to a maximum avoidance heading for each individual object. The maximum heading is $\pi/4$ as seen below

$$\psi_{oa}(t, c) = w_1 w_2 (\pi/4) \tag{3.27}$$

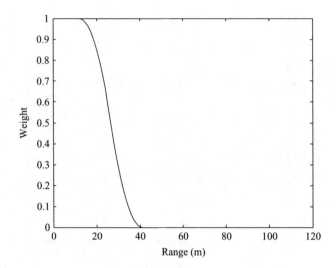

Figure 3.9 Range weighting function

where t is the time step and c is the obstacle being evaluated. The avoidance heading for all obstacles over a single time step (or one look) is

$$\psi_{\text{oalook}}(t) = \sum_{1}^{c} \psi_{\text{oa}}(t, c) \tag{3.28}$$

Following an evaluation of each obstacle at every time step, a final obstacle avoidance heading term is determined from the sum of the obstacle avoidance headings of each individual object within a specified bearing and range from the vehicle or

$$\psi_{\text{oatot}}(t) = \frac{\psi_{\text{oalook}}(t)}{cc} \tag{3.29}$$

where cc is the counter used to determine how many obstacles fall into this window. The counter is used to normalise this overall obstacle avoidance term to an average for all of the obstacles within the range above. This bearing and range of the window is determined through a rough evaluation of the weighting functions. In order to fall into the window, the gain factor must be equal to or exceed a value of $w_1 w_2 = 0.15$.

The obstacle avoidance term $\psi_{\text{oatot}}(t)$ is then incorporated into vehicle heading error as:

$$\tilde{\psi}_{\text{LOS}}(t) = \psi_{\text{track}}(t) - \psi_{\text{cont}}(t) - \arctan(\varepsilon(t)/\rho) + \psi_{\text{oatot}}(t) \tag{3.30}$$

This heading error drives the rudder commands to manoeuvre around detected objects in the track path.

The initial test performed on the two-dimensional sonar model was navigation around a single point obstacle. This is the simplest obstacle avoidance test for the two-dimensional model. Two variations of this test were run for the basic single point obstacle avoidance: a single point on the path and a single point off the path.

Figure 3.10(b) shows the rudder dynamics, vehicle heading and obstacle avoidance heading term for the duration of each vehicle run. The rudder action has a direct correlation with the obstacle avoidance heading and overall vehicle heading. The

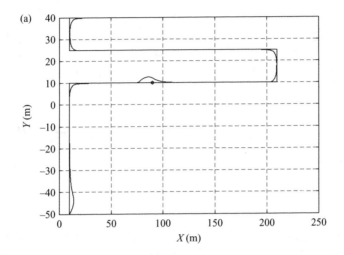

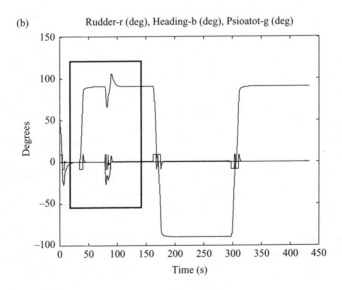

Figure 3.10 (a) Single point obstacle run centerline. (b) Single point obstacle run rudder/heading/ψ_{oa}. (c) Zoom-in of marked area in (b) showing rudder, turn rate and heading response details

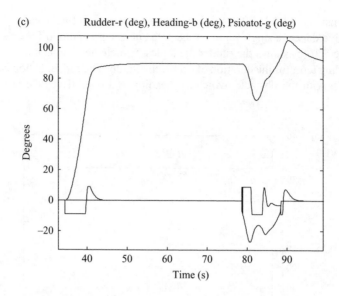

Figure 3.10   Continued

large angle motions of the heading are the 90° turns made to track the ordered vehicle path. There is an associated rudder action with each of these turns as seen by the corresponding rudder curve. These rudder curves show that the maximum programmable rudder deflection is 9°. For all dynamic behaviours, whether associated with a turn or obstacle avoidance manoeuvre, the rudder initiates the turn with this maximum value. In order to regain track, the rudder action may vary.

A single point obstacle avoidance model (Figure 3.11) is far simpler than a multiple point obstacle avoidance model not only in the manoeuvring of the vehicle, but also in maintaining the obstacle picture. For multiple point obstacle avoidance, it is necessary to have a model that reacts to obstacles in a certain proximity to its path rather than all possible obstacles seen by the sonar scan. Weighting functions allow for an accurate compilation of this obstacle picture. The *REMUS* model builds an obstacle counter for obstacles having a weighting function gain factor ($w_1 w_2$) greater than 0.15. This value allows for a maximum rudder and bearing weight of approximately 0.386, that is, the square root of 0.15. Referring to the membership functions in Figures 3.8 and 3.9, a value of 0.386 correlates to a bearing and range of approximately ±30° and 30 m, respectively.

As seen in Figures 3.12 and 3.13, *REMUS* successfully avoids multiple points and multiple point clusters in the same fashion it avoided a single point. The rudder dynamics are minimal during all avoidance manoeuvres for an efficient response.

As seen in Figure 3.14 , an obstacle appearing in the vehicle path causes the vehicle heading to deviate from its track path heading of 90°, approximately the same amount as the obstacle avoidance heading.

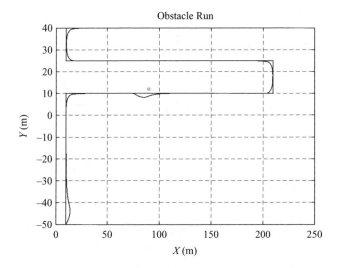

Figure 3.11   Single point obstacle runoff path

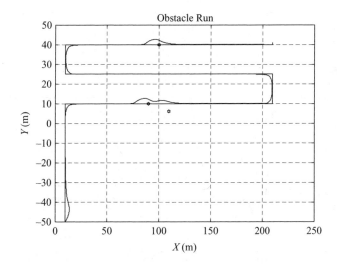

Figure 3.12   Multiple point obstacle run

3.4   Artificial potential functions

The use of artificial potential functions in vehicle guidance is useful in both directing and redirecting vehicle paths as well as in the avoidance of mapped obstacles [10,12]. Essentially, the local area in which the vehicle is being guided is mapped into a potential field such that heading commands are derived from the shaping of these potentials. In this manner, obstacles to be avoided are associated with high potential,

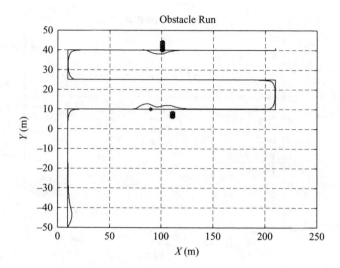

Figure 3.13   Multiple cluster obstacle

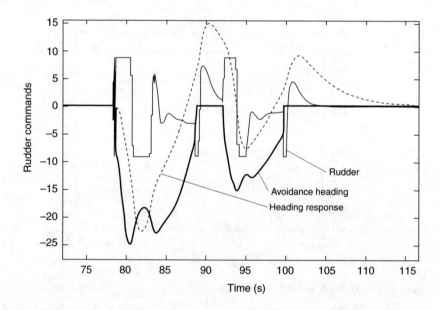

Figure 3.14   Vehicle heading comparison avoidance command and heading response

free spaces are represented by the valleys and the vehicle guidance seeks minima in the potential field. The motion of the vehicle is taken to be the motion of a particle moving along potential lines. In this work, we have considered the use of Gaussian potential functions to provide smooth path following where the paths derived are continuous with continuous derivatives and curvature. This is different from others such as Khatib [9], for instance, who used inverse radius potential functions with singularities at the obstacle locations. In all potential function based guidance laws, two potentials must be employed, one to follow the desired path or track and one to avoid any obstacles in the way. While the concepts are quite useful, in some cases instabilities can arise where potential functions combine with vehicle response lags to produce overall instability of path following. It should be intuitive since the potential and hence the guidance heading commands are derived from position based functions. Figure 3.1 shows the clear role of vehicle position in determining the heading command leading to unstable performance in the guidance loop if commands for reasonable turn rates are not considered carefully. That is, instabilities are possible and smooth paths with curvature limits must be generated. The guidance laws are derived by the following considerations.

3.4.1 Potential function for obstacle avoidance

Let us define a potential function within the local region Ω in which V has maxima located at the points of obstacles to be avoided. It assumes that a perception system has detected obstacles and located them so that the locations are known.

$$V(X, Y) > 0 \quad \forall (X, Y) \in \Omega \tag{3.31}$$

If there are obstacles within the region at x_i, y_i, then high potentials are assigned to those points, for example,

$$V(t) = \sum_i^N V_i \exp[((X(t) - x_i)^2 + (Y(t) - y_i)^2)/2\sigma_i^2] \tag{3.32}$$

A guidance law to minimise the potential function is obtained from seeking V to be always reducing along trajectories in (X, Y). We make

$$\frac{dV}{dt} = V_x' \dot{X} + V_y' \dot{Y} < 0 \quad \forall\, X, Y \in \Omega \tag{3.33}$$

leading to

$$\dot{X} = -\gamma V_x'; \quad \dot{Y} = -\gamma V_y'; \quad \gamma > 0 \tag{3.34}$$

and assuming that the vehicle path is commanded to follow tangent paths to the potential lines, a heading command for obstacle avoidance using the artificial potential field can be extracted using

$$\psi_{\text{oaapf}} = \tan^{-1}(V_y'/V_x') \tag{3.35}$$

Since V' is negative, the avoidance heading steers away from the obstacle. Also in Figure 3.15 it may be seen that the gradient is zero at the origin, which does not cause a problem since the origin is an unstable saddle point and paths always steer away from the obstacle.

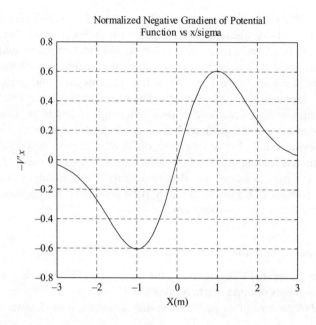

Figure 3.15 Negative potential gradient, $-V'_x$ versus normalised distance from centre, $\sigma = 1.0$

3.4.2 Multiple obstacles

With multiple obstacles present, each one has a similar potential that is summed to the total. This gives the more general form for the avoidance command.

$$\psi_{oaapf}(t)$$
$$= \tan^{-1} \frac{\sum_{i=1}^{N}(Y(t) - y_i)/\sigma_i^2 * V_i \exp[((X(t) - x_i)^2 + (Y(t) - y_i)^2)/2\sigma_i^2]}{\sum_{i=1}^{N}(X(t) - x_i)/\sigma_i^2 * V_i \exp[((X(t) - x_i)^2 + (Y(t) - y_i)^2)/2\sigma_i^2]}$$
(3.36)

An example of three obstacles in the field used in simulation from above is shown in Figure 3.16.

This is not particularly satisfactory alone since the vehicle is already following some defined track and that track has also to be maintained in the absence of obstacles. The problem is solved by the addition of a track following potential $V_{track(i)} = \beta(1 - S_i)$ in the direction of the current (i) track heading. The total guidance potential becomes the sum of the obstacle avoidance and the track following potentials.

$$V_{track(i)} = \begin{bmatrix} -\beta_x \\ -\beta_y \end{bmatrix}; \quad \beta_x = \beta \cos(\psi_{track(i)}); \quad \beta_y = \beta \sin(\psi_{track(i)})$$
(3.37)

so that the total potential is expressed as $V = V_{track} + V_{oa}$.

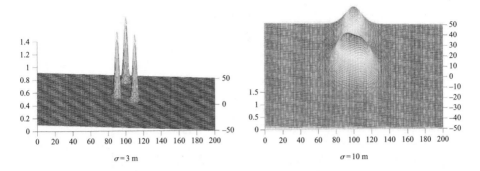

Figure 3.16 Potential function plotted vertically in the Y versus X field, $\sigma = 3$ and 10 m

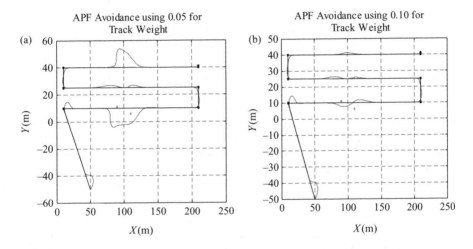

Figure 3.17 Path responses with $\beta = 0.05$ and 0.1 and $\sigma = 10$ m

This leads to the expression for heading command that includes the track following as well as the obstacle avoidance commands from the potential gradients as in the heading error for the autopilot $\tilde{\psi}(t)$

$$\tilde{\psi}(t) = (\psi_{\text{track}(j)} - \psi(t)) - \tan^{-1}(\varepsilon/\rho) - \psi_{\text{oaapf}}(t)$$

$$\psi_{\text{oaapf}}(t) = \tan^{-1}((V'_{\text{oay}} + \beta_y)/(V'_{\text{oax}} + \beta_x)) - \psi_{\text{track}(j)} \quad (3.38)$$

The resulting heading controller will judiciously avoid objects while maintaining track. Figure 3.17 shows the result for three obstacles placed around the planned tracks of the vehicle in which two choices for the parameter β are made.

Decreasing β in general allows weaker track following when presented with obstacles, increasing β, results in stricter track following and less avoidance. While there are no rules for the selection of β, it would be of interest to relate its choice to

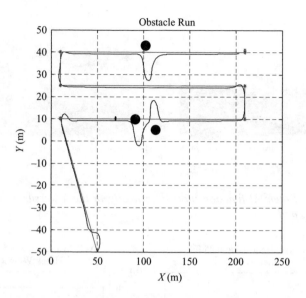

Figure 3.18   *Obstacle avoidance Gaussian potential function with* $\sigma = 3\,m$

the minimum avoidance range. In general, it can be said that β, a gradient, should be scaled according to the selection of the inverse of the standard deviation, σ in the Gaussian potentials. Since σ_l is a critical distance, it should be chosen with respect to the turning radius of the vehicle. With that said, a choice of β at 0.1 or 0.05 is in order, where σ_l is between 3 and 10 m. Since the potential gradient is used as the signal for heading change the sharper the gradient the sharper the avoidance turn so that small σ_l leads to sharper turns, and larger σ_l leads to softer turns.

Figure 3.18 shows the avoidance response using $\beta = 0.1$ and $\sigma = 3$ m, in which a responsive deviation is obtained. The minimum avoidance distance for this case is about 4 m. Varying σ produces deviations that increase the initial reaction to avoid with increasing σ. However, it is generally found that decreasing σ allows more overall path deviation (Figure 3.19). Thus, APFs can be tuned to optimise the path deviation while providing a minimum avoidance radius.

The results of the above simulation have indicated that APFs can be tuned. In the case of Gaussian APFs, the tuning parameters are the standard deviation and the track potential β. These methods can also be embedded into planned avoidance functions since the standard deviations and β values can be selected based on a balancing between path deviation and minimum avoidance distances.

3.5 Conclusions

In this chapter, guidance laws for track following, cross-track error control and obstacle avoidance using reactive, planned and APF methods have been discussed. While

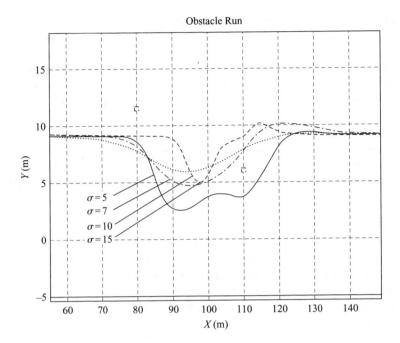

Figure 3.19 Deviation paths for σ = 5, 7, 10 and 15 m showing smoother paths with less deviation as σ increases

more work remains to be done, these methods of avoidance guidance form the basis of most of the needed algorithms. The use of APFs opens up the possibilities of an optimized avoidance plan once obstacles are detected.

3.6 Acknowledgements

The author would like to recognise the input of and discussions with Douglas P. Horner especially with respect to the use of APFs, and to Dr Tom Swean Office of Naval Research for financial support of the work in the Center for AUV Research at NPS.

References

1. Kaminer, I., A. Pascoal, E. Hallberg, and C. Silvestre (1998). Trajectory tracking for autonomous vehicles: an integrated approach to guidance and control. *AIAA Journal of Guidance and Dynamics*, 21(1), pp. 29–38.
2. Milam, M.B. (2003). Real time trajectory generation for constrained dynamics systems. Ph.D. thesis, Caltech.
3. Kanayama, Y. and B. Hartman (1997). Smooth local path planning for autonomous vehicle. *International Journal of Robotics Research*, 16(3), pp. 263–284.

4 Fodrea, L. and A.J. Healey (2003). Obstacle avoidance considerations for the REMUS autonomous underwater vehicle. In *Proceedings of the ASME OMAE Conference*, Paper # OMAE2003-37116.
5 Marco, D.B. and A.J. Healey (2001). Command, control and navigation: experimental results with the NPS ARIES AUV. *IEEE Journal of Oceanic Engineering* (Special Issue on Autonomous Ocean Sampling Networks), 26(4), pp. 466–477. http://web.nps.navy.mil/~me/healey/papers/IEEE_Marco_Healey.pdf
6 Healey, A.J. (2001). Dynamics and control of mobile robot vehicles. ME4823 Class Notes, Chapters 1–7. http://web.nps.navy.mil/~me/healey/ME4823/
7 Prestero, T. (2001). Verification of a six-degree of freedom simulation model for the REMUS autonomous underwater vehicle, M.S. thesis, Massachusetts Institute of Technology.
8 Papoulias, F.A. (1995). Non-linear dynamics and bifurcations in autonomous vehicle guidance and control. In *Underwater Robotic Vehicles: Design and Control* (Ed. J. Yuh). TST Press, Albuquerque, NM. ISBN 0-6927451-6-2.
9 Khatib, O. (1986). Real time obstacle avoidance for manipulators and mobile robots. *International Journal of Robotics Research*, 5(1), pp. 90–98.
10 Chuang, J.H. (1998). Potential based modeling of three dimensional workspace for obstacle avoidance. *IEEE Transactions on Robotics and Automation*, 14(5), pp. 778–785.
11 Fodrea, L.R. (2001). Obstacle avoidance control for the REMUS autonomous underwater vehicle. MSME thesis, Naval Postgraduate School, December. http://www.cs.nps.navy.mil/research/auv/theses/fodrea/Lynn_Fodrea/
12 Ge, S.S. and Y.J. Cui (2000). Dynamic motion planning for mobile robots using potential field method. In *Proceedings of the 8th Mediterranean Conference on Control and Automation* (MED 2000), Rio, Patras, Greece, 17–19 July.
13 Krogh, B.H. and C.E. Thorpe (1986). Integrated path planning and dynamic steering control for autonomous vehicles. In *Proceedings of the 1986 IEEE International Conference on Robotics and Automation*, pp. 1664–1669.
14 Lane, D.M., Y. Petillot, and I.T. Ruiz (2001). Underwater vehicle obstacle avoidance and path planning using a multi-beam forward looking sonar. *IEEE Journal of Oceanic Engineering*, 26(1), pp. 240–251.
15 Latrobe, J.C. (1991). *Robot Motion Planning*. Kluwer Academic Publishers, Norwell, MA.
16 Stentz, A. (1994). Optimal and efficient path planning for partially-known environments. In *Proceedings of the IEEE International Conference on Robotics and Automation (ICRA '94)*, vol. 4, pp. 3310–3317.
17 Warren, C.W. (1989). Global path planning using artificial potential fields. In *Proceedings of the IEEE Conference on Robotics and Automation*, pp. 316–321.

Chapter 4

Behaviour control of UUVs

M. Carreras, P. Ridao, R. Garcia and J. Batlle

4.1 Introduction

A control architecture can be defined as a multiple-component system whose main function is to provide the vehicle with a decision-making capability needed to perform a user-specified mission while ensuring the safety of the vehicle at all times. The vehicle becomes a physical agent which senses the environment through its sensors acting depending on its perception and its mission goals. Sometimes its acts are reflexes, probably with the goal of ensuring safety, sometimes they are the result of a reasoning process planned according to the state of the world the agent believes it understands, and according to its goals. The architecture must force its components to present a clear function and interface in order to make their interconnection simple with respect to a topology fixed by the architecture itself. Moreover, all this work has to be done in real time.

Three main approaches have been applied to the organisation of the control system. The first, hierarchical deliberative architectures, are strongly based on planning and on a world model, allowing reasoning and making predictions about the environment. This approach is used in the planning software architecture [1]. This is a hierarchical planner arranged in three homogeneous layers. The upper layer is in charge of the mission planning while the lower layer is responsible for issuing commands to the actuators. Barnett *et al.* [2] used a deliberative architecture called AUVC. It is organised in three hierarchical levels: planning level, control level and diagnostic level. The planner selects a predefined plan or creates a new plan based on the boundary conditions of the mission, and then the plan is transferred to the control layer. The algorithmic modules then execute it and invoke a replanning procedure if the deadlines or restrictions of the mission are violated. The Autonomous Benthic Explorer (ABE) control architecture [3] is organised in two layers. The lower layer deals with the low-level control of the robot's sensors and actuators while the higher layer provides mission planning and operational activity monitoring of the lower levels.

The second approach, behaviour-based architectures, also known as reactive architectures or heterarchies, are decomposed based on the desired behaviours for the robot. Normally, these missions are described as a sequence of phases with a set of active behaviours. The behaviours continuously react to the environment sensed by the perception system. The robot's global behaviour emerges from the combination of the elemental active behaviours. The real world acts as a model to which the robot reacts based on the active behaviours. As active behaviours are based on the sense–react principle, they are very robust and admirably suitable for dynamic environments. The work on behavioural architectures was started by Brooks [4] who proposed the subsumption architecture which divided the control system into a parallel set of competence-levels, linking the sensors with the actuators. It used priority arbitration through inhibition (one signal inhibits another) or suppression (one signal replaces another). Bellingham *et al.* [5] adapted the subsumption architecture to the *Sea Squirt* AUV. They isolated the actuator control and the sensor code from the control layer and avoided inter-layer communication to ease the scalability. They also extended the arbitration mechanism proposing a new method called masking whereby behaviour outputs are a sets of acceptable and unacceptable actions. The sets corresponding to all the enabled behaviours are combined and the most suitable is then chosen. Zheng [6] introduced the cooperation concept within a subsumption-like control architecture. The cooperation feature is closely connected to sharing actuators between behaviours. Two behaviours cooperate when they can be combined in a suitable way. He also used a layered sensing sub-system dealing with fault tolerance. Payton *et al.* [7] used a behavioural approach to build a distributed fault tolerant control architecture. When the system fails, instead of identifying all possible causes, they identify all possible actions. This is what is called 'do whatever works'. Boswell and Learney [8] applied a layered control architecture to the Eric underwater robot, introducing protected modules (modules which should never be subsumed), devices (modules which specify physical devices), and hormone emitters (modules which send a 'hormone level' affecting dependent behaviour modules). The distributed vehicle management architecture (DVMA) was applied to the *Twin Burger* AUV [9]. In their scheme, missions are defined as a sequence of different behaviours. The basic idea is that a behaviour can be created by a combination of specific functions given to the robot and a mission is accomplished by the robot performing sequentially appropriate behaviours. It has three levels: a hardware level, a function level and a behavioural level. Some years later, Rosenblatt *et al.* [10] applied the distributed architecture for mobile navigation (DAMN) to the *Oberon* ROV. Its main feature is the voting-based coordination mechanism used for arbitration.

The third approach, hybrid architectures, takes advantage of the two previous approaches, minimising their limitations. These are usually structured in three layers: (1) the deliberative layer, based on planning, (2) the control execution layer (turn on/turn off behaviours) and (3) a functional reactive layer. In the field of underwater robotics, Bonasso [11] presented the situated reasoning architecture applied to the *Hylas* underwater vehicle. Using a specially developed language GAPPS/REX, goals can be logically specified and compiled into runtime virtual circuits, mapping sensor information into actuator commands while allowing parallel execution. This architecture also includes a deliberative layer. Rock and Wang [12] described an architecture applied to *OTTER* AUV. It has a three level control structure including a task level. This level handles common tasks such as 'go there', 'track that object', etc., dealing with basic functions which have been planned as a sequence of orders. Actions were executed by means of a set of coordinated, parallel, finite state machines. Borrelly et al. [13] presented an open robot controller computer-aided design architecture (ORCCAD). They introduced the concepts of robot-tasks (RT) and module tasks (MT). A mission is built by sequencing RTs and defining event handlers. The automaton of the RTs is triggered by the main mission program, which in turn, triggers all the necessary MTs. Borges et al. [14] described the DCA architecture in order to permit the real-time parallel execution of tasks. This was based on a hierarchical structure consisting of three levels: organisation, coordination and functional levels. These were structured according to the principle of increasing precision with decreasing intelligence. Choi et al. [15] presented an architecture for the *ODIN II* AUV. It uses a supervisor to handle mission parameters on the basis of lower-level information and three separate blocks: sensory database, knowledge base and planner. Each block has multi-layers, demonstrating an increase in intelligence with the increase in the number of layers. Valavanis et al. [16] presented an AUV control system architecture as well as the concept of goal-oriented *modules*. The state-configured embedded control architecture is based on a two-level hybrid control architecture consisting of a supervisory control level and a functional control level. This architecture uses a master controller (MC) to coordinate the operation of the AUV by transferring control actions to several functionally independent modules. Finally, Ridao et al. [17] presented the intelligent task oriented control architecture (ITOCA), developed to be applied to the *SAUVIM* vehicle [18]. In this architecture, a set of parallel task modules and a coordinator task module are in charge of the execution of the mission subgoals while being supervised by a task supervisor. A planner supervisor is used for mission planning and supervision. For more information on control architectures, see Reference 19.

This chapter describes first an introduction to the principles of behaviour-based architectures in Section 4.2, as well as the behavioural layer of the hybrid control architecture of *URIS* UUV in Section 4.3. The set-up used for experimentation is reported in Section 4.4, while the applications are reported in Section 4.5. Conclusions are given in Section 4.6.

4.2 Principles of behaviour-based control systems

Behaviour-based robotics is a methodology for designing autonomous agents and robots. The behaviour-based methodology is a bottom-up approach inspired by biology, in which several behaviours act in parallel achieving goals. Behaviours are implemented as a control law mapping inputs to outputs. They can also store states constituting a distributed representation system. The basic structure consists of all behaviours taking inputs from the robot's sensors and sending outputs to the robot's

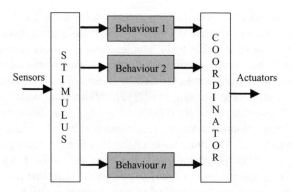

Figure 4.1   Structure of behaviour-based CA

actuators, see Figure 4.1. A coordinator is needed in order to send only one command at a time to the thrusters control systems.

The internal structure of a behaviour can also be composed of different modules interconnected by sensors, various other modules and finally, the coordinator [4]. However, behaviours must be completely independent of each other. The global structure is a network of interacting behaviours comprising low-level control and high-level deliberation abilities. The latter is performed by the distributed representation which can contain states and, consequently, change the behaviour according to their information.

The parallel structure of simple behaviours allows a real-time response with low computational cost. Autonomous robots using this methodology can be built easily at low cost. Behaviour-based robotics has demonstrated its reliable performance in standard robotic activities such as navigation, obstacle avoidance, terrain mapping, object manipulation, cooperation, learning maps and walking. For a more detailed review, see Reference 20.

There are a few basic principles which have been used by all researchers in behaviour-based robotics. These principles provide the keys to success of the methodology.

- *Parallelism.* Behaviours are executed concurrently.
- *Modularity.* The system is organised into different modules (behaviours). The important fact is that each module must run independently.
- *Situatedness/embeddedness.* The concept of 'Situatedness' means that a robot is situated in and surrounded by the real world. For this reason it must not operate using an abstract representation of reality, it must use the real perceived world. 'Embeddedness' refers to the fact that the robot exists as a physical entity in the real world. This implies that the robot is subjected to physical forces, damages and, in general, to any influence from the environment. This means that the robot should not try to model these influences or plan with them. Instead it should use this system–environment interaction to act and react with the same dynamics as the world.

- *Emergence*. This is the most important principle of behaviour-based robotics. It is based on the principles explained above and attempts to explain why the set of parallel and independent behaviours can arrive at a composite behaviour for the robot to accomplish the expected goals. Emergence is the property which results from the interaction between the robotic behavioural system and the environment. Due to emergence, the robot performs behaviours that were not pre-programmed. The interaction of behaviour with the environment generates new characteristics in the robot's behaviour which were not pre-designed.

4.2.1 Coordination

When multiple behaviours are combined and coordinated, the emergent behaviour appears. This is the product of the complexity between a robotic system and the real world. The two primary coordination mechanisms are:

- *Competitive methods*. The output is the selection of a single behaviour, see Figure 4.2(a). The coordinator chooses only one behaviour to control the robot. Depending on different criteria the coordinator determines which behaviour is best for the control of the robot. Preferable methods are suppression networks such as subsumption architecture, action-selection and voting-based coordination.
- *Cooperative methods*. The output is a combination function of all the active behaviours, see Figure 4.2(b). The coordinator applies a method which takes all the behavioural responses and generates an output which will control the robot. Behaviours that generate a stronger output will impose a greater influence on the final behaviour of the robot. Principal methods are vector summation such as potential fields and behavioural blending.

Basic behaviour-based structures use only a coordinator which operates using all the behaviours to generate the robot's response. However, there are more complex systems with different groups of behaviours coordinated by different coordinators. Each group generates an output and with these intermediate outputs, the robot's final response is generated through a final coordinator. These recursive structures are used in high-level deliberation. By means of these structures a distributed representation can be made and the robot can behave differently depending on internal states, achieving multi-phase missions.

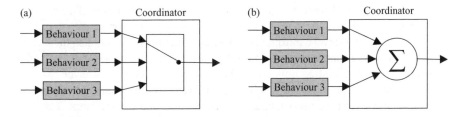

Figure 4.2 Coordination methodologies: (a) competition and (b) cooperation

4.2.2 Adaptation

One of the fields associated with behaviour-based robotics is adaptation. Intelligence cannot be understood without adaptation. If a robot requires autonomy and robustness it must adapt itself to the environment. The primary reasons for autonomous adaptivity are:

- The robot's programmer does not know all the parameters of the behaviour-based system.
- The robot must be able to perform in different environments.
- The robot must be able to perform in changing environments.

Many parameters can be adapted in a behaviour-based robotic system. At the moment there are only a few examples of real robotic systems which learn to behave and there is no established methodology to develop adaptive behaviour-based systems. The two approaches most commonly used are reinforcement learning and evolutionary techniques [20]. Both have interesting characteristics but also disadvantages like convergence time or the difficulties in finding a reinforcement or fitness function, respectively. In many cases, they are implemented over control architectures based on neural networks. Using the adaptive methodologies, the weights of the network are modified until an optimal response is obtained. The two approaches have demonstrated the feasibility of the theories in real robots.

4.3 Control architecture

The control architecture, or high-level controller, proposed in this section, is a hybrid control architecture. As explained in Section 4.1, the main advantages of hybrid architectures are reactive and fast responses due to a behaviour-based layer, and mission planning due to a deliberative layer. In this case, the deliberative layer has the goal of breaking down the mission to be accomplished into a set of tasks. The behaviour-based layer has the goal of carrying out each one of these tasks and is composed of a set of behaviours and a coordinator. The deliberative layer acts over the behaviour-based layer by configuring the particular set of behaviours and the priorities existing among them. It activates the best behaviour configuration for the current task. To decide if the task is being accomplished properly, it supervises what the robot perceives and also the actions that the behaviour-based layer is proposing. The behaviour-based layer acts over the low-level controller generating the actions to be followed. These actions depend directly on the current perception of the robot since behaviours are very reactive. Finally, the low-level controller acts over the actuators to accomplish these robot actions. The whole control system of the robot just described is depicted in Figure 4.3.

This section has centred only on the behaviour-based layer, that layer has been designed using the principles which behaviour-based control architectures propose. Therefore, the layer is composed of a set of behaviours and a coordinator. The most distinctive aspect of the layer is the coordination methodology, which is a hybrid approach between competitive and cooperative methodologies. Another distinctive

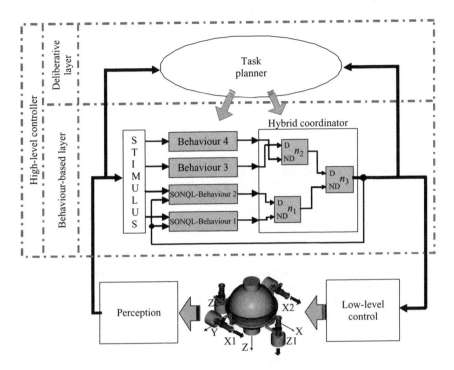

Figure 4.3 General schema of the control system conceived to control an autonomous robot

aspect can be found in the proposed behaviour-based layer. Due to the difficulty of manually tuning or designing each behaviour, some learning capabilities have been included in the behaviours to learn the internal state/action mapping which a reactive behaviour contains.

4.3.1 Hybrid coordination of behaviours

The main goal of this coordinator is to take advantage of competitive and cooperative methodologies. A competitive coordinator assures a good robustness as well as modularity and a short tuning time. On the other hand, a cooperative coordinator merges the knowledge of all the active behaviours, which, in normal situations, implies a rapid accomplishment of the task. The hybrid coordinator is able to operate as both a competitive and a cooperative method. As will be described later, coordination is achieved through a hierarchy among all the behaviours and an activation level. If higher-priority behaviours are fully activated, the coordinator will act as competitive. Alternatively, if higher-priority behaviours are partially activated, a merged control action will be generated.

The proposed hybrid coordination system was designed to coordinate a set of independent behaviours in a behaviour-based control architecture or control layer. The methodology allows the coordination of a large number of behaviours without the

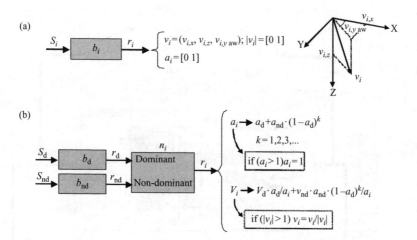

Figure 4.4 (a) The normalised robot control action v_i and the behaviour activation level a_i constitute the behaviour response r_i. (b) Hierarchical hybrid coordination node. The equations used to calculate the response of the node are shown

need of a complex designing phase or tuning phase. The addition of a new behaviour only implies the assignment of its priority with reference to other behaviours. The hybrid coordinator uses this priority and a behaviour activation level to calculate the resultant control action. Therefore, the response r_i of each behaviour is composed of the activation level a_i and the desired robot control action v_i, as illustrated in Figure 4.4. The activation level indicates the degree to which the behaviour wants to control the robot. This degree is expressed by a numerical value from 0 to 1.

The robot control action is the movement to be followed by the robot. There is a different movement for each degree of freedom (DOF). By movement, we mean the velocity the robot will achieve for a particular DOF. In the case of the underwater robot *URIS*, which has three controllable DOFs, the control action is a vector with three components. This vector is normalised and its magnitude cannot be greater than 1. Therefore, the units of the vector v_i do not correspond to any real units. After the coordination phase, this normalised vector will be re-escalated to the velocities of the vehicle.

The hybrid coordinator uses the behaviour responses to compose a final control action. This process is executed at each sample time of the high-level controller. The coordination system is composed of a set of nodes n_i. Each node has two inputs and generates a response which also has an activation level and a control action. The response of a node cannot be discerned from one of a behaviour. By using these nodes, the whole coordination process is accomplished. After connecting all the behaviour and node responses with other nodes, a final response will be generated to control the robot.

Each node has a dominant and a non-dominant input. The response connected to the dominant input will have a higher priority than the one connected to the non-dominant input. When the dominant behaviour is completely activated, $a_d = 1$, the response of the node will be equal to the dominant behaviour. Therefore, in this case, the coordination node will behave competitively. However, if the dominant behaviour is partially activated, $0 < a_d < 1$, the two responses will be combined. The idea is that non-dominant behaviours can modify the responses of dominant behaviours slightly when these are not completely activated. In this case, the node will behave cooperatively. Finally, if the dominant behaviour is not activated, $a_d = 0$, the response of the node will be equal to the non-dominant behaviour. These nodes are called hierarchical hybrid coordination nodes (HHCNs) as their coordination methodology changes depending on the activation level of the behaviours and the hierarchy between them.

Figure 4.4 shows the equations used to calculate the response of an HHCN. The activation level will be the sum of the activation levels of the input responses, in which the non-dominant activation level has been multiplied by a reduction factor. This factor, $(1 - a_d)k$, depends on the activation of the dominant behaviour and on the value of the integer parameter k. If $k = 1$, the activation level will linearly decrease as a_d increases. If more drastic reduction is desired, the value of k can be set at 2, 3, 4, ... This parameter does not have to be tuned for each node. The same value, for example, a quadratic reduction $k = 2$, can be applied to all the coordination nodes. Finally, if the new activation level is larger than 1, the level is saturated to 1.

The control action is calculated in the same way as the activation level. Vector v_i will be the sum of v_d and v_{nd}, applying the corresponding proportional factors. Therefore, each component of v_d will be taken in proportion of the activation level, a_d with respect to a_i and, each component of v_{nd} will be taken in proportion of the reduced activation level, and with respect to a_i. If the module of v_i is larger than 1, the vector will be resized to a magnitude equal to 1.

As commented on above, the hybrid coordinator is composed of a set of HHCNs which connect each pair of behaviours or nodes until a final response is generated. In order to build up this network of nodes, it is necessary to set up the hierarchy among the behaviours. This hierarchy will depend on the task to be performed. Once the priorities have been set, usually by the mission designer, the hybrid coordinator will be ready to use. Figure 4.3 shows an example of a set of three nodes which coordinate four behaviours.

4.3.2 Reinforcement learning-based behaviours

As described in Section 4.2, a commonly used methodology in robot learning is reinforcement learning (RL) [21]. RL is a class of learning algorithm where a scalar evaluation (reward) of the performance of the algorithm is available from the interaction with the environment. The goal of an RL algorithm is to maximise the expected reward by adjusting some value functions. This adjustment determines the control policy that is being applied. The evaluation is generated by a reinforcement function which is located in the environment. Most RL techniques are based on finite Markov decision processes (FMDP) causing finite state and action spaces. The main advantage of RL is that it does not use any knowledge database, as do most forms of machine learning, making this class of learning suitable for on-line robot learning. The drawbacks are the lack of generalisation among continuous variables and the difficulties in observing the Markovian state in the environment. A very widely used RL algorithm is the Q-learning [22] algorithm due to its good learning capabilities: on- and off-line policy.

Many RL-based systems have been applied to robotics over the past few years and most of them have attempted to solve the generalisation problem. To accomplish this, classic RL algorithms have usually been combined with other methodologies. The most commonly used methodologies are decision trees, cerebellar model articulation control (CMAC) function approximation [23], memory-based methods and neural networks (NN). These techniques modify the RL algorithms breaking in many cases their convergence proofs. Neural networks is a nonlinear method; however, it offers a high generalisation capability and demonstrated its feasibility in very complex tasks [24]. The drawback of NN is the interference problem. This problem is caused by the impossibility of generalising in only a local zone of the entire space. Interference occurs when learning in one area of the input space causes unlearning in another area [25].

In order to learn the internal state/action mapping of a reactive-behaviour of the behaviour-based control layer, the semi-online neural-Q-learning algorithm (SONQL) was developed [26]. This algorithm attempts to solve the generalisation problem combining the Q-learning algorithm with a NN function approximator. In order to solve the interference problem, the proposed algorithm introduces a database of the most recent and representative learning samples, from the whole state/action space. These samples are repeatedly used in the NN weight update phase, assuring the convergence of the NN to the optimal Q-function and, also, accelerating the learning process. The algorithm was designed to work in real systems with continuous variables. To preserve the real-time execution, two different execution threads, one for learning and another for output generation, are used. The use of the SONQL algorithm in the control architecture can be seen in Figure 4.3, and its internal structure in Figure 4.5.

4.4 Experimental set-up

4.4.1 URIS UUV

The *URIS* robot (see Figure 4.6(a)) was developed at the University of Girona with the aim of building a small-sized AUV. The hull is composed of a stainless steel sphere with a diameter of 350 mm, designed to withstand pressures of 3 atm (30 m depth). On the outside of the sphere there are two video cameras (forward and down-looking), two sonar beams (forward and down-looking) and a pressure vessel containing a MT9 MRU which estimates the attitude. A pressure sensor and a set of water leakage detectors complete the sensing package. The robot includes four thrusters (two in the X-direction and two in the Z-direction). Due to the stability of the vehicle in pitch and roll, the robot has four DOF: surge, heave, sway and pitch. Except for the sway, the other DOFs can be directly controlled. The vehicle has two onboard PC-104

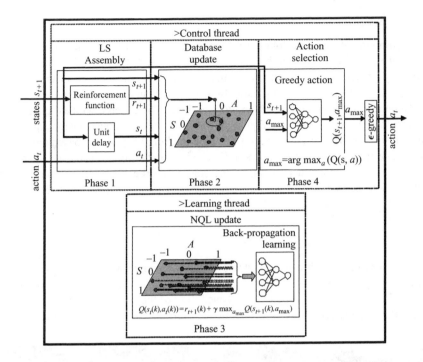

Figure 4.5 Diagram of the SONQL algorithm in its implementation for a robot behaviour. The control thread is used to acquire new learning samples and generate the control actions. The learning thread, with less priority, is used to update the NQL function

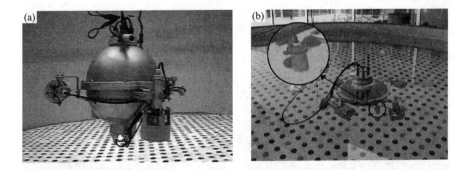

Figure 4.6 (a) URIS underwater vehicle. (b) Down-looking camera

computers. One runs the low level and high level controllers on a QNX real-time operating system. The other runs computer vision algorithms on a Linux operating system. Both computers are connected through an Ethernet network. An umbilical wire is used for communication, power and video signal transmissions.

4.4.2 Set-up

The Computer Vision and Robotics Group of the University of Girona (UdG) has an Underwater Robotics Research Facility (see Figure 4.7(a)) financed by the Spanish Ministry of Science and Technology, the Catalan Government and FEDER funds. The facility consists of a testing pool (18 × 8 × 5 m), an underwater control room, a robot hangar and several labs. The bottom of the pool is covered with a digital poster of a real underwater scene (see Figure 4.7(b,c)) allowing computer vision algorithms to be tested like image mosaicking [27], pipe tracking [28], vision-based navigation [29]. The underwater room allows a direct view of the robot during its operation, helping the development process. In the context on the PERSUB national project [30], it will be possible in the future to carry out remote experiments through the Internet.

4.4.3 Software architecture

When working with physical systems such as an underwater robot, a real-time operating system (OS) is usually required. The main advantage is better control of the CPU work. In a real-time OS, the scheduling of the processes to be executed by the CPU is done according to pre-emptive priorities. More priority processes will

Figure 4.7 (a) Water tank facility. (b) Digital image of a real scenario placed on the bottom of the water tank. (c) Virtual representation of the water tank used within NEPTUNE. (d) View of NEPTUNE during a simulation of URIS and GARBI UUVs

be executed first, as will advance processes which are already in execution. Using a correct priority policy it is possible to guarantee the frequency in which the control architecture has to be executed, which is very important to assure the controllability of the robot.

A software framework, based on a real-time operating system was specially designed for the URIS AUV. In particular, QNX OS was used. This framework is intended to assist the architecture designers to build the software architecture required to carry out a particular mission with the URIS AUV. The framework proposes the use of a set of distributed objects which represent the architecture. Each object represents a component of the robot (sensors or actuators), or a component of the control system (low- or high-level controllers). An interface definition language (IDL) is used to define the services offered by the objects. From the information contained in the IDL, an object skeleton is automatically generated. Each object has usually two threads of execution. One, the periodic thread, is executed at a fixed sample time and is used to perform internal calculations. The other thread, the requests thread, is used to answer requests from clients.

The priority of each object thread is set independently and, depending on this, the objects will be executed. If a sample time is not accomplished, a notification is produced. These notifications are used to redesign the architecture in order to accomplish the desired times. The software framework allows the execution of the objects in different computers without any additional work for the architecture designer. A server name is used in order to find the location of all the objects. Evidently, objects that are referred to as physical devices, such as sensors or actuators, have to be executed in the computer which has the interfaces for them. Communication between objects is performed in different ways depending on whether they are executed sharing the same logical space and if they are executed in the same computer. However, these variations are hidden by the framework, which only shows, to the architecture designer, a single communication system based on remote method invocation.

Although this software framework was developed to work under a real-time operating system, the execution of objects under other conventional OS is also supported. The main reason for that is the lack of software drivers of some devices for the QNX OS.

4.4.4 Computer vision as a navigation tool

All UUV missions require some method to sense vehicle position and localise it in the working environment. Normally, most tele-operated vehicles incorporate a video camera to provide visual feedback to the robot operator. From this viewpoint, the idea of using this sensor for positioning the vehicle is very attractive and cost-effective. A vision system for aiding a UUV's navigation is composed of a down-looking camera (see Figure 4.6(b)), a video digitiser, a host computer and, depending on the working depth, a source of light. Comparing these requirements with most on-board positioning sensors (e.g., accelerometers, gyros, Doppler velocity log, inertial navigation systems), it is obvious that computer vision emerges as a very low-cost sensing technology. Moreover, given the high resolution of digital imaging, measurement accuracy can be on the order of millimetres, while other sensing technologies obtain much less accurate resolutions. However, these measurements can only be used when the vehicle is close to the underwater terrain, and is limited by the visibility of the working area. Light attenuation in the water medium is one of the main problems the vision system has to face. Moreover, as the submersible increases the working depth, it has to carry artificial lights, increasing the power consumption of the vehicle. A simplified set-up for obtaining a drift-free localization of the vehicle exists for indoor applications [31]. This set-up, to be used in our indoor facility, is based on a coded pattern placed on the bottom of the water tank (see Figure 4.6).

For natural environments, a mosaicking system has been developed to obtain the localisation of the vehicle based on computer vision [27,29]. Its position and orientation can be calculated by integrating the apparent motion between consecutive images acquired by the down-looking camera. Knowledge of the pose at image acquisition instances can also be used to align consecutive frames to form a 'mosaic', that is, a composite image which covers the entire scene imaged by the submersible. As stated above, the vehicle motion can be recovered by means of these visual mosaics. This visual information allows the vehicle to localise itself on the mosaic map as it is being constructed (following the strategy known as simultaneous localisation and mapping – SLAM).

4.5 Results

This section presents an application in which a behaviour-based control layer is able to control the *URIS*' UUV to perform a pre-defined task. In this example, the positioning system was specifically designed to obtain a very good accuracy of the robot position and orientation. Since this positioning system is only suitable for structured environments, the results of our positioning system for natural environments are also reported. The goal pursued in the near future is to reproduce the experiments we have obtained in the structured environment, in a natural environment in which we will use the mosaicking techniques together with other conventional navigation sensors.

4.5.1 Target tracking task

The task consisted of following a target with the underwater robot *URIS* in a circular water tank (see Figure 4.8). Three basic aspects were considered. The first was to avoid obstacles in order to ensure the safety of the vehicle. In this case, the wall of the pool was the only obstacle. The second aspect was to ensure the presence of the target within the forward looking camera's field of view. The third aspect was to follow the target at a certain distance. Each one of these aspects was translated to a robot behaviour, hence, the behaviours were: the 'wall avoidance' behaviour, the 'target recovery' behaviour and the 'target following' behaviour. An additional behaviour was included, the 'tele-operation' behaviour, which allowed the robot to move according to the commands given by a human. This behaviour did not influence the outcome of the target following task, but was used to test the performance of the system, for example, by moving the vehicle away from the target.

Figure 4.8   Simplified experimental set-up. The water tank (a) and the URIS' *robot, while it was following the target (b), can be seen*

Due to the shallow water in the tank in which the experiments were performed, only the motions on the horizontal plane were considered. Therefore, to accomplish the target following task, only the surge and yaw control actions were generated by the behaviours. The other two controllable DOFs (heave and pitch) were not used. In the case of the heave movement, the low-level controller maintained the robot at an intermediate depth. Regarding the pitch orientation, a 0° set-point (normal position) was used. Note that the position and orientation of the robot was computed by a vision-based application which used the down-looking camera of the robot and a coded pattern placed on the bottom of the water tank [31].

After defining the behaviours present in this task, the next step was to set their priorities. To determine these priorities, the importance of each behaviour goal was ranked. The hierarchy used is as follows:

1. *'Wall avoidance' behaviour*. This was the highest priority behaviour in order to ensure the safety of the robot even if the task was not accomplished. The state/action mapping was manually programmed.
2. *'Tele-operation' behaviour*. This was given a higher priority than the next two behaviours in order to be able to drive the robot away from the target when desired. This behaviour used the commands provided by a human operator.
3. *'Target recovery' behaviour*. This was given a higher priority than the target following behaviour so as to be able to find the target when it was not detected. The state/action mapping was automatically learnt with the SONQL algorithm. Some experiments were specifically designed to assist the learning process.
4. *'Target following' behaviour*. This was the lowest priority behaviour since it should only control the robot when the other behaviour goals had been accomplished. The state/action mapping was manually programmed.

The establishment of the priorities allowed the composition of the behaviour-based control layer. Figure 4.9 shows the implementation of the four behaviours using three hybrid coordination nodes. Finally, the last step was the definition of the internal state/action mapping and the activation level function for each behaviour.

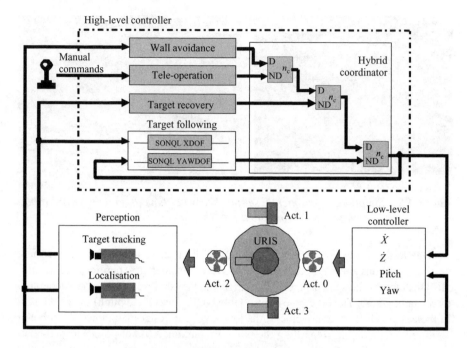

Figure 4.9 Control system architecture used in the target following task

The 'target following' behaviour was implemented with the SONQL algorithm. For more implementation details, see Reference 32.

The behaviour-based control layer and, in particular, the hybrid coordinator system showed a suitable methodology to solve a robot task. The architecture is simple and operates with robustness and good performance. In Figure 4.10 it can be seen how the robot followed the target at a certain distance. The target was moved manually around the water tank and, therefore, the robot trajectory was also a circumference. Note that the position of the target is approximate since there is no system to measure it.

4.5.2 Exploration and mapping of unknown environments

Several sea trials have been performed in the last few years in the coastal waters of Costa Brava. In each experimental run the pilot teleoperates the vehicle at a suitable range above the seabed. Then, as the vehicle moves, the acquired images are sent to the surface through the umbilical tether, where they are either stored to a tape or processed in real time. The sea trial reported in Figure 4.11 shows a trajectory performed by the vehicle in an area of the sea floor formed by rocks and algae. The trajectory starts at the bottom right and ends at the bottom left. It is obvious that the vehicle trajectory can be recovered directly from the computed mosaic. Moreover, it can be observed in this figure that image alignment is quite good, although the underwater terrain is not flat.

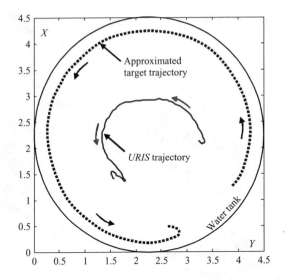

Figure 4.10   Trajectory of URIS *while following the target in the water tank*

4.6 Conclusions

A control architecture is a key component in the development of an autonomous robot. The control architecture has the goal of accomplishing a mission which can be divided into a set of sequential tasks. This chapter has presented behaviour-based control architectures as a methodology to implement this kind of controller. Its high interaction with the environment, as well as its fast execution and reactivity, are the keys to its success in controlling autonomous robots. The main attention has been given to the coordination methodology. Competitive coordinators assure the robustness of the controller, whereas cooperative coordinators determine the performance of the final robot trajectory. The structure of a control architecture for an autonomous robot has been presented. Two main layers are found in this schema; the deliberative layer which divides the robot mission into a set of tasks, and the behaviour-based layer which is in charge of accomplishing these tasks. This chapter has focused only on the behaviour-based layer. A behaviour coordination approach has been proposed, its main feature being a hybrid coordination of behaviours between competitive and cooperative approaches. A second distinctive part is the use of a learning algorithm to learn the internal mapping between the environment state and the robot actions.

Then, the chapter has introduced the main features of the *URIS* AUV and its experimental set-up. According to the navigation system we have shown that the estimation of the robot position and velocity can be effectively performed with a vision-based system using mosaicking techniques. This approach is feasible when the vehicle navigates near the ocean floor and is also useful to generate a map of the area. Finally, results on real data have been shown through an example in which the behaviour-based layer controlled *URIS* in a target following task. In these experiments

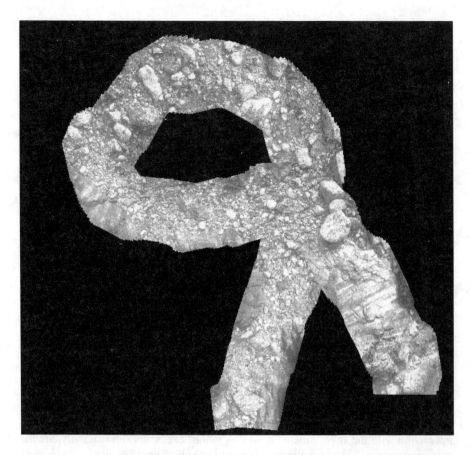

Figure 4.11   Part of a surveillance trajectory described by URIS *at Cala Margarida (Palamós-Girona)*

several behaviours were in charge of the control of the robot to avoid obstacles, to find the target and to follow it. One of these behaviours was automatically learnt demonstrating the feasibility of the learning algorithms. Finally, we have also shown the performance of the navigation system using the visual mosaicking techniques.

References

1 Hall, W.D. and M.B. Adams (1992). Autonomous vehicle taxonomy. In Proceedings of the IEEE International Symposium on Autonomous Underwater Vehicle Technology, Washington, DC.
2 Barnett, D., S. McClaran, E. Nelson, M. McDermott, and G. Williams (1996). Architecture of the Texas A&M autonomous underwater vehicle controller. In Proceedings of the IEEE Symposium on Autonomous Underwater Vehicles Technology, Monterrey, CA, pp. 231–237.

3 Yoerger, D.R., A.M. Bradley, and B. Waldem (1994). System testing of the autonomous benthic explorer. In Proceedings of the IARP 2nd Workshop on Mobile Robots for Subsea Environments, pp. 159–170.
4 Brooks, R. (1986). A robust layered control system for a mobile robot. *IEEE Journal of Robotics and Automation*, RA-2(1), pp. 14–23.
5 Bellingham, J.G., T.R. Consti, and R.M. Beaton (1990). Keeping layered control simple. In Proceedings of the IEEE Symposium on Autonomous Underwater Vehicles Technology, IEEE catalog no. 90CH2856-3, Washington, DC, pp. 3–8.
6 Zheng, X. (1992). Layered control of a practical AUV. In Proceedings of the IEEE Symposium on Autonomous Underwater Vehicles Technology, Washington, DC, pp. 142–147.
7 Payton, D., D. Keirsey, D. Kimble, J. Krozel, and J. Rosenblatt (1992). Do whatever works: a robust approach to fault-tolerant autonomous control. *Applied Intelligence*, 2(3), pp. 225–250.
8 Boswell, A.J. and J.R. Learney (1994). Using the subsumption architecture in an autonomous underwater robot: expostulations, extensions and experiences. In International Advanced Robotics Program, *Workshop on Mobile Robots for Subsea Environments*, Monterrey, CA.
9 Fujii, T. and T. Ura (1995). Autonomous underwater robots with distributed behaviour control architecture. In Proceedings of the IEEE International Conference on Robotics and Automation, New York, NY, pp. 1868–1873.
10 Rosenblatt, J., S. Williams, and H. Durrant-Whyte (2000). Behavior-based control for autonomous underwater exploration. In IEEE International Conference on Robotics and Automation, San Francisco, CA, pp. 920–925.
11 Bonasso, P. (1992). Reactive control of underwater experiments. *Applied Intelligence*, 2(3), pp. 201–204.
12 Rock, S.M. and H.H. Wang (1995). Task-directed precision control of the MBARI/Standford OTTER AUV. In Proceedings of the US/Portugal Workshop on Undersea Robotics and Intelligent Control, Lisbon, pp. 131–138.
13 Borrelly, J.-J., E. Coste-Maniere, B. Espiau, K. Kapellos, R. Pissard-Gibollet, D. Simon, and N. Turro (1998). The ORCCAD architecture. *International Journal of Robotics Research*, 17(4), pp. 338–359.
14 Borges, J., F. Lobo, and E. Pereira (1994). A dynamically configurable architecture for the control of autonomous underwater vehicles. In Proceedings of the IEEE Conference on Oceans, Brest, pp. 166–171.
15 Choi S.K., J. Yuh, and G.Y. Takashige (1995). Designing of an omni-directional intelligent navigator. In *Underwater Robotic Vehicles: Design and Control* (Ed. J. Yuh). TSI Press, Albuquerque, NM, chapter 11, pp 277–297. ISBN 0-962 7451-6-2.
16 Valavanis, P.K., D. Gracanin, M. Matijasevic, R. Kolluru, and G.A. Demetriou (1997). Control architectures for autonomous underwater vehicles. *IEEE Control Systems Magazine*, pp. 48–64.
17 Ridao, P., J. Yuh, J. Batlle, and K. Sugihara (2000). On AUV control architecture. In Proceedings of the IEEE International Conference on Intelligent Robots and Systems, Takamatsu, Japan.

18 Yuh, J. and S.K. Choi (1999). Semi-autonomous underwater vehicle for intervention missions. *Sea Technology*, 40(10), pp. 31–40.
19 Ridao, P., J. Batlle, J. Amat, and G.N. Roberts (1999). Recent trends in control architectures for autonomous underwater vehicles. *International Journal of Systems Science*, 30(9), pp. 1033–1056.
20 Arkin, R.C. (1998). *Behavior-Based Robotics*. MIT Press, Cambridge, MA.
21 Sutton, R. and A. Barto (1998). *Reinforcement Learning: An Introduction*. MIT Press, Cambridge, MA.
22 Watkins, C. and P. Dayan (1992). Q-learning. *Machine Learning*, 8, pp. 279–292.
23 Santamaria, J., R. Sutton, and A. Ram (1998). Experiments with reinforcement learning in problems with continuous state and action spaces. *Adaptive Behavior*, 6, pp. 163–218.
24 Tesauro, G. (1992). Practical issues in temporal difference learning. *Machine Learning*, 8(3/4), pp. 257–277.
25 Weaver, S., L. Baird, and M. Polycarpou (1998). An analytical framework for local feedforward networks. *IEEE Transactions on Neural Networks*, 9(3), pp. 473–482.
26 Carreras, M., P. Ridao and A. El-Fakdi (2003). Semi-online neural-Q-learning for real-time robot learning. In IEEE–RSJ International Conference on Intelligent Robots and Systems, Las Vegas, NY, 27–31 October.
27 Garcia, R., X. Cufi, and V. Ila (2003). Recovering camera motion in a sequence of underwater images through mosaicking. In Iberian Conference on Pattern Recognition and Image Analysis, *Lecture Notes in Computer Science*, no. 2652. Springer-Verlag, Berlin, pp. 255–262.
28 Antich, J., A. Ortiz, M. Carreras, and P. Ridao (2004), Testing the control architecture of a visually guided underwater cable tracker by using a UUV prototype. In 5th IFAC/EURON Symposium on Intelligent Autonomous Vehicles. IAV 04, Lisbon, Portugal.
29 Garcia, R., T. Nicosevici, P. Ridao, and D. Ribas (2003). Towards a real-time vision-based navigation system for a small-class UUV. In IEEE/RSJ International Conference on Intelligent Robots and Systems.
30 Spanish project reference number: MCYT (DPI2001-2311-C03-01).
31 Carreras, M., P. Ridao, R. Garcia, and T. Nicosevici (2003). Vision-based localization of an underwater robot in a structured environment. In IEEE International Conference on Robotics and Automation ICRA 03, Taipei, Taiwan.
32 Carreras, M., P. Ridao, J. Batlle, and T. Nicosevici (2002). Efficient learning of reactive robot behaviors with a neural-Q-learning approach. In IEEE/RSJ International Conference on Intelligent Robots and Systems, Lausanne, Switzerland, 30 September–4 October.

Chapter 5

Thruster control allocation for over-actuated, open-frame underwater vehicles

E. Omerdic and G.N. Roberts

5.1 Introduction

This chapter introduces a new hybrid approach associated with the thruster control allocation problem for over-actuated thruster-propelled open-frame underwater vehicles (UVs). The work presented herein is applicable to a wide class of control allocation problems, where the number of actuators is higher than the number of objectives. However, the application described here is focused on two remotely operated vehicles (ROVs) with different thruster configuration. The chapter expands upon the work previously reported by the authors [1–4].

Significant efforts have been undertaken in the research community over the last two decades to solve the control allocation problem for modern aircraft. Different methods were proposed such as direct control allocation [5,6], optimisation-based methods using l_2 norm [7–9] and l_1 norm [7,10,11], fixed-point method [12,13] and daisy-chain control allocation [14]. The problem of fault accommodation for UVs is closely related with the control allocation problem for aircraft. In both cases, the control allocation problem can be defined as the determination of the actuator control values that generate a given set of desired or commanded forces and moments.

For the unconstrained control allocation problem with a control energy cost function used as the optimisation criteria the optimal solution is pseudoinverse [15]. Pseudoinverse is a special case of general inverse (GI), which has the advantage of being relatively simple to compute and allowing some control in distribution of control energy among available actuators. However, in real applications actuator constraints must be taken into account, which leads to a constrained control allocation problem. Handling of constrained control is the most difficult problem for the GI approach. In some cases, the solution obtained by the generalised inverse approach is not feasible, that is, it violates actuator constraints. Durham [5] demonstrated that, except in certain degenerate cases, a general inverse cannot allocate controls inside a constrained control subset Ω that will map to the entire attainable command set Φ, that is, only a subset of Φ can be covered. Two methods are suggested to handle cases where attainable control inputs cannot be allocated [4]. The first approach [T-approximation (truncation)] calculates a GI solution and truncates any components of control vector which exceed their limits. The second approach [S-approximation (scaling)] maintains the direction of the desired control input command by scaling unfeasible pseudoinverse solution to the boundary of Ω [14].

The hybrid approach will be first explained using the low-dimensional control allocation problem with clear geometrical interpretation. After this the problem will be expanded for the higher-dimensional case. The problem formulation is presented in Section 5.2. Section 5.3 describes the nomenclature. The pseudoinverse is described in Section 5.4. A short review of the fixed-point iteration method is given in Section 5.5. Section 5.6 introduces the hybrid approach. Application to the control allocation problem for ROVs is given in Section 5.7. Section 5.8 provides some concluding remarks.

5.2 Problem formulation

The majority of modern aircraft and marine vessels represent over-actuated systems, for which it is possible to decompose the control design into the following steps [16]:

1. *Regulation task*: Design a control law, which specifies the total control effort to be produced (net force, moment, etc.).
2. *Actuator selection task*: Design a control allocator, which maps the total control effort (demand) onto individual actuator settings (thrust forces, control surface deflections, etc.).

Figure 5.1 illustrates the configuration of the overall control system. The control system consists of a control law (specifying the virtual control input, $\mathbf{v} \in \Re^k$) and a control allocator (allocating the true control input, $\mathbf{u} \in \Re^m$, where $m > k$, which distributes control demand among the individual actuators). In the system, the actuators generate a total control effort, $\mathbf{v}_{sys} \in \Re^k$, which is applied as the input to the system dynamics block and which determines the system behaviour.

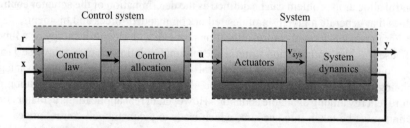

Figure 5.1 The overall control system architecture

Thruster control allocation for over-actuated, open-frame UVs

The main objective of the control allocation is to ensure that condition $\mathbf{v}_{sys} = \mathbf{v}$ is satisfied for all attainable $\mathbf{v}$. The standard constrained linear control allocation problem can be formulated as follows: for a given $\mathbf{v}$, to find $\mathbf{u}$ such that

$$\mathbf{Bu} = \mathbf{v} \tag{5.1}$$

$$\underline{\mathbf{u}} \leq \mathbf{u} \leq \overline{\mathbf{u}} \tag{5.2}$$

where the control effectiveness matrix $\mathbf{B}$ is a $k \times m$ matrix with rank k. Constraint (5.2) includes actuator position and rate constraints, where the inequalities apply component-wise. Consider the control allocation problem $\mathbf{Bu} = \mathbf{v}$, where

$$\mathbf{u} = \begin{bmatrix} u_1 \\ u_2 \\ u_3 \end{bmatrix} \in \mathfrak{R}^3 \quad (m = 3) \tag{5.3}$$

$$\mathbf{v} = \begin{bmatrix} v_1 \\ v_2 \end{bmatrix} \in \mathfrak{R}^2 \quad (k = 2) \tag{5.4}$$

$$\mathbf{B} = \begin{bmatrix} \frac{1}{2} & -\frac{1}{4} & -\frac{1}{4} \\ 0 & \frac{3}{5} & -\frac{2}{5} \end{bmatrix} \tag{5.5}$$

$$\underline{\mathbf{u}} = \begin{bmatrix} -1 \\ -1 \\ -1 \end{bmatrix} \leq \mathbf{u} = \begin{bmatrix} u_1 \\ u_2 \\ u_3 \end{bmatrix} \leq \overline{\mathbf{u}} = \begin{bmatrix} 1 \\ 1 \\ 1 \end{bmatrix} \tag{5.6}$$

Equation $\mathbf{Bu} = \mathbf{v}$ represents system of equations

$$\begin{aligned} \frac{1}{2}u_1 - \frac{1}{4}u_2 - \frac{1}{4}u_3 &= v_1 \\ \frac{3}{5}u_2 - \frac{2}{5}u_3 &= v_2 \end{aligned} \tag{5.7}$$

Each equations in (5.7) represents a plane in the true control space $\mathfrak{R}^3$. The intersection of these planes is line l:

$$l: \quad \mathbf{p} = \begin{bmatrix} \frac{104}{77}v_1 + \frac{10}{77}v_2 + \frac{1}{4}t \\ -\frac{40}{77}v_1 + \frac{85}{77}v_2 + \frac{1}{5}t \\ -\frac{60}{77}v_1 - \frac{65}{77}v_2 + \frac{3}{10}t \end{bmatrix} \tag{5.8}$$

where t is the parameter of the line. The constrained control subset Ω, which satisfies actuator constraints (5.6), is a unit cube in $\Re^3$. Geometric interpretation of the control allocation problem can be obtained by reformulating the problem as follows: *for a given $\mathbf{v}$, find intersection (solution set) $\Im$ of l and Ω.* Three cases are possible:

1. If $\Im$ is a segment, there is an infinite number of solutions (each point that lies on the segment is a solution).
2. If $\Im$ is a point, there is only one solution.
3. If $\Im$ is an empty set, no solution exists.

5.3 Nomenclature

5.3.1 Constrained control subset Ω

The following nomenclature is adopted for referring to Ω [5].

The boundary of Ω is denoted by $\partial(\Omega)$. A control vector belongs to $\partial(\Omega)$ if and only if at least one of its components is at a limit. Vertices are the points on $\partial(\Omega)$ where each control (component) receives a limit (min or max). In Figure 5.2, vertices are denoted as $\boxed{0}, \boxed{1}, \ldots, \boxed{7}$. In the general case, the number of vertices is equal to 2^m. Vertices are numerated using the following rule: if the vertex is represented in a binary form, then '0' in the kth position indicates that the corresponding control u_k is at a minimum $\underline{u}_k$, while '1' indicates it is at a maximum $\overline{u}_k$. Edges are lines that connect vertices and that lie on $\partial(\Omega)$. In Figure 5.2, edges are denoted as $\boxed{01}, \boxed{02}, \ldots, \boxed{67}$. They are generated by varying only one of the m controls, while the remaining $m-1$ controls are at their limits, associated with the two connected vertices. In the general

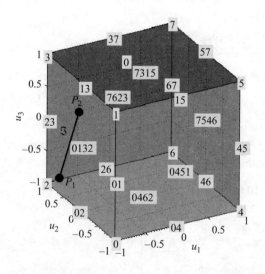

Figure 5.2   Constrained (admissible) control subset Ω

case, the number of edges is equal to $2^{m-1}\binom{m}{1}$. Two vertices are connected by an edge if and only if their binary representations differ in only one bit. Facets are plane surfaces on $\partial(\Omega)$ that contain two adjacent edges, that is, two edges that have a common vertex. In Figure 5.2, facets are denoted as 0132, 0451, ..., 7623. In the general case, the number of facets is equal to $2^{m-2}\binom{m}{2}$. A facet is defined as the set in the control space obtained by taking all but two controls at their limits and varying the two free controls within their limits.

5.3.2 Attainable command set Φ

The control effectiveness matrix **B** performs a linear transformation from the true control space $\mathfrak{R}^m$ to the virtual control space $\mathfrak{R}^k$. The image of $\Omega \subset \mathfrak{R}^m$ is called the attainable command set and denoted by Φ. Φ is a subset of the virtual control space $\Phi_v = \{\mathbf{v} : \|\mathbf{v}\|_\infty \leq 1\}$ and represents a convex polyhedron, whose boundary $\partial(\Phi)$ is the image of the facets of Ω. It is important to emphasise that not all facets of Ω are mapped on the boundary $\partial(\Phi)$; most of these facets are mapped to the interior of Φ. If any k columns of **B** are linearly independent (non-coplanar controls), then mapping **B** is one-to-one on $\partial(\Phi)$. The attainable command set Φ for the control allocation problem (5.3)–(5.6) is shown in Figure 5.3.

Images of vertices from $\partial(\Omega)$ are called vertices (if they lie on $\partial(\Phi)$) or nodes (if they lie in the interior of Φ). In Figure 5.3, 1, 2, 3, 4, 5 and 6 are vertices, while 0 and 7 are nodes. In a similar way, images of edges from $\partial(\Omega)$ are called edges (if they lie on $\partial(\Phi)$) or connections (if they lie in the interior of Φ). In Figure 5.3, 13, 23, 26, 46, 45 and 15 are edges, while 01, 02, 04, 37, 57 and 67 are connections. Images of facets that lie on $\partial(\Omega)$ are called facets (if they lie on $\partial(\Phi)$) or faces (if they lie in the interior of Φ). For the problem (5.3)–(5.6), Φ is two-dimensional and there are no faces or facets. If Φ is three-dimensional, facets or faces are parallelograms. Let $\mathbf{v}_d = [-0.5 \quad 0.6]^T$ be the desired virtual control input (Figure 5.3). Substituting $v_1 = -0.5$ and $v_2 = 0.6$ in (5.8) yields

$$l: \quad \mathbf{p} = \begin{bmatrix} -\dfrac{46}{77} + \dfrac{1}{4}t \\ \dfrac{71}{77} + \dfrac{1}{5}t \\ -\dfrac{9}{77} + \dfrac{3}{10}t \end{bmatrix} \tag{5.9}$$

The intersection $\Im$ of l and Ω is a segment $\Im = P_1 P_2$ (see Figure 5.2), where $P_1 = [-1 \quad 3/5 \quad -3/5]^T$ (for $t = t_1 = -123/77$) and $P_2 = [-1/2 \quad 1 \quad 0]^T$ (for $t = t_2 = 30/77$). The point P_1 belongs to the 0132 facet, while P_2 belongs to the 7623 facet. Each point on the segment $\Im$ is a solution. In order to extract a unique, 'best' solution from $\Im$, it is necessary to introduce criteria, which are minimised by the chosen solution. By introducing criteria, the problem can be reformulated as

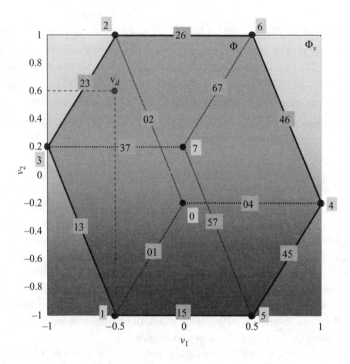

Figure 5.3 Attainable command set Φ

a constrained optimisation problem

$$\min_{\mathbf{u}} \|\mathbf{u}\|_p \tag{5.10}$$

subject to (5.1) and (5.2). The solution depends on the choice of norm used. The most suitable criteria for underwater applications is a control energy cost function, for which case $p = 2$.

5.4 Pseudoinverse

The unconstrained minimum norm allocation problem (5.10) subject to (5.1) has an explicit solution given by

$$\mathbf{u} = \mathbf{B}^+ \mathbf{v} \tag{5.11}$$

where

$$\mathbf{B}^+ = \mathbf{B}^T (\mathbf{B} \mathbf{B}^T)^{-1} \tag{5.12}$$

is the pseudoinverse of $\mathbf{B}$ [15]. For the constrained control allocation problem, where the constraint (5.2) is required to be satisfied, the solution (5.11) may become

Thruster control allocation for over-actuated, open-frame UVs   93

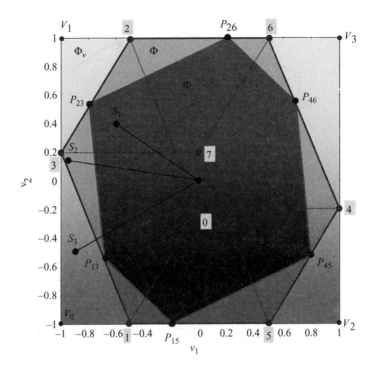

Figure 5.4   *Three typical cases for position of virtual control inputs relative to Φ_p and Φ*

unfeasible, depending on the position of **v** inside Φ_v. The virtual control space (square $\Phi_v = V_0 V_1 V_3 V_2$; Figure 5.4) is mapped by the pseudoinverse to a parallelogram $\Omega_v = U_0 U_1 U_3 U_2$ (Figure 5.5(a)). The intersection of the parallelogram Ω_v with the cube Ω is a convex polygon $\Omega_p = R_{13} R_{15} R_{45} R_{46} R_{26} R_{23}$, where the vertex R_{ij} lies on the edge *ij* of Ω. The subset $\Phi_p \subset \Phi_v$ such that $\mathbf{B}^+(\Phi_p) = \Omega_p$ is a convex polygon $P_{13} P_{15} P_{45} P_{46} P_{26} P_{23}$, whose vertex P_{ij} lies on the edge *ij* of Φ (Figure 5.4). The pseudoinverse image of Φ is a polygon Ω_e. Let $S = [v_1 \quad v_2]^T$ denote an arbitrary point from Φ_v. When a point S moves inside Φ_v, the corresponding line l moves in the true control space. For a given S, the pseudoinverse will select the solution from $\Im = l \cap \Omega$ where the line l intersects the parallelogram Ω_v. Three characteristic cases are considered, regarding the position of S relative to the partitions of Φ_v (see Figure 5.4).

In the first case, point S_1 lies inside Φ_p and the solution set is a segment $\Im_1 = P_1 P_2 = l_1 \cap \Omega$, $P_1 = [-1 \quad 14/25 \quad -4/25]^T$, $P_2 = [-9/20 \quad 1 \quad 1/2]^T$ (Figure 5.5(a)). The pseudoinverse solution $T_1 = \mathbf{B}^+(S_1) = [-0.7584 \quad 0.7532 \quad 0.1299]^T$ is the point where the segment $P_1 P_2$ intersects the parallelogram Ω_v. This solution is feasible, since it belongs to Ω. From all solutions in $\Im$, the solution T_1, selected by pseudoinverse, is optimal in the l_2

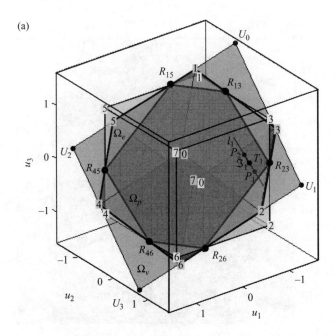

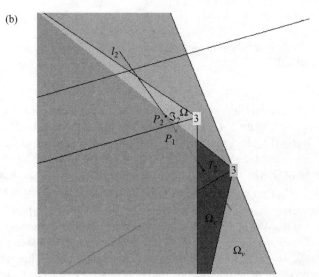

Figure 5.5 *Pseudoinverse solution for three typical cases: (a) Case 1: $S_1 \in \Phi_p \Rightarrow T_1 = l_1 \cap \Omega_v \in \Omega_p \subset \Omega$, (b) Case 2: $S_2 \in \Phi \backslash \Phi_p \Rightarrow T_2 = l_2 \cap \Omega_v \notin \Omega$ and (c) Case 3: $S_3 \in \Phi_v \backslash \Phi \Rightarrow T_3 = l_3 \cap \Omega_v \notin \Omega$*

(c)

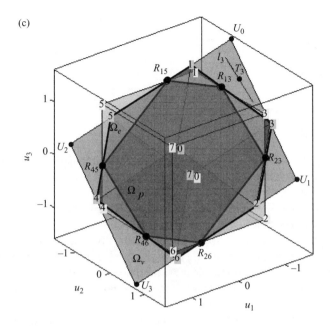

Figure 5.5 Continued

sense. In the second case, the solution set is a segment $\Im_2 = P_1 P_2 = l_2 \cap \Omega$, $P_1 = [-1 \quad 43/50 \quad 89/100]^T$, $P_2 = [-9/20 \quad 1 \quad 1/2]^T$ (Figure 5.5(b)). The pseudoinverse solution $T_2 = \mathbf{B}^+(S_2) = [-1.2455 \quad 0.6636 \quad 0.5955]^T$ represents the point where the line l_2 intersects the parallelogram Ω_v.

This solution is unfeasible, since it lies on the line l_2 outside $\Im_2$ and does not belong to Ω. Finally, in the last case the line l_3 does not intersect Ω and the exact solution of the control allocation problem does not exist (Figure 5.5(c)). However, the intersection of l_3 and Ω_v is a point $T_3 = \mathbf{B}^+(S_3) = [-1.2805 \quad -0.0844 \quad 1.1234]^T$. This pseudoinverse solution is unfeasible, since it lies outside Ω.

5.5 Fixed-point iteration method

One of the methods for solving the problem with l_2 norm is the fixed-point method [13,16,17]. This method finds the control vector **u** that minimises

$$J(\mathbf{u}) = (1 - \varepsilon)\|\mathbf{W}_v(\mathbf{Bu} - \mathbf{v})\|_2^2 + \varepsilon \|\mathbf{W}_u \mathbf{u}\|_2^2 \tag{5.13}$$

subject to

$$\underline{\mathbf{u}} \leq \mathbf{u} \leq \overline{\mathbf{u}} \tag{5.14}$$

where $|\varepsilon| < 1$. The algorithm proceeds by iterating on the equation

$$\mathbf{u}_{k+1} = \operatorname{sat}\lfloor(1 - \varepsilon)\eta \mathbf{B}^T \mathbf{Q}_1 \mathbf{v} - (\eta \mathbf{H} - \mathbf{I})\mathbf{u}_k\rfloor \tag{5.15}$$

where

$$Q_1 = W_v^T W_v \tag{5.16}$$

$$Q_2 = W_u^T W_u \tag{5.17}$$

$$H = (1-\varepsilon)B^T Q_1 B + \varepsilon Q_2 \tag{5.18}$$

$$\eta = 1/\|H\|_2 \tag{5.19}$$

and sat(**u**) is the saturation function that clips the components of the vector **u** to their limits. The condition for stopping the iteration process could be, for example, $|J(\mathbf{u}_{k+1}) - J(\mathbf{u}_k)| < \text{tol}$.

The fixed-point algorithm is very simple, and most computations need to be performed only once before iterations start. Remarkably, the algorithm also provides an exact solution to the optimisation problem and it is guaranteed to converge. To improve the efficiency, Burken et al. [12,13] suggest selecting the initial point $\mathbf{u}_0$ as the true control input calculated at the previous time sample, that is, $\mathbf{u}_0(t) = \mathbf{u}(t-T)$. The main drawback is that convergence of the algorithm can be very slow and strongly depends on the problem. In addition, the choice of the parameter ε is delicate, since it affects the trade-off between the primary and the secondary optimisation objectives, as well as the convergence of the algorithm.

5.6 Hybrid approach

The first step in the hybrid approach algorithm is the calculation of the pseudoinverse solution **u** (5.11). If **u** satisfies constraints (5.2), that is, if $\mathbf{v} \in \Phi_p$ (Case 1), the obtained solution is feasible, optimal in the l_2 sense, and algorithm stops. Otherwise, the fixed-point iteration method is activated, that is able to find the exact (feasible) solution $\mathbf{u}_f^*$ for $\mathbf{v} \in \Phi \setminus \Phi_p$ (Case 2) and good (feasible) approximation for $\mathbf{v} \in \Phi_v \setminus \Phi$ (Case 3). The algorithm will be explained for Case 2 in more detail in the following paragraph.

The number of iterations depends on the desired accuracy and the choice of the initial point (solution), which must be feasible. Since the (unfeasible) pseudoinverse solution $\mathbf{u}_2 = T_2$ is already found in the first step of the algorithm, it is natural to choose the feasible approximation of this solution as the initial point for iteration. Two choices are available: the T-approximation (truncation) $\mathbf{u}_{2t}^* = T_{2t}^* = [-1 \quad 0.6636 \quad 0.5955]^T$ is obtained from T_2 by truncating (clipping) components that exceed their limits. Another choice is the S-approximation (scaling) $\mathbf{u}_{2s}^* = T_{2s}^* = [-1 \quad 0.5328 \quad 0.4781]^T$ that is obtained by scaling T_2 by factor $f_2 = 0.8029$ such that $T_{2s}^* = \mathbf{u}_{2s}^* \in \partial(\Omega_p) \subset \partial(\Omega)$. Approximate solutions T_{2t}^* and T_{2s}^*, and corresponding virtual control inputs S_{2t}^* and S_{2s}^* are shown in Figure 5.6(a) and (b), respectively.

In order to compare different approximations, it is necessary to define criteria for comparison.

(a)

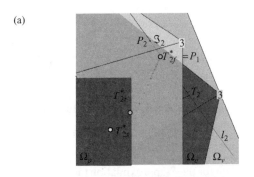

(b)

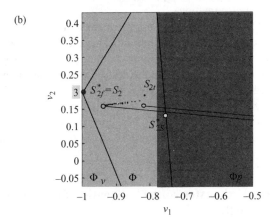

Figure 5.6   *The fixed-point iteration method is able to find the exact solution for* $v_2 \in \Phi \setminus \Phi_p$: *(a) if* $v_2 \in \Phi \setminus \Phi_p$, $T_{2f}^* = P_1 \in \mathfrak{I}_2$ *is exact solution, optimal in* l_2 *sense and (b) if* $v_2 \in \Phi \setminus \Phi_p$, *then* $S_{2f}^* = S_2$

Definition 1 (Approximation error)   *The approximation error is defined as* $e = v - v^*$, *where* v *is the virtual control input,* $v^* = Bu^*$ *is an approximation of* v *and* u^* *is an approximation of* $u = B^+ v$. *In order to be able to compare different approximations, two scalar errors are introduced: direction error* $\theta = \mathrm{a\,cos}(v^T \cdot v^* / \|v\|_2 \|v^*\|_2)$ *and magnitude error* $\|e\|_2 = \|v - v^*\|_2$. *The direction error represents the angle between* v *and* v^*, *while the magnitude error represents the module of the approximation error vector* e *(Figure 5.7). In the case when* $\theta = 0$, *the approximation* v^* *preserves the direction of the original vector* v.

The vector v_{2t}^* in Figure 5.6(a) is not colinear with v_2 and the direction error is $\theta_{2t} = \mathrm{a\,cos}(v_2^T \cdot v_{2t}^* / \|v_2\|_2 \|v_{2t}^*\|_2) = 1.4249°$, while the magnitude error is $\|e_{2t}\|_2 = \|v_2 - v_{2t}^*\|_2 = 0.1227$. On the other side, vector v_{2s}^* is colinear with v_2 and the direction error is $\theta_{2s} = \mathrm{a\,cos}(v_2^T \cdot v_{2s}^* / \|v_2\|_2 \|v_{2s}^*\|_2) = 0.0°$, while the magnitude error is $\|e_{2s}\|_2 = \|v_2 - v_{2s}^*\|_2 = 0.1874$. In general, the T-approximation has lower magnitude error than the S-approximation. However, the direction error is higher for

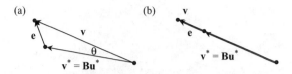

Figure 5.7   *Approximation error* $\mathbf{e} = \mathbf{v} - \mathbf{v}^* = \mathbf{v} - \mathbf{Bu}^*$: *(a) case* $\theta \neq 0$ *and (b) case* $\theta = 0$

the T-approximation than for the S-approximation (actually, direction error is always equal to zero for S-approximation).

In the second step of the algorithm, approximate solutions T_{2t}^* and T_{2s}^* are used as initial points for fixed-point iterations. Individual iterations are shown in Figure 5.6(a) as black dots, if they start from the T-approximation (T_{2t}^*), and as grey dots, if they start from the S-approximation (T_{2s}^*). If the desired virtual control input $\mathbf{v}_2 = S_2$ lies in $\Phi \setminus \Phi_p$, the fixed-point algorithm converges toward the exact solution $T_{2f}^* = P_1$ (regardless of the choice of initial point), which lies on the solution set $\Im_2$ and has lower l_2 norm than any other point in $\Im_2$. The corresponding sequence in the virtual control space converges toward the desired $S_{2f}^* = S_2$ (see Figure 5.6(b)).

5.7 Application to thruster control allocation for over-actuated thruster-propelled UVs

Two ROVs (*FALCON*, SeaEye Marine Ltd. and *URIS*, University of Girona, Figure 5.8) with different thruster configurations are used to demonstrate the performance of the proposed hybrid approach.

The normalised control allocation problem for motion in the horizontal plane is defined by [4,18]

Thruster control matrix (*FALCON*): $\underline{\mathbf{B}} = \begin{bmatrix} \frac{1}{4} & \frac{1}{4} & \frac{1}{4} & \frac{1}{4} \\ \frac{1}{4} & -\frac{1}{4} & \frac{1}{4} & -\frac{1}{4} \\ \frac{1}{4} & -\frac{1}{4} & -\frac{1}{4} & \frac{1}{4} \end{bmatrix}$

Thruster control matrix (*URIS*): $\underline{\mathbf{B}} = \begin{bmatrix} \frac{1}{2} & \frac{1}{2} & 0 & 0 \\ 0 & 0 & \frac{1}{2} & \frac{1}{2} \\ \frac{1}{4} & -\frac{1}{4} & \frac{1}{4} & -\frac{1}{4} \end{bmatrix}$

Thruster control allocation for over-actuated, open-frame UVs

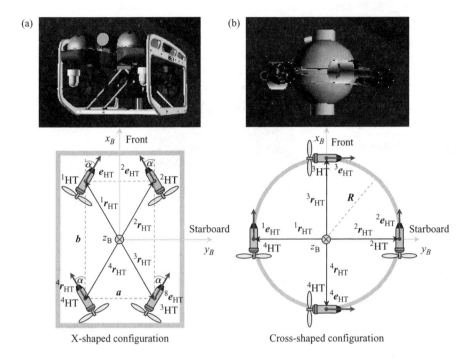

Figure 5.8 Two common configurations of the horizontal thrusters: (a) FALCON and (b) URIS

The virtual control input: $\underline{\tau} = \begin{bmatrix} \underline{\tau}_X \\ \underline{\tau}_Y \\ \underline{\tau}_N \end{bmatrix}$ $(k = 3)$ (5.22)

The true control input: $\underline{\mathbf{u}} = \begin{bmatrix} \underline{u}_1 \\ \underline{u}_2 \\ \underline{u}_3 \\ \underline{u}_4 \end{bmatrix}$ $(m = 4)$ (5.23)

The actuator constraints: $\begin{bmatrix} -1 \\ -1 \\ -1 \\ -1 \end{bmatrix} \leq \begin{bmatrix} \underline{u}_1 \\ \underline{u}_2 \\ \underline{u}_3 \\ \underline{u}_4 \end{bmatrix} \leq \begin{bmatrix} +1 \\ +1 \\ +1 \\ +1 \end{bmatrix}$ (5.24)

The nomenclature for the constrained control subset $\underline{\Omega}$ and the attainable command set $\underline{\Phi}$ was introduced in section 5.3. The shape of $\underline{\Omega}$ is the same for both configurations, while the shape of $\underline{\Phi}$ varies with configuration. Unfortunately, $\underline{\Omega}$ is the four-dimensional hypercube and cannot be easily visualised, in contrast to $\underline{\Phi}$ that is

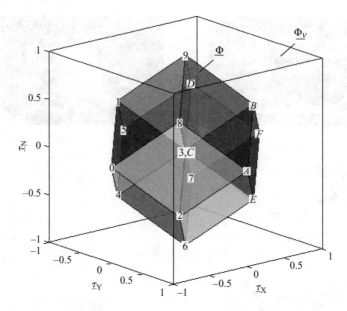

Figure 5.9 Attainable command set $\underline{\Phi}$ for the X-shaped thruster configuration

shown in Figure 5.9 (for the X-shaped thruster configuration) and Figure 5.10 (for the cross-shaped thruster configuration).

Using the MATLAB command `convhulln`, it was found that a convex hull of $\underline{\Phi}$ consists of all but 3 and C vertices for the X-shaped configuration, and 6 and 9 vertices for the cross-shaped configuration. Hence, nodes 3 and C (6 and 9) lie inside $\underline{\Phi}$ for the X-shaped (cross-shaped) configuration.

Since every 3×3 partition of **B** in (5.20) and (5.21) is non-singular, the equation $\mathbf{Bu} = \underline{\tau}$ has a unique solution $\underline{\mathbf{u}} \in \partial(\underline{\Omega})$ on the boundary $\underline{\tau} \in \partial(\underline{\Phi})$.

In the general case, the character of the solution depends on the position of the vector $\underline{\tau}$ relative to $\underline{\Phi}$. Three cases are possible:

1. If $\underline{\tau}$ lies inside $\underline{\Phi}$, then the solution set $\underline{\Im}$ has an infinite number of points and the control allocation problem has an infinite number of solutions.
2. If $\underline{\tau}$ lies on the boundary $\partial(\underline{\Phi})$, then solution set $\underline{\Im}$ is a single point, which is a unique solution of the control allocation problem.
3. Finally, if $\underline{\tau}$ lies outside $\underline{\Phi}$, that is, if $\underline{\tau} \in \underline{\Phi}_v \backslash \underline{\Phi}$, then solution set $\underline{\Im}$ is an empty set, that is, no exact solution exists.

As stated before, in order to extract a unique, 'best' solution from a solution set for case 1, it is necessary to introduce criteria, which are minimised by the chosen solution. The most suitable criterion for underwater applications is a control energy cost function, since minimising this criterion means maximising operational battery life, which is a very important issue for future development of UVs. The hybrid

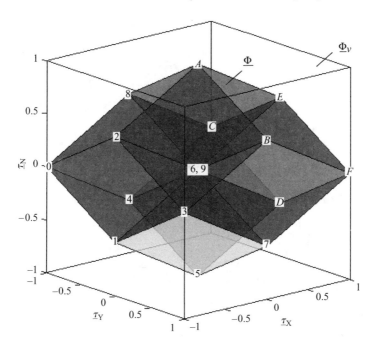

Figure 5.10 Attainable command set $\underline{\Phi}$ for the cross-shaped thruster configuration

approach for control allocation is able to find the exact solution of the problem for cases 1 and 2, optimal in the l_2 sense, and a good approximate solution for case 3.

The input command vector $\underline{\tau}_d = [\tau_X \quad \tau_Y \quad \tau_N]^T$ for motion in the horizontal plane lies in the virtual control space $\underline{\Phi}_v$, that is, the unit cube in $\Re^3$:

$$\underline{\Phi}_v = \{\underline{\tau} \in \Re^3 \mid \|\underline{\tau}\|_\infty \leq 1\} \subset \Re^3 \tag{5.25}$$

For the constrained control allocation problem, where the constraint $\underline{u} \in \underline{\Omega}$ is required to be satisfied, the pseudoinverse solution may become unfeasible, depending on the position of $\underline{\tau}_d$ inside $\underline{\Phi}_v$. The virtual control space $\underline{\Phi}_v$ can be partitioned into characteristic regions, as indicated in Figure 5.11. The two characteristic regions inside $\underline{\Phi}_v$ are $\underline{\Phi}_p$ (the feasible region for the pseudoinverse) and $\underline{\Phi} \supset \underline{\Phi}_p$ (the attainable command set). The shape of $\underline{\Phi}$ is already given in Figure 5.9 for *FALCON* and Figure 5.10 for *URIS*. It should be emphasised that, for the general constrained control allocation problem, there is an infinite number of exact solutions for $\underline{\tau}_d \in \underline{\Phi}$, while no exact solution exists for $\underline{\tau}_d \in \underline{\Phi}_v \backslash \underline{\Phi}$. The pseudoinverse is able to find the exact feasible solution of the control allocation problem, optimal in the l_2 sense, only if $\underline{\tau}_d \in \underline{\Phi}_p$. Otherwise, for $\underline{\tau}_d \in \underline{\Phi}_v \backslash \underline{\Phi}_p$, the solution obtained by pseudoinverse is unfeasible. However, as previously demonstrated, the fixed-point iteration method (activated in the second stage of the hybrid algorithm) is able to find the exact solution, optimal in the l_2 sense, for cases $\underline{\tau}_d \in \underline{\Phi} \backslash \underline{\Phi}_p$.

(a)

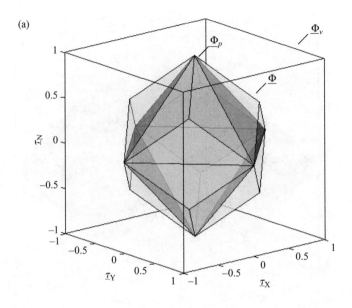

(b)

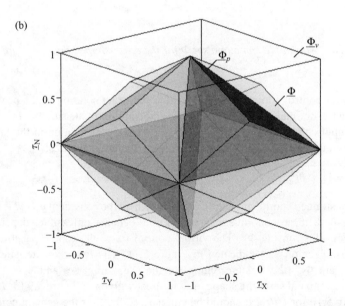

Figure 5.11 Partitions of the virtual control space $\underline{\Phi}_v$: (a) FALCON and (b) URIS

In the following, the hybrid approach will be used to find the exact solution for case $\underline{\tau}_d \in \underline{\Phi}\setminus\underline{\Phi}_p$. Let $\underline{\tau}_d = [0.70\ 0.20\ 0.25]^T$ for *FALCON*. The pseudoinverse solution $\underline{u} = [1.15\ 0.25\ 0.65\ 0.75]^T$ (5.11) is unfeasible, since $\underline{u}_1 > 1$. The T-approximation is: $\underline{u}_t^* = [1\ 0.25\ 0.65\ 0.75]^T$ and the S-approximation

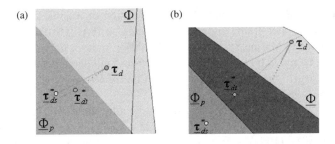

Figure 5.12 Fixed-point iterations in the virtual control space: (a) FALCON and (b) URIS

is: $\underline{\mathbf{u}}_s^* = [1\ 0.2174\ 0.5652\ 0.6522]^T$. These lead to approximate solutions $\underline{\boldsymbol{\tau}}_{dt}^* = [0.6625\ 0.1625\ 0.2125]^T$ and $\underline{\boldsymbol{\tau}}_{ds}^* = [0.6087\ 0.1739\ 0.2174]^T$, respectively (Figure 5.12(a)). Individual fixed-point iterations converge in the true control space toward the exact solution $\underline{\mathbf{u}}_f = [1.00\ 0.10\ 0.80\ 0.90]^T$, optimal in the l_2 sense. Corresponding iterations in the virtual control space are shown in Figure 5.12(a). It can be seen that they converge toward $\underline{\boldsymbol{\tau}}_d$. A similar case for URIS is shown in Figure 5.12(b) for $\underline{\boldsymbol{\tau}}_d = [0.975\ -0.025\ 0.475]^T$. In most cases the number of iterations to achieve desired accuracy is smaller for initial point $\underline{\mathbf{u}}_t^*$ than $\underline{\mathbf{u}}_s^*$, that is, the T-approximation is a better choice for the initial iteration than the S-approximation.

5.8 Conclusions

A hybrid approach for control allocation is described in this chapter. The approach integrates the pseudoinverse and fixed-point iteration method. The standard pseudoinverse method allocates feasible solution only on a subset of the attainable command set. By introducing fixed-point iterations, the feasible region is extended to the entire set and the obtained solution is optimal in the l_2 sense. A clear geometrical interpretation of the approach is demonstrated on the low-dimensional control allocation problem. The hybrid approach is used to solve the control allocation problem for two ROVs with different thruster configuration, overactuated for motion in the horizontal plane. The results confirm that the hybrid approach is able to allocate the entire attainable command set in an optimal way.

References

1 Omerdic, E. and G.N. Roberts (2003). Thruster fault accommodation for underwater vehicles. In 1st IFAC Workshop on Guidance and Control of Underwater Vehicles GCUV'03, 9–11 April, 2003, Newport, South Wales, UK, pp. 221–226.

2 Omerdic, E., G.N. Roberts, and P. Ridao (2003). Fault detection and accommodation for ROVs. In 6th IFAC Conference on Manoeuvring and Control of Marine Craft (MCMC 2003), Girona, Spain.

3 Omerdic, E. and G.N. Roberts (2004). Extension of feasible region of control allocation for open-frame underwater vehicles. In IFAC Conference on Control Applications in Marine Systems (CAMS 2004), Ancona, Italy.
4 Omerdic, E. and G.N. Roberts (2004). Thruster fault diagnosis and accommodation for open-frame underwater vehicles. *Control Engineering Practice*, 12(12), pp. 1575–1598.
5 Durham, W.C. (1993). Constrained control allocation. *Journal of Guidance, Control, and Dynamics*, 16(4), pp. 717–725.
6 Durham, W.C. (1994). Constrained control allocation: three moment problem. *Journal of Guidance, Control, and Dynamics*, 17(2), pp. 330–336.
7 Enns, D. (1998). Control allocation approaches. In AIAA Guidance, Navigation, and Control Conference and Exhibit, Boston, MA, pp. 98–108.
8 Snell, S.A., D.F. Enns, and W.L. Garrard (1992). Nonlinear inversion flight control for a supermaneuverable aircraft. *Journal of Guidance, Control, and Dynamics*, 15(4), pp. 976–984.
9 Virnig, J.C. and D.S. Bodden (1994). Multivariable control allocation and control law conditioning when control effectors limit. In AIAA Guidance, Navigation and Control Conference and Exhibit, Scottsdale.
10 Lindfors, I. (1993). Thrust allocation method for the dynamic positioning system. In 10th International Ship Control Systems Symposium (SCSS'93), Ottawa, pp. 3.93–3.106.
11 Ikeda, Y. and M. Hood (2000). An application of l_1 optimisation to control allocation. In AIAA Guidance, Navigation and Control Conference and Exhibit, Denver, CO.
12 Burken, J., P. Lu, and Z. Wu (1999). Reconfigurable Flight Control Designs with Applications to the X-33 Vehicle. NASA/TM-1999-206582.
13 Burken, J., P. Lu, Z. Wu, and C. Bahm (2001). Two reconfigurable flight-control design methods: robust servomechanism and control allocation. *Journal of Guidance, Control, and Dynamics*, 24(3), pp. 482–493.
14 Bordignon, K.A. (1996). Constrained control allocation for systems with redundant control effectors. Ph.D. thesis, Virginia Polytechnic Institute and State University, Blacksburg, VA.
15 Fossen, T.I. (2002). *Marine Control Systems*. Marine Cybernetics AS, Trondheim, Norway.
16 Härkegård, O. (2003). Backstepping and control allocation with application to flight control. Ph.D. thesis, Department of Electrical Engineering, Linköping University, Sweden.
17 Bodson, M. (2002). Evaluation of optimization methods for control allocation. *Journal of Guidance, Control and Dynamics*, 24(4), pp. 703–711.
18 Omerdic, E. (2004). Thruster fault diagnosis and accommodation for overactuated open-frame underwater vehicles. Ph.D. thesis, Mechatronics Research Centre, University of Wales, Newport, UK.

Chapter 6

Switching-based supervisory control of underwater vehicles

G. Ippoliti, L. Jetto and S. Longhi

6.1 Introduction

Great effort is currently being devoted to the development of underwater robots with self-governing capabilities, able to reliably perform complex tasks in different environments and load conditions. Depending on the experimental situation, the different possible vehicle configurations may or may not be known in advance. However, in general, it is neither *a priori* known when the operating conditions are changed nor which is the new vehicle configuration after the change.

Traditional control techniques can be rather inadequate for the control of these kinds of mode-switch processes, therefore it is useful to develop advanced control techniques in order to supply the underwater vehicle with the necessary 'intelligence' for achieving some degree of self-governing capability. In this regard, different control strategies have been developed and efficient implementations of such controllers on real environments have been proposed. Significant solutions based on adaptive control, robust control, variable structure control and Lyapunov-based control have been recently investigated (see, e.g., References 1–9). The purpose of this chapter is to propose two different methods, both based on switching control. The main feature of switching control strategies is that one builds up a bank of alternative candidate controllers and switches among them according to a suitably defined logic [10–32]. This makes the approach particularly suited to deal with large parametric variations and/or uncertainties. The switching logic is driven by a specially designed supervisor that uses the measurements to assess the performance of the candidate controller currently in use and also the potential performance of alternative controllers. At each time instant, the supervisor decides which candidate controller should be put in the feedback loop with the process. Switching control is also used in different methods and implementations such as, for example, gain scheduling [33, 34] and sliding mode control [35, 36]. All these switching control schemes can be considered as examples of hybrid dynamical systems [37, 38]. Supervisory switching controllers can be divided into two categories: those based on process estimation [10–16, 18, 21, 23–25, 27–30] and those based on a direct performance evaluation of each candidate controller [17, 19, 20, 22, 26, 31, 32].

Estimator-based supervisors continuously compare the behaviour of the process with the behaviour of several admissible process models to determine which model is more likely to describe the actual process. This model is regarded as an 'estimate' of the actual process. In real time, the supervisor places in the loop the candidate controller that is more adequate for the estimated model.

Performance-based supervision is characterised by the fact that the supervisor attempts to assess directly the potential performance of every candidate controller, without trying to estimate the model of the process. The supervisor computes suitable indices that provide a measure of the controller performance and applies the controller for which the corresponding indices are small.

In this chapter, both categories of switching control schemes are investigated and two different control strategies are proposed. The first refers to the case of a poor knowledge of the different possible vehicle configurations, the second concerns the case when adequate *a priori* information is available. Reference is made to the remotely operated vehicle (ROV) developed by the ENI Group (Italy) for the exploitation of combustible gas deposits at great sea depths. The proposed switching controllers have been applied to the position control problem of this underwater vehicle that, during its tasks, is subjected to different load configurations, which introduce considerable variations of its mass and inertial parameters. On the basis of the performed numerical simulations, satisfactory performance of the proposed control systems seem to be really attainable.

The chapter is organised in the following way. In Section 6.2, basic definitions and concepts of supervisory control based on multiple models are introduced. The estimation-based switching control and the hierarchically supervised switching control are described in Sections 6.3 and 6.4 respectively. In Section 6.5, the stability analysis for the proposed switching-based supervisory controllers is reported; some details on the vehicle dynamics and on its linearised model are recalled in Section 6.6. The results of the numerical simulations are reported in Section 6.7.

6.2 Multiple models switching-based supervisory control

In this section, basic definitions and concepts of supervisory control based on multiple models are introduced. This approach has been developed in recent years for improving the performance of dynamical systems operating in rapidly varying environments. The main reason for using multiple models is to ensure the existence of at least one model sufficiently close to the unknown plant at each time instant [18, 29].

The approach based on a multiple model logic-based switching control law has been considered here for developing two different solutions to the vehicle control problem. The first method is an estimation-based switching control (EBSC), which consists of an adaptive control policy improved by the connection with a supervised switching logic. The EBSC algorithm is applied when the different possible vehicle configurations are not known. The second method is a hierarchically supervised switching control (HSSC), which consists of the connection of a multiple bank of fixed candidate controllers with two supervisors operating at two different hierarchical levels. The HSSC algorithm is applied when an adequate information on the different possible vehicle configurations is available.

In both cases, N identification models M_i, $i = 1, \ldots, N$, are used in parallel with the given plant, the objective is to determine which among them is closest (according to some criterion) to the plant at any given instant.

In the EBSC approach, for each possible plant model M_i, $i = 1, \ldots, N$, a corresponding controller C_i, $i = 1, \ldots, N$, is designed [18, 29]. In general, some of these controllers are fixed while the others are adaptively updated to the incoming information on the plant dynamics. The structure of the control system is shown in Figure 6.1.

The performance of each controller can be evaluated only after it is used because, at any time instant, only one control input can be chosen. On the contrary, a parallel evaluation of all the identification models is possible. Therefore, the output responses $\hat{y}_i(\cdot)$, $i = 1, 2, \ldots, N$, of the multiple identification models are used to estimate the performance of the corresponding controller. At every instant, the identification errors $\hat{e}_i(\cdot) := \hat{y}_i(\cdot) - y(\cdot)$, where $y(\cdot)$ is the plant output, are determined and suitable performance indices $J_i(\cdot)$, $i = 1, 2, \ldots, N$, functions of $\hat{e}_i(\cdot)$ are computed. The model corresponding to $\min_i J_i(\cdot)$, $i = 1, \ldots, N$, is chosen by the supervisor S to determine the plant control input at that time instant.

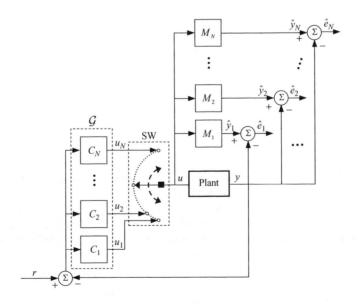

Figure 6.1 Estimation-based switching controller

108 Advances in unmanned marine vehicles

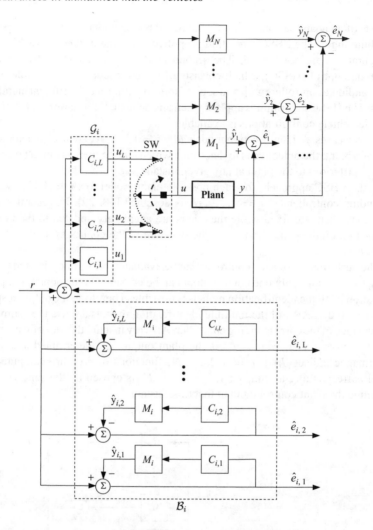

Figure 6.2 Hierarchically supervised switching controller

In the HSSC approach, a family of fixed controllers is designed for each plant model M_i, $i = 1, \ldots, N$. The HSSC consists of the connection of an estimation-based switching logic with a predicted performance-based switching control. This approach has the advantage of not requiring any sequential adaptation of controller parameters but can be applied only if the possible different operating conditions of the plant are known *a priori*. The structure of the proposed controller is shown in Figure 6.2 where it is realised through the hierarchical connection of two supervisors S_1 and S_2.

In brief, the logic of the overall control law can be explained as follows. A set of linearised models M_i, $i = 1, \ldots, N$, of the plant is defined. Each model describes the plant dynamics corresponding to a particular operating condition. For each M_i, a proper set $\mathcal{G}_i$ of differently tuned controllers $C_{i,j}$, $j = 1, \ldots, L$, is designed. For each M_i, the hierarchically lower supervisor S_2 selects the most appropriate $C_{i,j}$ among the elements of $\mathcal{G}_i$. The selection is performed minimising a suitably defined 'distance' between the predicted and the desired output response. The task of hierarchically higher supervisor S_1 is to govern the switching among the different families of controllers $\mathcal{G}_i$, $i = 1, \ldots, N$, recognising which is the linearised model M_i with output $\hat{y}_i(\cdot)$ corresponding to the current operating condition. This last operation is accomplished minimizing the performance indices $J_i(\cdot)$, $i = 1, 2, \ldots, N$, functions of the identification errors $\hat{e}_i(\cdot) := \hat{y}_i(\cdot) - y(\cdot)$, like in the EBSC approach.

The need to define a family $\mathcal{G}_i$ of controllers for each possible M_i is motivated by the consideration that if the transient specifications are too strict, a single time-invariant controller does not yield fully satisfactory closed-loop performance. For example, a regulator producing a step response with a short rise time is likely to produce a large overshoot and/or poorly damped oscillations. On the other hand, a smooth behaviour of the step response is often coupled with too long rise and settling times. To reduce the aforementioned inconveniences, it appears quite natural to design a time-varying controller which is suitably modified, from time to time, according to the characteristics of the produced output response [17].

6.3 The EBSC approach

The structure of the control system is defined assuming N linear discrete-time models M_i, $i = 1, \ldots, N$, of the plant, where the l-vector of model parameters $\boldsymbol{\theta}_i$ takes value in the box $\Theta \subset \mathbb{R}^l$ defined by:

$$\Theta := \{\boldsymbol{\theta} \in \mathbb{R}^l, \; \theta_i \in [\theta_i^-, \theta_i^+], \; i = 1, 2, \ldots, l\} \qquad (6.1)$$

Models M_i represent the different plants corresponding to N possible different operating conditions. In particular, autoregressive moving average with an exogenous variable (ARMAX) models are assumed and generalised minimum variance regulators C_i are designed for controlling the corresponding models M_i, $i = 1, \ldots, N$. These controllers guarantee the minimisation of the following cost function I_i [39]:

$$I_i = E\{\hat{y}_i^2(k+1) + r_i u^2(k)\}, \qquad i = 1, \ldots, N, \quad k \in \mathbb{Z}^+ \qquad (6.2)$$

where r_i is a positive scalar, $\hat{y}_i(k)$ is the output of model M_i and $u(k)$ is the plant input (see Figure 6.1)

At every time instant k, only one controller of the set C_i, $i = 1, \ldots, N$, is applied to the plant and a parallel architecture of N models M_i, $i = 1, \ldots, N$, forced by the same control input of the plant, yields the outputs $\hat{y}_i(k)$, $i = 1, \ldots, N$, which are used to identify the model which best approximates the plant.

The outputs of the multiple models are used to obtain an estimate of the controllers' performance. Defining the identification errors as:

$$\hat{e}_i(k) := \hat{y}_i(k) - y(k), \qquad i = 1, \ldots, N \qquad (6.3)$$

the following performance index is defined:

$$J_i(k) := \alpha \hat{e}_i^2(k) + \beta \sum_{\tau=k_0}^{k-1} \lambda^{(k-\tau-1)} \hat{e}_i^2(\tau), \quad i = 1, \ldots, N, \quad k \geq 0, \quad k \in \mathbb{Z} \tag{6.4}$$

where $\alpha > 0$, $\beta > 0$ and $0 < \lambda < 1$ are design parameters and k_0 is the initial time instant [27]. Parameters α and β determine the relative weight given by instantaneous and long-term measures respectively, while λ determines the memory of the index.

At every time instant k, the performance indices $J_i(k)$, $i = 1, 2, \ldots, N$, are monitored by the supervisor S which acts on the switching box (SW) to select the controller $C_{\bar{i}}$ corresponding to the model $M_{\bar{i}}$ with the minimum functional $J_{\bar{i}}(k)$ (see Figure 6.1).

The switching from the actual controller C_i towards $C_{\bar{i}}$ is performed only if the relative functionals satisfy the following hysteresis condition [27]:

$$J_{\bar{i}}(k) < J_i(k)(1 - q) \tag{6.5}$$

where $0 < q < 1$ is a free design parameter and J_i is the value of the functional corresponding to the actual C_i.

6.3.1 An implementation aspect of the EBSC

For the above control architecture to be efficient, neither the choice of all adaptive models nor that of all fixed models are advisable. In fact, while using adaptive models requires many update equations at every time instant (one equation for each parameter vector $\boldsymbol{\theta}_i$) entailing a significant overhead in computation, fixed models require a large value for N (number of models) to assure good performance. Therefore, a control structure composed by a combination of adaptive and fixed controllers is proposed in order to obtain a zero steady error by adaptive controllers and small transient errors by fixed controllers with the best tradeoff between performance and computation efficiency [27]. In particular, in the implementation of the EBSC approach, the following design issue has been addressed where the convergence of the minimum variance self-tuning regulator has been improved by including a re-initialized adaptive model M_N (in addition to the free running adaptive model M_{N-1}) and the corresponding re-initialized performance index [27]. If there is a switch to a fixed model M_j at the instant k, the parameter vector $\boldsymbol{\theta}_N(k)$ of M_N is reset to $\boldsymbol{\theta}_j(k)$; M_N is left to adapt from this value until a different fixed model is chosen [27]. A well-known fact about adaptive systems is that convergence is fast when the initial parametric error is small. This is precisely what is accomplished by using switching between the fixed models to determine good initial conditions for the adaptation of M_N. In the developed solution for controlling the considered ROV, this combination of $N - 2$ fixed models and two adaptive models is implemented. This solution uses switching to rapidly obtain a rough initial estimate followed by tuning to improve accuracy at each plant transition.

6.4 The HSSC approach

The structure of the control system is defined with reference to a vehicle working at N different known operating conditions. A bank of L different PID regulators is designed for each of the N operating conditions. At every time instant, the selected controller $C_{i,j}$, $i = 1, \ldots, N$, $j = 1, \ldots, L$, is applied to the plant. As in the EBSC, a parallel architecture of N models M_i, $i = 1, 2, \ldots, N$, is defined, where models M_i represent the different plants corresponding to N possible different operating conditions.

To meet the requirements on the transient output response, N families $\mathcal{G}_i$, $i = 1, 2, \ldots, N$, of classical stabilising PID controllers $C_{i,j}$, $j = 1, 2, \ldots, L$, have been designed, each one for each possible configuration M_i of the plant. The purpose is to define, for each M_i, an overall time-varying control strategy picking up the best features of each single $C_{i,j}$ (see Figure 6.2). An advantage of using PID controllers is their robustness with respect to some degree of uncertainty in the parameter values.

Denote by $\Sigma_{i,j}$ and $S_{i,j}$, $i = 1, \ldots, N$, $j = 1, \ldots, L$, the closed-loop systems given by the feedback connection of $C_{i,j}$ with the related plant model M_i and with the true plant, respectively. By the procedure described in Reference 40, for a chosen i, controllers $C_{i,j}$ have been designed so that the related systems $\Sigma_{i,j}$, $j = 1, 2, \ldots, L$, have the distributions of chosen closed-loop poles to guarantee different dynamic characteristics. The models M_i are forced by the same control input of the plant and yield the outputs $\hat{y}_i(\cdot)$, $i = 1, \ldots, N$, which are used to identify the model which best approximates the plant (see Figure 6.2). To this purpose, the same performance index $J_i(\cdot)$ defined by (6.4) has been adopted.

At every time instant, the performance index $J_i(\cdot)$ is monitored by supervisor $\mathcal{S}_1$ to deduce the controller bank $\mathcal{G}_{\bar{i}}$ to be applied to the plant, that is, the bank $\mathcal{G}_{\bar{i}}$ of L different PID controllers corresponding to the model $M_{\bar{i}}$ with the minimum index $J_{\bar{i}}$ (see Figure 6.2). The switching from the actual controller bank $\mathcal{G}_i$ towards $\mathcal{G}_{\bar{i}}$ is performed only if the relative functionals satisfy the hysteresis condition (6.5).

Supervisor $\mathcal{S}_2$ selects the PID regulator $C_{i,j}$, $j = 1, 2, \ldots, L$, according to the switching policy outlined in the following subsection.

6.4.1 The switching policy

Let $\mathcal{G}_i$ be the controller bank chosen by $\mathcal{S}_1$ and let $C_{i,j}$, $j = 1, 2, \ldots, L$, be the corresponding PID controller chosen by $\mathcal{S}_2$ and connected to the plant. The performance of the different possible controllers $C_{i,j}$ are evaluated on the basis of the predicted output of the corresponding closed loop systems $\Sigma_{i,j}$, $j = 1, 2, \ldots, L$ (see Figure 6.2). Defining the output prediction errors as:

$$\hat{e}_{i,j}(k+h) := r(k+h) - \hat{y}_{i,j}(k+h), \qquad h = 1, 2, \ldots, p, \quad j = 1, 2, \ldots, L \tag{6.6}$$

where $r(k)$ is the external reference and $\hat{y}_{i,j}(k+h)$ is the predicted output of system $\Sigma_{i,j}$ computed by setting system $\Sigma_{i,j}$ to the same initial condition of $S_{i,j}$ at time instant k. Then the following functionals are defined:

$$J_{i,j}(k) := \alpha_i \hat{e}_{i,j}^2(k) + \beta_i \sum_{h=1}^{p} \lambda_i^h [\hat{e}_{i,j}^2(k+h)(1+n_j v_j)] \qquad j = 1, 2, \ldots, L$$

where $\alpha_i > 0$, $\beta_i > 0$ and $0 \leq \lambda_i < 1$ are free design parameters which determine the relative weight of instantaneous, future long-term errors and the memory of the index respectively, n_j is a non-negative integer which is increased by 1 whenever a sign change of $\hat{e}_{i,j}(k+h)$ is observed, and v_j is a fixed percentage of $\hat{e}_{i,j}(k+h)$. The terms n_j and v_j have been introduced to penalise excessive oscillations.

At each time instant k, performance indices $J_{i,j}(k)$ are monitored by supervisor S_2 to deduce the controller to be applied to the plant, that is, through the SW, supervisor S_2 imposes the controller $C_{i,\bar{j}}$ of the bank $\mathcal{G}_i$ producing the minimum index $J_{i,\bar{j}}(k)$. The switching from the actual controller $C_{i,j}$ towards $C_{i,\bar{j}}$ is performed only if $J_{i,\bar{j}}(k) < J_{i,j}(k)(1-q_i)$, where $0 < q_i < 1$ is a free design parameter and $J_{i,j}(k)$ is the value of the functional corresponding to the actual $C_{i,j}$.

6.5 Stability analysis

For linear plant, the stability analysis for the above switching-based supervisory controllers is reported in the following subsections. Application to the considered nonlinear problem is justified by the possibility of performing an accurate linearisation of the ROV dynamics as discussed in Reference 41.

6.5.1 Estimation-based supervisory control

Consider a minimum variance self-tuning control of an unknown SISO linear time-invariant system described by an ARMAX model. The parameters of the minimum variance controller depend on the unknown plant parameters. In an adaptive control scheme [39, 42, 43] these parameters can be tuned through an estimate of the plant parameters (minimum variance self-tuning regulator). Denoting with $\widehat{\theta}(k)$ the estimate of plant parameters θ at time k obtained by the weighted extended recursive least squares (ERLS) estimation algorithm [42], the stability of the closed-loop system is guaranteed if:

$$\lim_{k \to \infty} \widehat{\theta}(k) \longrightarrow \theta \quad \text{a.s.} \tag{6.7}$$

(where a.s. stands for almost surely).

The considered ERLS parameter-estimation method satisfies the above condition (6.7) if the convergence conditions stated in Reference 42 are met.

Consider now the proposed EBSC, implemented with $N-1$ fixed minimum variance regulators, one for each fixed model M_i, $i = 1, \ldots, N-1$, and one self-tuning minimum variance regulator based on the adaptive model M_N. At every time instant the regulator corresponding to the model with the minimum performance index (6.4) is applied to the unknown plant. If the convergence conditions for the ERLS parameter-estimation method are met [42], the parameters of the adaptive model asymptotically converge a.s. to the true values. Therefore the adaptive regulator R_N is stabilising for the unknown plant [42] and moreover the steady-state prediction error:

$$\bar{e}_N(k) := \lim_{k_0 \to -\infty} \hat{e}_N(k), \quad k \in \mathbb{Z}, \tag{6.8}$$

is zero a.s., entailing the zeroing of the steady-state performance index $\bar{J}_N(k) = \lim_{k_0 \to -\infty} J_N(k)$, where k_0 is the initial time instant. Hence, in the steady-state condition, the control algorithm can switch on the adaptive controller C_N or a fixed controller C_j characterised by:

$$\bar{J}_j(k) = \lim_{k_0 \to -\infty} J_j(k) = 0 \tag{6.9}$$

Condition (6.9) implies that the steady-state prediction error $\bar{e}_j(k)$ is zero and by the technical Lemma stated in Reference 18, it follows that controller C_j is stabilising for the unknown plant.

Concluding, in the steady-state condition the EBSC applies the adaptive controller C_N or the fixed controller C_j a.s. and both stabilising for the unknown plant.

The hysteresis algorithm (6.5) is used for reducing the frequency of switching phenomena and for stopping the switching in a finite time to the controller for which the prediction error tends to zero. The same stability conditions also hold for the EBSC implemented as in Section 6.3.1.

6.5.2 Hierarchically supervised switching control

With reference to the ideal linear case, the stability analysis of the HSSC relies on the consideration that, while a fixed PID regulator $C_{i,j}$ is acting, the closed-loop system $\Sigma_{i,j} \equiv S_{i,j}$ behaves like a stable, linear, time-invariant system. Hence, closed-loop stability follows if a sufficiently long dwell-time is allowed, between any two intervals where switching occurs. In fact, the asymptotic stability of each $\Sigma_{i,j}$, $i = 1,\ldots,N$, $j = 1,\ldots,L$, does not imply the stability of the resulting time-varying switched closed-loop system, unless some conditions on the switching sequence are imposed. In particular, it is here assumed that each time interval of a fixed length T, where switching may occur, is followed by a suitably defined dwell-time of length $T_{D,k}$, $k = 0, 1, \ldots$, where switching is forbidden. The length of each dwell time interval which guarantees internal stability depends on the particular fixed stabilizing controller where switching stopped at the end of each switching interval T and can be computed in the following way.

Let $[t_{\ell-1}, t_\ell)$ be the time interval over which the plant is in the fixed configuration of model M_i and the time-varying controller is switching inside family $\mathcal{G}_i$. Denote by $A_{i,j}$, $j = 1, 2, \ldots, L$, the dynamic matrix of the time-invariant $\Sigma_{i,j}$ at the end of each switching interval T, when the switching stopped on a particular $C_{i,j} \in \mathcal{G}_i$, moreover let $\Phi_i(t_\ell - 1, t_{\ell-1})$ be the state transition matrix of the closed-loop switched system over $[t_{\ell-1}, t_\ell)$. Let Δ_i be defined as $\Delta_i \triangleq \max_j \|A_{i,j}\|$, then, for any $A_{i,j}$, $j = 1, 2, \ldots, L$, the associated T_{D_j} is chosen as:

$$T_{D,j} = \min t : \{\|A_{i,j}^t\|\Delta_i^T < K_M < 1\} \tag{6.10}$$

Condition (6.10) guarantees $\|\Phi_i(t_\ell - 1, t_{\ell-1})\| < 1$, $i = 1, 2, \ldots, N$, $\ell = 1, 2, \ldots$. The value of K_M has to be chosen sufficiently smaller than unity to compensate the possible closed-loop instability occurring after a plant transition and before supervisor S_1 identifies the new appropriate controller bank. The time interval T has to be chosen on the basis of the estimated transient duration corresponding to a step change of the external reference. Once the transient settles, switching can be stopped without problems because any fixed $C_{i,j}$ is able to adequately track a constant signal. Of course, during the dwell time $T_{D,j}$ chosen according to (6.10), no step change of the external reference should occur. For the considered application this is not a strict requirement.

6.6 The ROV model

The ROV considered in this chapter is equipped with four thrusters and connected with the surface vessel by a supporting cable (see Figure 6.3).

The control system is composed of two independent parts, the first monitoring the vehicle depth, placed on the surface vessel, and the second monitoring the position and orientation of the vehicle, placed on the vehicle itself.

Let us consider the inertial frame $R(0, x, y, z)$ and the body reference frame $R_a(0_a, x_a, y_a, z_a)$ [2]. The ROV position with respect to R is expressed by the origin of the system R_a, while its orientation by roll, pitch and yaw angles ψ, θ and ϕ, respectively. As the depth z is controlled by the surface vessel, the ROV is considered to operate on surfaces parallel to the x–y plane. The controllable variables are x, y and the yaw angle ϕ. The roll and pitch angles ψ and θ will not be considered in the dynamic model; their values have been proved to be negligible in a wide range of load conditions and with different intensities and directions of the underwater current [2, 6], as confirmed by experimental tests [41]. In fact, the vehicle has been designed with the centre of buoyancy coincident with the centre of mass. The vehicle is not buoyant, its weight in water is very high and it is suspended from the surface vessel by a cable connected to the top of the vehicle. The centre of mass is always under the hanging point and the umbilical and the suspending cable are hinged to the vehicle structure allowing the vehicle to rotate freely with respect to the cable. With this mechanical configuration the vehicle is able to maintain its trim independently of surface vessel position and underwater current. Moreover, due to the vehicle mass the velocity and acceleration are always low implying a negligible influence on pitch and roll variations.

The following equations, describing the ROV motion, are obtained by the analysis of the applied forces [2, 6]:

$$p_1\ddot{x} + (p_2|\cos(\phi)| + p_3|\sin(\phi)|)V_x|V| + p_4 x - p_5 V_{cx}|V_c| = T_x \quad (6.11)$$

$$p_1\ddot{y} + (p_2|\sin(\phi)| + p_3|\cos(\phi)|)V_y|V| + p_4 y - p_5 V_{cy}|V_c| = T_y \quad (6.12)$$

$$p_6\ddot{\phi} + p_7\dot{\phi}|\dot{\phi}| + p_8|V_c|^2 \sin\left(\frac{\phi - \phi_c}{2}\right) + p_9 = M_z \quad (6.13)$$

(a)

(b)

Figure 6.3 (a) The remotely operated vehicle (ROV). (b) The surface vessel

where $V_c = [V_{cx}, V_{cy}]^T$ is the time-invariant underwater current velocity and $V = [V_x, V_y]^T = [(\dot{x} - V_{cx}), (\dot{y} - V_{cy})]^T$; the quantities T_x, T_y and M_z are the decomposition of the thrust and torque provided by the four propellers along the axes of R [6]. Coefficients p_i, ($i = 1, \ldots, 9$) of the above equations depend on the considered three different load configurations [2, 6] and on the environment conditions, so that their values significantly vary during the ROV tasks. The interactions produced by the cable on the vehicle position are considered in the parameters p_4 and p_5 which are functions of the cable diameter and length [2, 6].

6.6.1 The linearised model

Defining the following state, input and output variables:

$$x = [x_1, x_2, x_3, x_4, x_5, x_6]^T = [x - x_0, y - y_0, \phi - \phi_0, \dot{x} - \dot{x}_0, \dot{y} - \dot{y}_0, \dot{\phi} - \dot{\phi}_0]^T$$
$$u = [T_x - T_{x0}, T_y - T_{y0}, M_z - M_{z0}]^T$$
$$y = [x - x_0, y - y_0, \phi - \phi_0]^T$$

the linearisation of (6.11)–(6.13) around an operating point $[x_0, y_0, \phi_0, \dot{x}_0, \dot{y}_0, \dot{\phi}_0, T_{x0}, T_{y0}, M_{z0}]$ results in a 3×3 transfer matrix $W(s)$ [2, 41, 6]. The further simplifying assumption of a diagonal $W(s)$ can be motivated as in Reference 41 and has been introduced here for testing the applicability of the proposed switching control scheme.

6.7 Numerical results

The switching controllers have been designed assuming a second order model for each diagonal element $W_\ell(s)$, $\ell = 1, 2, 3$, of the transfer matrix $W(s)$ [41]. Element $W_\ell(s)$ can be expressed in the form:

$$W_\ell(s) = \frac{K_\ell \omega_{n\ell}^2}{s^2 + 2\zeta_\ell \omega_{n\ell} s + \omega_{n\ell}^2}, \quad \ell = 1, 2, 3 \quad (6.14)$$

As a consequence of the simplified diagonal form of $W(s)$, the controller is composed of three independent SISO controllers for the x, y and ϕ variables. For the sake of brevity, the numerical tests reported here only refer to the first component of the controlled variables vector, that is the x variable. Namely, they refer to the decoupled Eq. (6.11). A sample period of 0.5 s is considered. For the constructive characteristics of the thruster system, the constraint on the control effort $|T_x| < 1.5 \times 10^4$ N is imposed. The simulation tests have been carried out considering different load conditions at a depth of 200 m and an external reference given by a piecewise constant function with step variations of its amplitude. The vehicle output (the x variable) is affected by sample values of Gaussian white noise with zero mean and variance σ_m^2. By the analysis of sensors installed on the vehicle, σ_m^2 is chosen as 0.5×10^{-4} for the translation movement.

For the implementation of the EBSC, it is assumed that the different possible operating conditions are not known and that the corresponding parameters of the plant model vary in the following compact set: $K_1 \omega_{n_1}^2 \in [0.1139 \times 10^4, 0.78927 \times 10^4]$, $2\zeta_1 \omega_{n_1} \in [0.0301, 0.1103]$ and $\omega_{n_1}^2 \in [0.0329, 0.0428]$. The following models, uniformly distributed over Θ, are used: 28 fixed linear discrete-time models, one free running adaptive model (model 15) and one re-initialised adaptive model (model 30).

For the implementation of the HSSC, it is assumed that the three different possible operating conditions are known. The corresponding parameters of the plant model, that is, the static gain K_1, the natural frequency ω_{n_1} and the relative damping ζ_1, have been identified as in Reference 40 for the three different operating conditions and they assume the values: $K_1 = [2.66 \times 10^{-4}, 24 \times 10^{-4}, 5.77 \times 10^{-4}]$, $\omega_{n_1} = [0.21, 0.18, 0.19]$ and $\zeta_1 = [0.14, 0.19, 0.20]$. For each of the three different load configurations, the corresponding family $\mathcal{G}_i$, $i = 1, 2, 3$, of controllers is composed of three PID regulators $C_{i,j}$, $i, j = 1, 2, 3$. The corresponding third-order, closed-loop systems given by the feedback connection of each $C_{i,j}$ with the related plant model (6.14) are denoted by $\Sigma_{i,j}$, $i, j = 1, 2, 3$. Following the procedure described in Reference 40, controllers $C_{i,j}$ have been designed so that the related systems $\Sigma_{i,j}$ have one real pole $s_i^{(j)}$ and a pair of complex conjugate poles $\alpha_i^{(j)} \pm j \beta_i^{(j)}$. The poles of the nine closed-loop systems $\Sigma_{i,j}$ are the following:

$$s_i^{(j)} = 0.5 \zeta_i^{(j)} \omega_{n_i}^{(j)}$$

$$\alpha_i^{(j)} \pm j \beta_i^{(j)} = \zeta_i^{(j)} \omega_{n_i}^{(j)} \pm \sqrt{(1 - \zeta_i^{(j)2})} \omega_{n_i}^{(j)}, \quad i, j = 1, 2, 3$$

where $\zeta_i^{(j)} = 0.5$, $\omega_{n_i}^{(j)} = \omega_{n_1} + (j - 1) 0.5 \omega_{n_1}$, ω_{n_1} being the natural frequency of the linearised model (6.14) of Eq. (6.11). The distributions of closed-loop poles have been chosen to obtain three system families with different dynamic characteristics. The family of closed-loop systems $\Sigma_{i,1}$, $i = 1, 2, 3$, is expected to exhibit a smooth step response with a little overshoot but, as a counterpart, with a possible long rise time. System family $\Sigma_{i,3}$, $i = 1, 2, 3$ represents the opposite situation, it is expected to exhibit a fast dynamics, characterised by a short rise time but, as a counterpart, by a possible large overshoot. System family $\Sigma_{i,2}$, $i = 1, 2, 3$ represents an intermediate situation. The discretisation of the nine different PID $C_{i,j}$, $i, j = 1, 2, 3$, has been performed according to the procedure described in Reference 40, assuming a sampling period of 0.5 s.

For both the switching controllers, namely, the EBSC and the HSSC, several simulations have been executed considering different operative conditions for the ROV. In the sample reported in Figures 6.4 and 6.5 for the EBSC and in Figures 6.6 and 6.7 for the HSSC, the ROV works at a depth of 200 m with underwater current velocities $V_{cx} = V_{cy} = 0.15$ m/s. The vehicle task is to carry the flow-line frame (the main component of the underwater structure to be assembled) and the relative installation module to the set-point at the time instant $k = 200$ s (see Figures 6.4 and 6.6 for the EBSC and the HSSC, respectively), to maintain the position after the load has been discharged at the time instant $k = 400$ s and to reach without load two different set-points at time instants $k = 600$ s and $k = 800$ s (see Figures 6.5 and 6.7 for the EBSC and the HSSC, respectively).

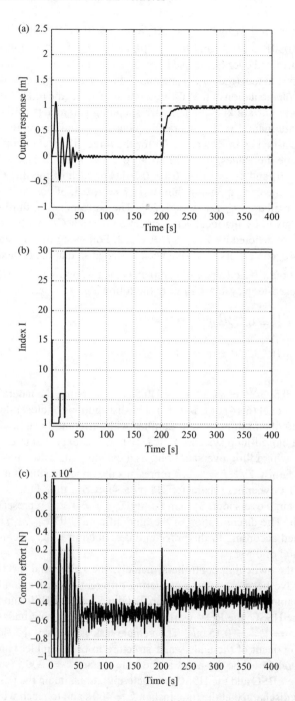

Figure 6.4 Vehicle carrying the load. (a) Output response. (b) Switching sequence of S. (c) Control effort

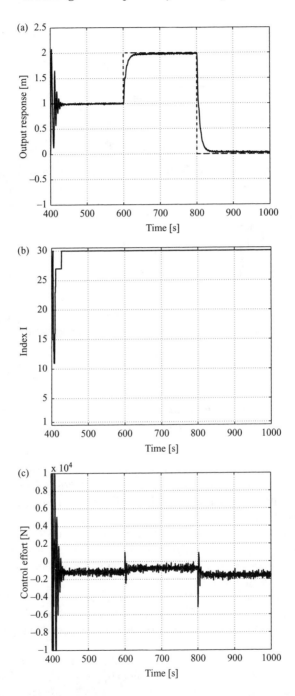

Figure 6.5 Vehicle without load. (a) Output response. (b) Switching sequence of S. (c) Control effort

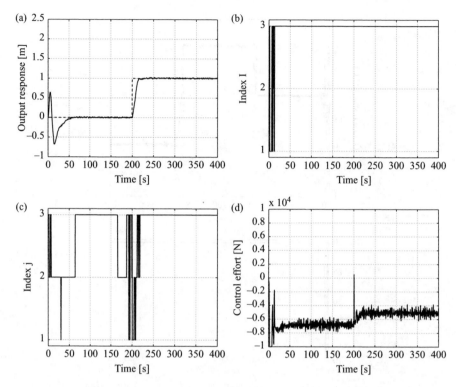

Figure 6.6 Vehicle carrying the load. (a) Output response. (b) Switching sequence of S_1. (c) Switching sequence of S_2. (d) Control effort

Figures 6.4(a), 6.5(a), 6.6(a) and 6.7(a) show the output response produced by the EBSC and the HSSC, respectively. Figures 6.6(b), 6.7(b) 6.6(c) and 6.7(c) show the switching sequences of supervisors S_1 and S_2, respectively, and Figures 6.4(b) and 6.5(b) show the switching sequences of supervisor S.

Numerical results evidence a better performance of the HSSC with respect to the EBSC in terms of set-point tracking capability when unexpected set-point changes occur. This is an expected consequence of the fact that EBSC operates with a lower amount of *a priori* available information.

Figure 6.6(a) at time instant 200 s and Figure 6.7(a) at time instants 600 and 800 s show a step response with a short rise time and with an improved steady state set-point tracking with respect to the step response of the EBSC reported in Figures 6.4(a) and 6.5(a).

Before S or S_1 identify the correct ROV load configuration (see Figures 6.4(b), 6.5(b) and Figures 6.6(b), 6.7(b) for the switching sequence of the EBSC and of the HSSC, respectively), the output response shows a significative transient phase (Figures 6.4(a) and 6.6(a) at time instant 0 s for the EBSC and the HSSC, respectively and Figures 6.5(a) and 6.7(a) at time instant 400 s for the EBSC and the HSSC,

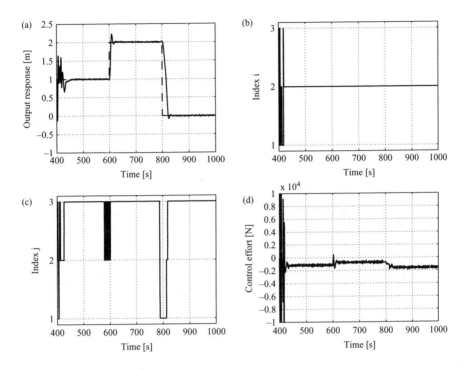

Figure 6.7 Vehicle without load. (a) Output response. (b) Switching sequence of S_1. (c) Switching sequence of S_2. (d) Control effort

respectively). This transient behaviour could be improved in a normal operating condition but is almost unavoidable in the present situation which depicts the worst possible reference scenario for a set-point tracking controller: neither the change instant nor the new operating condition is known *a priori*. Using a single family of PID controllers for the HSSC (namely dropping S_1) or using a single adaptive controller instead of the EBSC produced very poor results when a change of the vehicle configuration occurred and the output response showed an unacceptable oscillatory behaviour. These results are not reported here for brevity.

6.8 Conclusions

Underwater vehicles operate in dynamical environments where sudden changes of the working conditions occur from time to time. The need for an effective control action calls for refined techniques with a high degree of robustness with respect to large parametric variations and/or uncertainties. Supervised switching control gives the theoretical framework where appropriate control strategies can be developed. Both the switching algorithms proposed here are based on a multiple models approach to describe the different operating conditions.

The EBSC algorithm consists of the connection of an estimation-based switching logic with a bank of N minimum variance controllers, two of the which are adaptively adjusted on line. The introduction of a supervised switching logic was able to greatly improve the adaptation capability of the resulting control law, thus enhancing the autonomy of the controlled vehicle. The EBSC method is particularly well suited when the different environments where the vehicle operates are not well known.

The HSSC algorithm is given by the connection of a bank of fixed controller families with two supervisors operating at two different levels of priority. The hierarchical higher supervisor S_1 recognises if the acting control family is appropriate for the actual configuration of the vehicle. This task is accomplished by means of an identification procedure similar to that of the EBSC. The hierarchical lower supervisor S_2 drives the switching logic inside each family of fixed controllers for improving the transient response features. As all the single controllers are designed off-line, the HSSC algorithm is more appropriate when the different possible operating conditions of the vehicle are known *a priori*.

The main merits of the proposed techniques are in terms of their simplicity and robustness with respect to different operating conditions. On the basis of numerical simulations, satisfactory performance of these control systems are shown to be attainable.

References

1 Conte, G. and A. Serrani (1994). H_∞ control of a remotely operated underwater vehicle. In Proceedings ISOPE'94, Osaka.
2 Conter, A., S. Longhi, and C. Tirabassi (1989). Dynamic model and self-tuning control of an underwater vehicle. In Proceedings of the 8th International Conference on Offshore Mechanics and Arctic Engineering, The Hague, pp. 139–146.
3 Corradini, M.L. and G. Orlando (1997). A discrete adaptive variable structure controller for mimo systems, and its application to an underwater ROV. *IEEE Transactions on Control Systems Technology*, 5(3), pp. 349–359.
4 Cristi, R., A.P. Fotis, and A.J. Healey (1990). Adaptive sliding mode control of autonomous underwater vehicles in the dive plane. *IEEE Journal of Oceanic Engineering*, 15(3), pp. 152–160.
5 T.I. Fossen (1994). *Guidance and Control of Ocean Vehicles*. John Wiley & Sons, New York, USA, 1994.
6 Longhi, S. and A. Rossolini (1989). Adaptive control for an underwater vehicle: Simulation studies and implementation details. In Proceedings of the IFAC Workshop on Expert Systems and Signal Processing in Marine Automation, The Technical University of Denmark, Lyngby, Copenhagen, Denmark, pp. 271–280.
7 Longhi, S., G. Orlando, A. Serrani, and A. Rossolini (1994). Advanced control strategies for a remotely operated underwater vehicle. In Proceedings of the First World Automation Congress (WAC'94), Maui, HI, USA, pp. 105–110.

8 Kaminer, I., A.M. Pascoal, C.J. Silvestre, and P.P. Khargonekar (1991). Control of an underwater vehicle using H_∞ synthesis. In Proceedings of the 30th IEEE Conference on Decision and Control, Brighton, England, pp. 2350–2355.
9 Aguiar, A.P. and A.M. Pascoal (2001). Regulation of a nonholonomic autonomous underwater vehicle with parametric modeling uncertainty using lyapunov functions. In Proceedings of the 40th IEEE Conference on Decision and Control, Orlando, FL, USA, pp. 4178–4183.
10 Baldini, M., M.L. Corradini, L. Jetto, and S. Longhi (1999). A multiple-model based approach for the intelligent control of underwater remotely operated vehicles. In Proceedings of the 14th Triennial World Congress of IFAC, Vol. Q, Beijing, P.R. China, pp. 19–24.
11 Borrelli, D., A.S. Morse, and E. Mosca (1998). Discrete-time supervisory control of families of two degrees-of-freedom linear set-point controllers. *IEEE Transactions on Automatic Control*, 44(1), pp. 178–181.
12 D'Amico, A., G. Ippoliti, and S. Longhi (2001). Adaptation and learning in neural networks multiple models based control of mobile robots. In Proceedings of the IFAC Workshop on Adaptation and Learning in Control and Signal Processing (ALCOSP 2001), Villa Erba, Cernobbio-Como, Italy.
13 Hespanha, J. (2001) Tutorial on supervisory control. Lecture notes for the workshop control using logic and switching. In 40th IEEE Conference on Decision and Control (CDC01), Orlando, FL, USA.
14 Hespanha, J.P., D. Liberzon, and A.S. Morse (1999). Logic-based switching control of a nonholonomic system with parametric modeling uncertainty. *Systems & Control Letters*, 38(3), pp. 167–177 (Special Issue on Hybrid Systems).
15 Hespanha, J.P., D. Liberzon, A.S. Morse, B.D.O. Anderson, T.S. Brinsmead, and F. de Bruyne (2001). Multiple model adaptive control, part 2: switching. *International Journal of Robust and Nonlinear Control*, 11(5), pp. 479–496 (Special Issue on Hybrid Systems in Control).
16 Hockerman-Frommer, J., S.R. Kulkarni, and P.J. Ramadge (1998). Controller switching based on output prediction errors. *IEEE Transactions on Automatic Control*, 43(5), pp. 596–607.
17 Ippoliti, G., L. Jetto, and S. Longhi (2003). Improving PID control of underwater vehicles through a hybrid control scheme. In Proceedings of the 1st IFAC Workshop on Guidance and Control of Underwater Vehicles (GCUV '03), Newport, South Wales, UK, pp. 19–24.
18 Ippoliti, G., and S. Longhi (2004). Multiple models for adaptive control to improve the performance of minimum variance regulators. IEE Proceedings – Control Theory and Applications, 151(2), pp. 210–217.
19 Ippoliti, G., L. Jetto, S. Longhi, and V. Orsini (2004). Adaptively switched set-point tracking PID controllers. In Proceedings of the 12th Mediterranean Conference on Control and Automation (MED'04), Kusadasi, Aydin, Turkey.
20 Ippoliti, G., L. Jetto, and S. Longhi (2004). Hierarchical switching scheme for PID control of underwater vehicles. In Proceedings of the IFAC Conference on Control Applications in Marine Systems (CAMS 2004), Ancona, Italy, pp. 439–446.

21 Karimi, A., I. D. Landau, and N. Motee (2001). Effects of the design paramters of multimodel adaptive control on the performance of a flexible transmission system. *International Journal of Adaptive Control and Signal Processing*, 15(3), pp. 335–352 (Special Issue on Switching and Logic).
22 Kosut, R.L. (2004). Iterative unfalsified adaptive control: analysis of the disturbance-free case. In Proceedings of the 1999 American Control Conference pp. 566–570.
23 Morse, A.S. (1995). Control using logic-based switching. In *Trends in Control: A European Perspective*, A. Isidori, Ed. Springer-Verlag, London, UK, pp. 69–113.
24 Morse, A.S. (1996). Supervisory control of families of linear set-point controllers – Part I: exact matching. *IEEE Transactions on Automatic Control*, 41(10), pp. 1413–1431.
25 Morse, A.S. (1997). Supervisory control of families of linear set-point controllers – Part 2: robustness. *IEEE Transactions on Automatic Control*, 42(11), pp. 1500–1515.
26 Mosca, E., F. Capecchi, and A. Casavola (2001). Designing predictors for mimo switching supervisory control. *International Journal of Adaptive Control and Signal Processing*, 15(3), pp. 265–286 (Special Issue on Switching and Logic).
27 Narendra, K.S., J. Balakrishnan, and M.K. Ciliz (1995). Adaptation and learning using multiple models, switching, and tuning. *IEEE Control Systems Magazine*, 15(3), pp. 37–51.
28 Narendra, K.S. and J. Balakrishnan (1997). Adaptive control using multiple models. *IEEE Transactions on Automatic Control*, 42(2), pp. 171–187.
29 Narendra, K.S. and C. Xiang (2000). Adaptive control of discrete-time systems using multiple models. *IEEE Transactions on Automatic Control*, 45(9), pp. 1669–1686.
30 Narendra, K.S. and O.A. Driollet (2001). Stochastic adaptive control using multiple models for improved performance in the presence of random disturbances. *International Journal of Adaptive Control and Signal Processing*, 15(3), pp. 297–317 (Special Issue on Switching and Logic).
31 Safonov, M.G. and T.-C. Tsao (1994). The unfalsified control concept and learning. In Proceedings of the 33rd Conference on Decision and Control, pp. 2819–2824.
32 Woodley, B.R., J.P. How, and R. L. Kosut (1999). Direct unfalsified controller design solution via convex optimization. In Proceedings of the 1999 American Control Conference, pp. 3302–3306.
33 Blanchini, F. (2000). The gain scheduling and the robust state feedback stabilization problems. *IEEE Transactions on Automatic Control*, 45(11), pp. 2061–2070.
34 Rugh, W.J. and J.S. Shamma (2000). Research on gain scheduling. *Automatica*, 36, pp. 1401–1425.
35 Corradini, M.L. and G. Orlando (1998). Variable structure control of discretized continuous-time systems. *IEEE Transactions on Automatic Control*, 43(9), pp. 1329–1334.
36 Utkin, V.I. (1992). *Sliding Modes in Control and Optimization*. Springer-Verlag, New York, USA.

37 Antsaklis, P.J. and A. Nerode (1998). Hybrid control systems: an introductory discussion to the special issue. *IEEE Transactions on Automatic Control*, 43(4), pp. 457–460.
38 Morse, A.S., C.C. Pantelides, S.S. Sastry, and J.M. Schumacher (1999). Introduction to the special issue on hybrid systems. *Automatica*, 35, pp. 347–348.
39 Isermann, R., K.H. Lachmann, and D. Matko (1992). *Adaptive Control Systems*. Prentice-Hall, UK.
40 Astrom, K.J. and T. Hagglund (1995). *PID Controllers: Theory, Design, and Tuning*, 2nd edition. ISA – The Instrumentation, Systems, and Automation Society.
41 Ippoliti, G., S. Longhi, and A. Radicioni (2002). Modelling and identification of a remotely operated vehicle. *Journal of Marine Engineering and Technology*, A(AI), pp. 48–56.
42 Goodwin, G.C. and K.S. Sin (1984). *Adaptive Filtering Prediction and Control*. Prentice-Hall, Englewood Cliffs, NJ, USA.
43 Middleton, R.H., G.C. Goodwin, D.J. Hill, and D.Q. Mayne (1988). Design issues in adaptive control. *IEEE Transactions on Automatic Control*, 33(1), pp. 442–463.

Chapter 7

Navigation, guidance and control of the *Hammerhead* autonomous underwater vehicle

D. Loebis, W. Naeem, R. Sutton, J. Chudley and A. Tiano

7.1 Introduction

The development of autonomous underwater vehicles (AUVs) for scientific, military and commercial purposes in applications such as ocean surveying [1], unexploded ordnance hunting [2] and cable tracking and inspection [3] requires the corresponding development of navigation, guidance and control (NGC) systems, which should work in accord with each other for proper operation. Navigation systems are necessary to provide knowledge of vehicle position and attitude. The guidance systems manipulate the output of the navigation systems to generate suitable trajectories to be followed by the vehicle. This takes into account the target and any obstacles that may have been encountered during the course of a mission. The control systems are responsible for keeping the vehicle on course as specified by the guidance processor. In the *Hammerhead* AUV, this is achieved through manipulating the rudder and the hydroplanes (canards) of the vehicle. The need for accuracy in NGC systems is paramount. Erroneous position and attitude data in navigation systems can lead to a meaningless interpretation of the collected data, which in turn affects the accuracy of the corresponding guidance and control systems. This, if not contained properly may lead to a catastrophic failure of an AUV during a specific mission. The integrated NGC of the *Hammerhead* AUV is depicted in Figure 7.1.

Hammerhead, shown in Figure 7.2, was developed from a deep mobile target (DMT) torpedo of 3 m length and 30 cm diameter that was purchased by Cranfield University (CU). Initial modifications were made to transform the torpedo into a PC controlled AUV [4]. Subsequently, research teams from the University of Plymouth (UP) and CU have successfully developed an integrated NGC system for the vehicle. In this collaborative work, CU developed a navigation subsystem based on a laser stripe illumination methodology developed previously [5]. Interested readers on the

128 *Advances in unmanned marine vehicles*

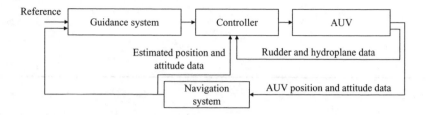

Figure 7.1 *Navigation, guidance and control for the* Hammerhead *AUV*

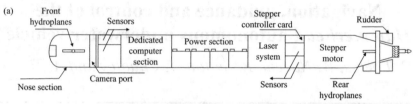

Figure 7.2 *(a) The schematic of the* Hammerhead. *(b) The* Hammerhead *strapped on its trailer*

details of this subsystem are referred to Reference 6 (also see Chapter 13). The focus of this chapter is on the NGC work undertaken by the UP research team. The navigation system is based on a GPS/INS integrated system (GPS, global positioning system; INS, inertial navigation system) equipped with adaptive Kalman filtering (KF) techniques. The proposed guidance laws, namely the pure pursuit and hybrid guidance systems are formulated for cable/pipeline inspection task. Three approaches are suggested for autopilot design: linear quadratic Gaussian controller with loop transfer recovery (LQG/LTR), a model predictive controller (MPC) optimised using a genetic algorithm (GA) and a fuzzy-based GA-MPC.

7.2 The *Hammerhead* AUV navigation system

A growing number of research groups around the world are developing integrated navigation systems utilising INS and GPS [7–9]. However, few of these works make explicit the essential need for fusion of several INS sensors that enable the users to maintain the accuracy or even to prevent a complete failure of this part of navigation system, before being integrated with the GPS. Kinsey and Whitcomb [10], for example, use a switching mechanism to prevent a complete failure of the INS. Although simple to implement, the approach may not be appropriate to use to maintain a certain level of accuracy.

Several estimation methods have been used in the past for multisensor data fusion and integration purpose in AUVs [11]. To this end, simple Kalman filters (SKF) and extended Kalman filters (EKF) and their variants have been popular methods in the past and interest in developing the algorithms has continued to the present day. However, a significant difficulty in designing an SKF/EKF can often be traced to incomplete *a priori* knowledge of the process covariance matrix (**Q**) and measurement noise covariance matrix (**R**). In most practical applications, these matrices are initially estimated or even unknown. The problem here is that the optimality of the estimation algorithm in the SKF/EKF setting is closely connected to the quality of *a priori* information about the process and measurement noise [12]. It has been shown that insufficiently known *a priori* filter statistics can reduce the precision of the estimated filter states or introduces biases to their estimates. In addition, incorrect *a priori* information can lead to practical divergence of the filter [13]. From the aforementioned it may be argued that the conventional SKF/EKF with fixed (**Q**) and/or (**R**) should be replaced by an adaptive estimation formulation as discussed in the next section.

7.2.1 Fuzzy Kalman filter

In this section, an on-line innovation-based adaptive scheme of the KF to adjust the **R** matrix employing the principles of fuzzy logic is presented. The fuzzy logic is chosen mainly because of its simplicity and closeness to human reasoning. These enable a satisfactory performance being developed empirically in practice without complicated mathematics. These have motivated the interest in the topic, as testified by related articles which have been appearing in the literature [14–16].

The fuzzy logic Kalman filter (FKF) proposed herein is based on an innovation adaptive estimation (IAE) approach using a technique known as covariance-matching [12]. The idea behind the technique is to make the actual value of the covariance of the innovation sequences match its theoretical value.

The actual covariance is defined as an approximation of the innovation of time k (Inn_k) sample covariance through averaging inside a moving estimation window of size M [17] which takes the following form:

$$\hat{\mathbf{C}}_{\text{Inn}_k} = \frac{1}{M} \sum_{j=j_0}^{k} \text{Inn}_k \cdot \text{Inn}_k^T \tag{7.1}$$

where $j_0 = k - M + 1$ is the first sample inside the estimation window. An empirical heuristic experiment is conducted to choose the window size M that is adequate to capture the dynamic of the Inn_k actual covariance. From experimentation it was found that a good size for the moving window in Eq. (7.1) used in this work is 15. The value of M is dependent on the dynamic of the Inn_k and therefore can vary for different types of applications.

The theoretical covariance of the innovation sequence is defined as [12]:

$$\mathbf{S}_k = \mathbf{H}_k \cdot \mathbf{P}_k^- \cdot \mathbf{H}_k^T + \mathbf{R}_k \tag{7.2}$$

The logic of the adaptation algorithm using covariance matching technique can be qualitatively described as follows. If the actual covariance value $\hat{\mathbf{C}}_{\text{Inn}_k}$ is observed, whose value is within the range predicted by theory $\mathbf{S}_k$ and the difference is very near to zero, this indicates that both covariances match almost perfectly and only a small change is needed to be made on the value of $\mathbf{R}$. If the actual covariance is greater than its theoretical value, the value of $\mathbf{R}$ should be decreased. On the contrary, if $\hat{\mathbf{C}}_{\text{Inn}_k}$ is less than $\mathbf{S}_k$, the value of $\mathbf{R}$ should be increased. This adjustment mechanism lends itself very well to being dealt with using a fuzzy-logic approach based on rules of the kind:

$$\textbf{IF } \langle \text{antecedent} \rangle \textbf{ THEN } \langle \text{consequent} \rangle \tag{7.3}$$

To implement the above covariance matching technique using the fuzzy logic approach, a new variable called $\textbf{delta}_k$, is defined to detect the discrepancy between $\hat{\mathbf{C}}_{\text{Inn}_k}$ and $\mathbf{S}_k$. The following fuzzy rules of the kind of Eq. (7.3) are used [14]:

$$\textbf{IF } \langle \textbf{delta}_k \cong 0 \rangle \textbf{ THEN } \langle \mathbf{R}_k \text{ is unchanged} \rangle \tag{7.4}$$

$$\textbf{IF } \langle \textbf{delta}_k > 0 \rangle \textbf{ THEN } \langle \mathbf{R}_k \text{ is decreased} \rangle \tag{7.5}$$

$$\textbf{IF} \langle \textbf{delta}_k < 0 \rangle \textbf{ THEN } \langle \mathbf{R}_k \text{ is increased} \rangle \tag{7.6}$$

Thus $\mathbf{R}$ is adjusted according to

$$\mathbf{R}_k = \mathbf{R}_{k-1} + \Delta \mathbf{R}_k \tag{7.7}$$

where $\Delta \mathbf{R}_k$ is added or subtracted from $\mathbf{R}$ at each instant of time. Here $\textbf{delta}_k$ is the input to the fuzzy inference system (FIS) and $\Delta \mathbf{R}_k$ is the output.

On the basis of the above adaptation hypothesis, the FIS can be implemented using three fuzzy sets for $\textbf{delta}_k$: N = Negative, Z = Zero and P = Positive. For $\Delta \mathbf{R}_k$ the fuzzy sets are specified as I = Increase, M = Maintain and D = Decrease.

7.2.2 Fuzzy logic observer

To monitor the performance of an FKF, another FIS called the fuzzy logic observer (FLO) [14] is used. The FLO assigns a weight or degree of confidence denoted as c_k,

Table 7.1 Fuzzy rule based FLO

$|\text{delta}|_k$	$\mathbf{R}_k$		
	Z	S	L
Z	G	G	AV
S	G	AV	P
L	AV	P	P

a number on the interval [0,1], to the FKF state estimate. The FLO is implemented using two inputs: the values of $|\mathbf{delta}_k|$ and $\mathbf{R}_k$. The fuzzy labels for the membership functions: Z = Zero, S = Small and L = Large. Three fuzzy singletons are defined for the output c_k and are labelled as G = Good, AV = Average and P = Poor with values 1, 0.5 and 0, respectively. The basic heuristic hypothesis for the FLO is as follows: if the value of $|\mathbf{delta}_k|$ is near to zero and the value of $\mathbf{R}_k$ is near to zero, then the FKF works almost perfectly and the state estimate of the FKF is assigned a weight near 1. On the contrary if one or both of these values increases far from zero, it means that the FKF performance is degrading and the FLO assigns a weight near 0. Table 7.1 gives the complete fuzzy rule base of each FLO.

7.2.3 Fuzzy membership functions optimisation

GAs in single- and multi-objective mode are used here to optimise the membership functions of the FKF. To translate the FKF membership functions to a representation useful as genetic material, they are parameterised with real-valued variables. Each of these variables constitutes a gene of the chromosomes for the multi-objective genetic algorithm (MOGA). Boundaries of chromosomes are required for the creation of chromosomes in the right limits so that the MOGA is not misled to some other area of search space. The technique adopted in this chapter is to define the boundaries of the output membership functions according to the furthest points and the crossover points of two adjacent membership functions. In other words, the boundaries of FKF consist of three real-valued chromosomes (Chs), as in Figure 7.3.

The trapezoidal membership functions' two furthest points, $-0.135\,(D_1)$, $-0.135\,(D_2)$ and $0.135\,(I_3)$, $0.135\,(I_4)$ of FKF, remain the same in the GA's description to allow a similar representation as the fuzzy system's definition. As can be seen from Figure 7.3, D_3 and M_1 can change value in the 1st Ch boundary, D_4, M_2 and I_1 in the 2nd Ch boundary, and finally, M_3 and I_2 in 3rd Ch. Table 7.2 shows the encoding used for optimisation of the membership functions.

7.2.4 Implementation results

This section discusses the implementation of the FKF optimised using MOGA discussed earlier for fusing heading data acquired during a real-time experiment conducted in the Roadford Reservoir, Devon, UK. The *Hammerhead* AUV model

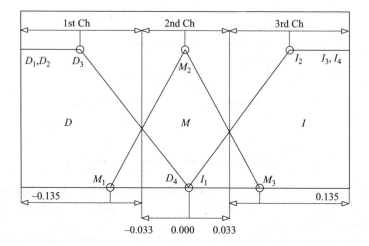

Figure 7.3 Membership function and boundaries of $\mathbf{R}_k$

Table 7.2 FKF boundaries

Limit	Parameter		
	D_3, M_1	D_4, M_2, I_1	M_3, I_2
Upper limit	−0.135	−0.033	0.033
Lower limit	−0.033	0.033	0.135

used herein was derived using SI techniques [18], which will be discussed in Section 7.3. The system matrix (**A**), input matrix (**B**) and output vector (**C**) are:

$$\mathbf{A} = \begin{bmatrix} 0 & 1 \\ -0.98312 & 1.9831 \end{bmatrix}, \quad \mathbf{B} = \begin{bmatrix} -0.003196 \\ -0.0036115 \end{bmatrix}, \quad \mathbf{C} = \begin{bmatrix} 1 & 0 \end{bmatrix}$$

with yaw (x_1) and delayed yaw (x_2) as the states of the system. It is assumed in this model that the forward velocity of the vehicle is constant at 1 m/s and the vehicle is not at an angle of roll and pitch. Process and measurement noise components are both zero mean white noise. Input to the system (indicated by δ_{r_k}) is rudder deflection. This model is assumed to be sufficiently accurate to represent the dynamics of the vehicle, and for this reason, any output produced by the model after being excited by an input, can be considered as an actual output value. This assumption also motivates the use of the model output as a reference in measuring the performance of the FKF-MOGA algorithm.

To test the FKF-MOGA algorithms, real data obtained from a TCM2 electronic compass and an inertial measurement unit (IMU), as a response to the input shown in

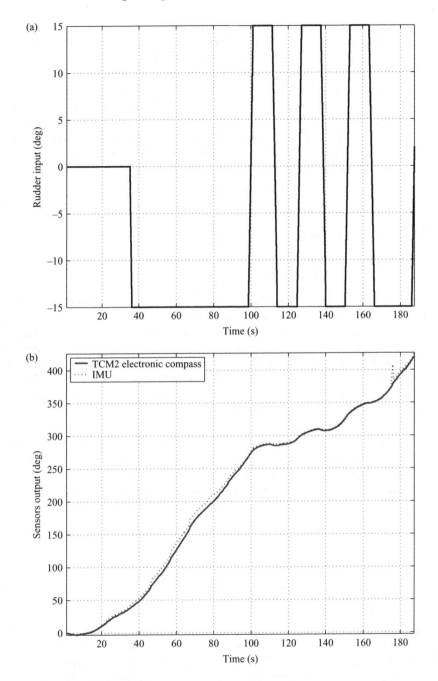

Figure 7.4 (a) Rudder input. (b) TCM2 electronic compass and IMU output

Figure 7.4(a), are fused together with two sets of simulated data. To produce the simulated data, the noise in Figure 7.5(a) and (b) are simply added to the TCM2 electronic compass and IMU real data, respectively. In this particular scenario, the second TCM2 electronic compass (sensor-3) is located in close proximity to the propeller DC motor of the vehicle, whose internal temperature increases with time and affects the sensor ambient temperature. A similar scenario can also be considered to occur when the second IMU (sensor-4) is located in close proximity to the laser unit used in the VNS whose initial internal temperature is high and settles down after sometime. This particular scenario can result in the noise characteristic shown in Figure 7.5(b).

The initial conditions are

$$x_0 = \begin{bmatrix} 0 \text{ rad} \\ 0 \text{ rad} \end{bmatrix}; \quad \mathbf{P}_0 = \begin{bmatrix} 0.01 \text{ rad}^2 & 0 \\ 0 & 0.01 \text{ rad}^2 \end{bmatrix} \quad (7.8)$$

and $\mathbf{Q}_k$ is made constant as

$$\mathbf{Q}_k = \begin{bmatrix} 0 \text{ rad}^2 & 0 \\ 0 & 0.1725 \times 10^{-7} \text{ rad}^2 \end{bmatrix} \quad (7.9)$$

The values of $\mathbf{P}_0$ and $\mathbf{Q}_k$ are determined heuristically. In real-time applications, the $\mathbf{Q}_k$ values are dependent on temporal and spatial variations in the environment such as sea conditions, ocean current, and local magnetic variations and therefore, appropriate adjustments to the initial values of $\mathbf{Q}$ also need to be undertaken. However, given the fact that the *Hammerhead* AUV mostly operates in a stable environment, the problem with the $\mathbf{Q}$ adjustment is reserved for future work. The actual value of $\mathbf{R}$ is assumed unknown, but its initial value is selected according to the heading accuracy of the sensors, that is, 1 deg^2 standard deviation.

The covariance matching technique discussed previously is then implemented to maintain the performance of the estimation process. Subsequent optimisation of $\Delta \mathbf{R}_k$ membership functions using MOGA is done using the parameters shown in Table 7.3. Trade-off graphs of this particular search are shown in Figure 7.6.

Table 7.4 shows the performance of the sensors, indicated by J_{zv} and J_{ze}, where

$$J_{zv} = \sqrt{\frac{1}{n} \sum_{k=1}^{n} (za_k - z_k)^2} \quad (7.10)$$

$$J_{ze} = \sqrt{\frac{1}{n} \sum_{k=1}^{n} (za_k - \hat{z}_k)^2} \quad (7.11)$$

Here, za_k is the actual value of the yaw, z_k is the measured yaw, $\hat{z}_k$ is the estimated yaw at an instant of time k and n = number of samples. A close look on the J_{zv} and J_{ze} of each sensor indicates that the FKF with GA (single objective optimisation) has improved the accuracy of the heading information of sensor-1 to sensor-4. However, the result of fusing the estimated sensor data has shown a slightly inferior performance, indicated by $J_{ze} = 0.2487$ rad, compared to the performance of sensor-1, indicated by $J_{zv} = 0.2340$ rad. This can be understood as a direct result of fusing a relatively

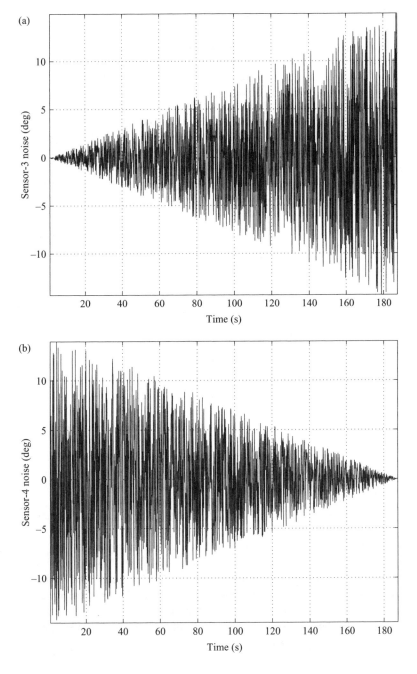

Figure 7.5 (a) *Sensor-3 noise.* (b) *Sensor-4 noise*

Table 7.3   MOGA parameters

Parameters	Values
Number of objective functions	5
Number of generation	25
Number of individual per generation	10
Generation gap in selection operation	0.95
Rate in recombination operation	0.8
Rate in mutation operation	0.09

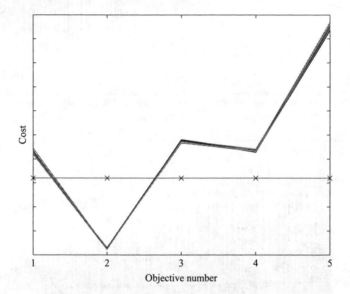

Figure 7.6   Trade-off graphs for the FKF search

accurate sensor-1, with other sensors that are less accurate. A further comparison is made between individual sensor performance of GA and MOGA case. It is clear that the individual sensor performance of the MOGA case, with the exception of sensor-1, has produced some improvements, with sensor-2 as the most noticeable one. It is clear that the improvement on sensor-2 has brought about an overall significant improvement on the quality of the estimation of the MOGA fused sensor, which is indicated by $J_{ze} = 0.2088$ rad.

7.2.5   GPS/INS navigation

Here, the fused estimated yaw obtained previously is treated as a single imaginary yaw sensor and used by other INS sensors to transform data from a body coordinate to

Table 7.4 Comparison of performance

Sensor	J_{zv}(rad)	J_{ze}(rad)	
		Non-MOGA	MOGA
Sensor-1	0.2340	0.2090	0.2094
Sensor-2	0.2960	0.3047	0.2761
Sensor-3	0.6558	0.4131	0.4130
Sensor-4	0.3852	0.2552	0.2551
Fused		0.2487	0.2088

a geographical (North–East–Down(NED)) coordinate frame where integration with converted GPS data is performed using a combination of FKF and EKF techniques and can be referred to as fuzzy extended Kalman filter (FEKF). Two GPS/INS scenarios are considered. The first scenario is where the vehicle performs a surface mission. The second scenario is where the vehicle performs an underwater mission.

7.2.5.1 Two-dimensional (2D)/surface mission

A continuous time model of the vehicle motion appropriate to this problem is taken to be

$$\dot{X}(t) = \mathbf{F}(X(t)) + W(t) \tag{7.12}$$

$$Z(t) = \mathbf{H}(X(t)) + V(t) \tag{7.13}$$

Denoted by $X(t) = [X_{\text{NED}}(t)\ \ Y_{\text{NED}}(t)\ \ \psi_{\text{im}}(t)\ \ r(t)\ \ u(t)\ \ v(t)]^{\text{T}}$ are the model states. $X_{\text{NED}}(t)$ and $Y_{\text{NED}}(t)$ are the longitude and latitude of the AUV position converted from deg min s in an Earth-centred Earth-fixed coordinate frame into metres in the NED coordinate frame, $\psi_{\text{im}}(t)$ is the yaw angle obtained from the imaginary yaw sensor, $r(t)$ is yaw rate, $u(t)$ and $v(t)$ are the surge and sway velocity, respectively. In this system model, $\mathbf{F}(\cdot)$ and $\mathbf{H}(\cdot)$ are both continuous function, continuously differentiable in $X(t)$. The $W(t)$ and $V(t)$ are both zero mean white noise for the system and measurement models respectively. The model states are related through the following kinematically based set of functions ($\mathbf{F}(X(t))$) in Eq. (7.12):

$$\dot{u}(t) = 0 \tag{7.14}$$

$$\dot{v}(t) = 0 \tag{7.15}$$

$$\dot{\psi}_{\text{im}}(t) = r(t) \tag{7.16}$$

$$\dot{r}(t) = 0 \tag{7.17}$$

$$\dot{X}_{\text{NED}}(t) = u(t) \cos \psi_{\text{im}}(t) - v(t) \sin \psi_{\text{im}}(t) \tag{7.18}$$

$$\dot{Y}_{\text{NED}}(t) = u(t) \sin \psi_{\text{im}}(t) + v(t) \cos \psi_{\text{im}}(t) \tag{7.19}$$

To obtain an EKF with an effective state prediction equation in a simple form, the continuous time model of Eqs. (7.14)–(7.19) have been linearised about the current state estimates, producing:

$$\mathbf{F}_{\text{2D-linearised}}(t) = \begin{bmatrix} 0 & 0 & -u(t)\sin\psi_{\text{im}}(t) - v(t)\cos\psi_{\text{im}}(t) & 0 & \cos\psi_{\text{im}}(t) & -\sin\psi_{\text{im}}(t) \\ 0 & 0 & u(t)\cos\psi_{\text{im}}(t) - v(t)\sin\psi_{\text{im}}(t) & 0 & \sin\psi_{\text{im}}(t) & \cos\psi_{\text{im}}(t) \\ 0 & 0 & 0 & 1 & 0 & 0 \\ 0 & 0 & 0 & 0 & 0 & 0 \\ 0 & 0 & 0 & 0 & 0 & 0 \\ 0 & 0 & 0 & 0 & 0 & 0 \end{bmatrix}$$

(7.20)

The output measurements are related through the states by the following matrix:

$$\mathbf{H}_{\text{2D-linearised}} = \begin{bmatrix} 0 & 0 & 0 & 0 & 1 & 0 \\ 0 & 0 & 0 & 0 & 0 & 1 \\ 0 & 0 & 1 & 0 & 0 & 0 \\ 0 & 0 & 0 & 1 & 0 & 0 \\ 1 & 0 & 0 & 0 & 0 & 0 \\ 0 & 1 & 0 & 0 & 0 & 0 \end{bmatrix}$$

(7.21)

when GPS signal is available, and when it is not,

$$\mathbf{H}_{\text{2D-linearised}} = \begin{bmatrix} 0 & 0 & 0 & 0 & 1 & 0 \\ 0 & 0 & 0 & 0 & 0 & 1 \\ 0 & 0 & 1 & 0 & 0 & 0 \\ 0 & 0 & 0 & 1 & 0 & 0 \end{bmatrix}$$

(7.22)

where $\mathbf{F}_{\text{2D-linearised}}$ and $\mathbf{H}_{\text{2D-linearised}}$ are, respectively, equivalent to $\mathbf{A}$ and $\mathbf{C}$ in linear dynamic system. Subsequent discretisation with period $T = 0.125$ s of the linearised model results in the EKF algorithm.

The initial conditions are $X_0 = 0\mathbf{I}_{6\times 6}$ and $\mathbf{P}_0 = 0.01\mathbf{I}_{6\times 6}$, and $\mathbf{Q}$ is made constant as

$$\begin{bmatrix} 10\,\text{m}^2 & 0 & 0 & 0 & 0 & 0 \\ 0 & 10\,\text{m}^2 & 0 & 0 & 0 & 0 \\ 0 & 0 & 0.0175\text{rad}^2 & 0 & 0 & 0 \\ 0 & 0 & 0 & 0.1\,(\text{rad/s})^2 & 0 & 0 \\ 0 & 0 & 0 & 0 & 0.1\,(\text{m/s})^2 & 0 \\ 0 & 0 & 0 & 0 & 0 & 0.1\,(\text{m/s})^2 \end{bmatrix}$$

(7.23)

The actual value of **R** is assumed unknown but its initial value is selected as

$$\begin{bmatrix} 1000 \text{ m}^2 & 0 & 0 & 0 & 0 & 0 \\ 0 & 1000 \text{ m}^2 & 0 & 0 & 0 & 0 \\ 0 & 0 & 0.0873 \text{ rad}^2 & 0 & 0 & 0 \\ 0 & 0 & 0 & 0.0175 \text{ (rad/s)}^2 & 0 & 0 \\ 0 & 0 & 0 & 0 & 2\text{(m/s)}^2 & 0 \\ 0 & 0 & 0 & 0 & 0 & 2\text{(m/s)}^2 \end{bmatrix}$$

(7.24)

The FEKF algorithm is then implemented on the diagonal element of $\mathbf{R}_k$.

Figure 7.7(a) shows the *Hammerhead* AUV trajectory obtained using GPS, dead reckoning using INS sensors (through double integration of the accelerometer data with respect to time) and integrated GPS/INS. As the initial value of **R** for both $X_{NED}(t)$ and $Y_{NED}(t)$ is 1000 m², the standard EKF algorithm puts less weight on the position obtained by GPS and more on the prediction of position obtained from dead reckoning method (using INS sensor data). Figure 7.7(b) shows that the matrix has been adjusted accordingly and more weight is given to the GPS data, and therefore the estimated trajectory in the integrated GPS/INS is 'pulled' a little bit further to the GPS trajectory. However, discrepancies can still be observed between the integrated GPS/INS estimate with respect to the GPS fixes. There are several explanations to this erratic behaviour. The first possibility is that it is caused by the poor level of accuracy of the low-cost GPS being used in this particular application. It is important to note that the proposed algorithm has detected a persistent high actual covariance ($\hat{\mathbf{C}}_{Inn_k}$) for both X_{NED} and Y_{NED} throughout the trajectory. This results in insufficient weight being given to the GPS fixes in the FEKF and more on the position obtained by the dead reckoning. The second possibility is that the GPS receiver did not lock into a sufficient number of satellites with a sufficiently small value of position dilution of precision (PDOP) that can provide the required level of accuracy. The use of a differential global positioning system (DGPS) receiver or a GPS receiver with a wide area augmentation system (WAAS) or a European geostationary navigation overlay service (EGNOS) capability can be considered as a way forward to alleviate this problem.

7.2.5.2 Three-dimensional (3D)/surface–depth mission

Many missions performed by AUVs require the vehicle to operate not only on the surface of the sea, but also at a particular depth. Examples of such AUVs and their specific missions can be found in Reference 11. The *Hammerhead* AUV is also designed to be able to dive to a certain depth and perform a particular mission, such as tracking underwater cables for maintenance purposes or landmark recognition for an underwater absolute positioning system as proposed in Reference 19. To carry out these missions, the *Hammerhead* AUV is equipped with underwater image acquisition techniques [20], coupled with a laser stripe illumination (LSI) methodology developed previously by Cranfield University [5] to provide an enhanced viewing of the seabed

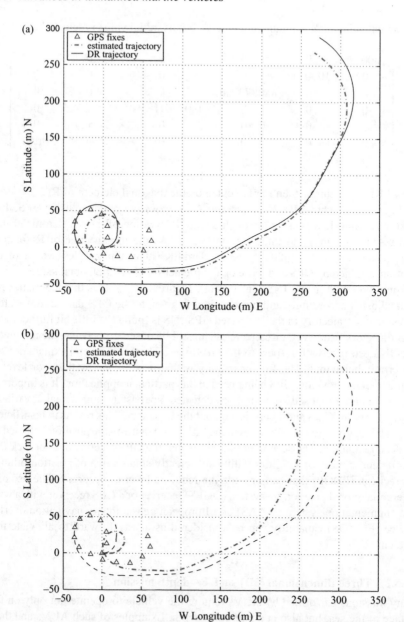

Figure 7.7 (a) AUV trajectory obtained using GPS, INS sensors (dead reckoning method) and GPS/INS using EKF without adaptation. (b) AUV trajectory obtained using GPS, INS sensors (dead reckoning method) and GPS/INS using EKF with adaptation

below the vehicle, and a depth controller developed by Naeem [21], which will be discussed in the section to come about guidance and control.

The concept of 3D navigation system enhanced by the proposed techniques is demonstrated in this section. The real data used herein are those generated by the individual TCM2 electronic compass and IMU, their respective simulated counterparts, and their overall fused values. Further real-time experiments are considered to be imperative and must be conducted before a full-scale pseudo real-time implementation of the proposed techniques can be undertaken. This, however, due to the amount of time required to do so and to analyse the data produced thereby, is considered to be suitable for the future work of the *Hammerhead* AUV.

The mission scenario adopted in this section is designed to mimic the actual cable-tracking or landmark recognition that will be performed in the future by the *Hammerhead* vehicle. This involves acquiring GPS/INS data on the surface and subsequently finding the estimated trajectory before sending the vehicle to a certain depth. Once the vehicle is under the water, the GPS signals are completely lost and the GPS/INS navigation system is replaced by a pure dead reckoning navigation system. During this period, the underwater image acquisition algorithms continuously observe the area beneath the vehicle to find a cable to be tracked or underwater landmarks to be identified and used as underwater absolute position fixes. In conditions where sufficient illumination is available in identifying those objects, produced either by the LSI or natural ambient light, the vehicle is then controlled to maintain its current depth. Otherwise, the depth controller algorithm will act accordingly and send the vehicle further down until sufficient illumination is obtained. After a certain period of time the vehicle is sent back to the surface to obtain GPS fixes that are used to reset the drift or the accumulated error produced by the dead reckoning navigation system.

The surge, sway and heave of the vehicle are obtained by integrating body coordinate frame acceleration data. The true values of the surge and sway are, respectively, defined as 1.3 m/s and ±0.1 m/s. The heave values are defined into five parts. The first is the heave of the vehicle when it is operating on the surface, that is, true values are assumed to be 0 m/s. The second part is the heave of the vehicle as it is descending to a certain depth, defined here as -0.1 m/s. Once the vehicle reaches this, the depth controller is employed to maintain the depth of the vehicle. Consequently, the true heave during this period is defined to be 0 m/s. The vehicle is sent back to the surface, and the heave during ascending period is defined to be 0.1 m/s. Finally, the vehicle is back to the surface and the heave once again is defined to be 0 m/s. It is clear that the errors added to these true values will contribute to the total drift suffered by the dead reckoning navigation system in finding the position of the vehicle when it is operating under the water.

$$X(t) = [X_{NED}(t)\ Y_{NED}(t)\ Z_{NED}(t)\ \theta(t)\ q(t)\ \psi_{im}(t)\ r(t)\ u(t)\ v(t)\ w(t)]^T$$

is the state vector for a continuous time model of the vehicle motion appropriate to this problem, which is taken to be as in Eqs. (7.12) and (7.13), where $X_{NED}(t)$ and $Y_{NED}(t)$ are the longitude and latitude of the AUV position converted from deg min s in the Earth-centred Earth-fixed coordinate frame into metres in the NED coordinate frame, $Z_{\text{NED}}(t)$ is the depth of the vehicle, $\theta(t)$ is the pitch, $q(t)$ is the pitch rate, $\psi_{\text{im}}(t)$ is the yaw angle obtained from the imaginary yaw sensor, $r(t)$ is yaw rate, $u(t), v(t)$ and $w(t)$ are the surge, sway and heave velocity, respectively. The model states are related through the following kinematically based set of functions:

$$\dot{X}_{\text{NED}}(t) = u(t)\cos\psi_{\text{im}}(t)\cos\theta(t) - v(t)\sin\psi_{\text{im}}(t)$$
$$+ w(t)\cos\psi_{\text{im}}(t)\sin\theta(t) \tag{7.25}$$

$$\dot{Y}_{\text{NED}}(t) = u(t)\sin\psi_{\text{im}}(t)\cos\theta(t) + v(t)\cos\psi_{\text{im}}(t) + w(t)\sin\psi_{\text{im}}(t)\sin\theta(t) \tag{7.26}$$

$$\dot{Z}_{\text{NED}}(t) = -u(t)\sin\theta(t) + w(t)\cos\theta(t) \tag{7.27}$$

$$\dot{\theta}(t) = q(t) \tag{7.28}$$

$$\dot{q}(t) = 0 \tag{7.29}$$

$$\dot{\psi}_{\text{im}}(t) = r(t) \tag{7.30}$$

$$\dot{r}(t) = 0 \tag{7.31}$$

$$\dot{u}(t) = a_{x_{\text{BODY}\to\text{NED}}}(t) \tag{7.32}$$

$$\dot{v}(t) = a_{y_{\text{BODY}\to\text{NED}}}(t) \tag{7.33}$$

$$\dot{w}(t) = a_{z_{\text{BODY}\to\text{NED}}}(t) \tag{7.34}$$

where $a_{x_{\text{BODY}\to\text{NED}}}(t)$, $a_{y_{\text{BODY}\to\text{NED}}}(t)$ and $a_{z_{\text{BODY}\to\text{NED}}}(t)$ are the acceleration of the vehicle acquired in the body coordinate frame and transformed subsequently to the NED coordinate frame.

The output measurements are related through the states by an identity matrix $\mathbf{I}_{10\times 10}$ when the vehicle is operating on the surface. When the vehicle is operating under the water, pure dead reckoning is used. Linearisation about the current estimates of the continuous time model of Eqs. (7.25)–(7.34), producing the $\mathbf{F}(t)$ of the system, which in this case is defined as $\mathbf{F}_{\text{3D-linearised}}(t)$, with the corresponding $\mathbf{H}_{\text{3D-linearised}}(t) = \mathbf{I}_{10\times 10}$. The FEKF algorithm is then implemented after subsequent discretisation with period $T = 0.125$ s. The initial conditions are $X_0 = 0\mathbf{I}_{10\times 10}$ and $\mathbf{P}_0 = 0.01\mathbf{I}_{10\times 10}$, and $\mathbf{Q}$ is made constant as with the following components:

$$\mathbf{Q}_{[1,1]} = 0.01\text{m}^2$$

$$\mathbf{Q}_{[2,2]} = 0.01\text{m}^2$$

$$\mathbf{Q}_{[3,3]} = 0.01\text{m}^2$$

$$\mathbf{Q}_{[4,4]} = 0.000001\text{rad}^2$$

$$\mathbf{Q}_{[5,5]} = 0.01(\text{rad/s})^2$$

$$\mathbf{Q}_{[6,6]} = 0.000001\text{rad}^2$$

$\mathbf{Q}_{[7,7]} = 0.01\,(\text{rad/s})^2$

$\mathbf{Q}_{[8,8]} = 0.01\,(\text{m/s})^2$

$\mathbf{Q}_{[9,9]} = 0.01\,(\text{m/s})^2$

$\mathbf{Q}_{[10,10]} = 0.01\,(\text{m/s})^2$

The initial value of **R** is selected as:

$\mathbf{R}_{[1,1]} = 10\,\text{m}^2$

$\mathbf{R}_{[2,2]} = 10\,\text{m}^2$

$\mathbf{R}_{[3,3]} = 5\,\text{m}^2$

$\mathbf{R}_{[4,4]} = 0.000001\,\text{rad}^2$

$\mathbf{R}_{[5,5]} = 0.000001\,(\text{rad/s})^2$

$\mathbf{R}_{[6,6]} = 0\,\text{rad}^2$

$\mathbf{R}_{[7,7]} = 0\,(\text{rad/s})^2$

$\mathbf{R}_{[8,8]} = 2\,(\text{m/s})^2$

$\mathbf{R}_{[9,9]} = 2\,(\text{m/s})^2$

$\mathbf{R}_{[10,10]} = 2\,(\text{m/s})^2$

Figures 7.8 and 7.9 show the result of implementing the proposed FEKF algorithm to the 3D/surface–depth mission described herein using the yaw produced by an individual sensor (here represented by sensor-4) and fused sensor, respectively. Readers interested in the complete results and analysis are referred to Reference 22.

It is clear from Figure 7.8 that the initial GPS/INS surface trajectory using the yaw produced by sensor-4 contains an unexpected drift in vertical direction. This is a direct result of assuming the measurement noise in this direction as being higher than its corresponding process noise. The values of the measurement and process covariance matrices are indicated, respectively, as $\mathbf{R}_{[3,3]} = 5\,\text{m}^2$ and $\mathbf{Q}_{[3,3]} = 0.01\,\text{m}^2$. Consequently, the EKF algorithm puts more confidence on the process, that is, integrating the $\dot{Z}_{[\text{NED}]}$, than the measurement of depth from the pressure transducer. It is clear, as indicated by Eq. (7.27), that integrating the value of $\dot{Z}_{[\text{NED}]}$ consequently integrates the noise in the $u(t)$, ψ_{im}, w and θ. This in turn produces an accumulation of error and needs to be reset to $Z_{\text{NED}} = 0\,\text{m}$, immediately before the vehicle dives. Once the vehicle is below the surface, the depth controller and the underwater image acquisition algorithms will work side by side to find objects of interest and to maintain a constant depth thereafter for a specific period of time. The vehicle is then sent back to the surface to obtain GPS fixes used to reset the drift produced by the dead reckoning process during the underwater mission. A similar case of dead reckoning error also occurs at this stage. Although the depth has been reset to 0 m, the EKF algorithm soon puts more confidence on the vertical dead reckoning process

144 Advances in unmanned marine vehicles

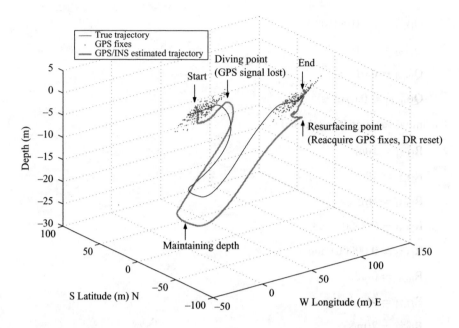

Figure 7.8 *True trajectory, GPS fixes and GPS/INS using yaw produced by sensor-4 only*

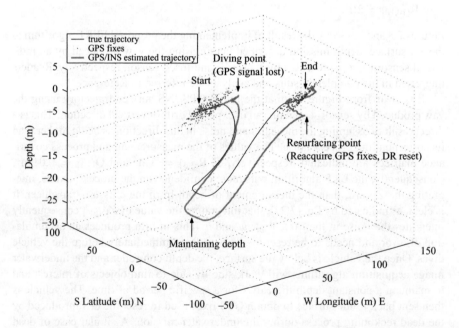

Figure 7.9 *True trajectory, GPS fixes and GPS/INS using yaw produced by fused sensor*

and consequently produces an estimate of depth larger than 0 m. This also happens to the horizontal (X_{NED} and Y_{NED}) estimation process. As the measurement covariance matrices for both the longitude and latitude are $\mathbf{R}_{[1,1]} = \mathbf{R}_{[2,2]} = 10\,\text{m}^2$, the estimation process puts more weight on the dead reckoning processes, which are assumed to have $\mathbf{Q}_{[1,1]} = \mathbf{Q}_{[2,2]} = 0.01\,\text{m}^2$ process covariance matrices.

Figure 7.9 shows the trajectory produced using the fused yaw sensor and with the values of $\mathbf{R}$ adjusted by FEKF. It is clear as presented, that as the assumed values of $\mathbf{R}$ for the longitude and latitude are quite low, $10\,\text{m}^2$, compared to the true ones, which are simulated to be $225\,\text{m}^2$ and $100\,\text{m}^2$, the FEKF estimation process put initial weight more on the GPS fixes than the dead reckoning solutions. However, as the filter learns the true nature of $\mathbf{R}$ of these quantities, the FEKF makes an appropriate adjustment by putting more weight on the dead reckoning solution than on the GPS fixes. It can also be observed how the filter learns the true value of $\mathbf{R}_{[3,3]}$, which is simulated to be $0.0001\,\text{m}^2$. This time the vehicle is not estimated to have depth larger than 0 m, as in the case with the trajectory using yaw data produced by sensor-4. As before, once the vehicle is below the surface, the depth controller and the underwater image acquisition algorithms will work side by side to find objects of interest and to maintain a constant depth thereafter for a specific period of time. The vehicle is then sent back to the surface to obtain GPS fixes used to reset the drift produced by the dead reckoning process during the underwater mission. It is also clear here how the FEKF has learned the true nature of the $\mathbf{R}$ values. It can be observed from Figure 7.9 how the FEKF algorithm puts extra confidence on the GPS fixes right after the vehicle reaches the surface. Soon afterwards however, the algorithm recognises the high level of noise inherent in the acquired GPS signals and puts less confidence thereon. Small discrepancies still exist between the true end and the estimated end of the mission. However, it is clear that without the FEKF, the estimated end could easily coincide with the last GPS fix and cause a significant position error.

7.3 System modelling

All controller designs are based on a model of the physical system to be controlled. This gives the modelling process utmost importance before any real time controller can be developed. It is imperative that the designer gain significant depth into system behaviour via extensive simulations using a model of the process as an alternative to the physical system. Clearly, this requires a model that can replicate the system's dynamic behaviour as closely as possible.

Modelling an underwater vehicle is a complex task because of the nonlinear nature of the vehicle dynamics and the degrees of freedom of vehicle movement. In addition, cross-coupling effects make the controller design even more intricate. Fortunately, there is a plentiful amount of literature available on the mathematical modelling of underwater vehicles and it is generally applicable to all types of underwater vessels. However, a major difficulty in using these generalised models is the evaluation of hydrodynamic coefficients which require tank tests on a full-scale physical model of the vehicle provided the test facility is available.

An alternate route to modelling an AUV using SI is thus suggested and used herein. An SI approach is useful in providing reliable and accurate models in a short time without relying too much on mathematical modelling techniques. This feature therefore is attractive for the underwater vehicle manufacturers, where a vehicle configuration changes frequently to suit the mission requirements.

AUV modelling using SI approaches has been investigated before [23–26], but most of the work involved has been done on identifying a model by generating data from a mathematical model of the vehicle. However, for *Hammerhead*, the SI is performed on input-output data obtained from actual in-water experiments. SI theory is well established and the reader is referred to Reference 18 for a comprehensive treatment of the subject.

7.3.1 Identification results

Trials for SI have been performed at South West Water's Roadford Reservoir, Devon, and at Willen Lake in Milton Keynes. Experiments were designed that could obtain the best possible data for model development. Ideally, the requirement is to have completely noise free data which is impossible in a real world environment. The *Hammerhead* is a low-speed AUV that swims at approximately 2 knots. This gives some insight about the sampling period to be chosen. Clearly, too high a sampling rate in this case will give no advantage whatsoever. A sampling rate of 1 Hz is thus chosen iteratively which is adequate to obtain ample dynamical information about the system. By the same token, the frequency for the input signal is chosen as 0.1 Hz which was deemed sufficient to excite the interesting modes of the system.

Some common types of excitation signals used are the uniformly distributed random numbers, pseudo random binary sequence, and its variants such as multistep, and doublet input. For depth channel identification, it was found that the vehicle would hit the bed without the depth autopilot in the loop for long duration inputs. Therefore, it was decided to perform several short duration experiments with a different multi-step input being applied to the hydroplane in each experiment. The experiments can easily be merged for model identification using the SI toolbox in MATLAB. Please note that all available measurements were pre-filtered and resampled at 1 Hz before any model parameters could be identified. Due to this, most of the high-frequency contamination was eliminated and hence it was decided to extract an autoregressive with exogeneous (ARX) or state space model without modelling the noise separately.

7.3.1.1 Rudder–yaw channel

The input to this channel are the rudder deflections and the output is the vehicle's yaw or heading angle. The heading information is available from an onboard TCM2 compass and IMU. However, the vehicle response was obtained from TCM2 in these particular trials. It has been observed that there is no strong cross coupling present between the yaw and depth channels. Moreover, the roll data remain unaltered with respect to the change in vehicle's heading. A single-input single-output (SISO) model has thus been developed for this channel from the data and is given by:

$$G(q^{-1}) = \frac{-0.04226q^{-1} + 0.003435q^{-2}}{1 - 1.765q^{-1} + 0.765q^{-2}} \quad (7.35)$$

where q^{-1} is the delay operator. This model has been verified by independent data sets that were not used in the modelling process. In addition, correlation based tests were also carried out to gauge the quality of the model, and these were found to be adequate.

7.3.1.2 Hydroplane–depth channel

The input to this channel is the hydroplane deflections whilst the output is the depth of vehicle taken from a pressure transducer. It has been mentioned that only multistep inputs were employed to excite the depth dynamics of the vehicle due to the reasons explained before. The vehicle was allowed to swim freely in six degrees of freedom, however, only a single input (hydroplane) was manipulated and the heading, depth, roll and pitch data were analysed and recorded.

The *Hammerhead* data obtained from the depth trials reveal some cross coupling effects between the depth and heading angle. A multivariable model is therefore the ideal choice. However, exploiting the fact that the heading angle does not vary significantly when the vehicle is fully submerged, a SISO model involving only hydroplane deflections and depth has been developed. Several data sets containing these parameters were collected and suitable data were averaged, resampled and then merged to estimate the model coefficients.

A fourth-order ARX model was chosen iteratively which gives the best fit between measured and model predicted outputs. Equation (7.36) below presents the ARX(441) model of the depth dynamics of the *Hammerhead*

$$G(q^{-1}) = \frac{0.002681q^{-1} - 0.00327q^{-2} - 0.0007087q^{-3} + 0.001322q^{-4}}{1 - 3.6773q^{-1} + 5.0839q^{-2} - 3.1348q^{-3} + 0.72826q^{-4}} \quad (7.36)$$

The depth dynamic model has also been verified by analysing the correlation tests and time domain cross validation.

7.4 Guidance

For *Hammerhead*, two novel guidance strategies have been proposed. These guidance laws were specifically designed for underwater cable inspection tasks for cruising type vehicles. The first guidance strategy presented has been borrowed from airborne systems and is termed as pure pursuit guidance. This scheme generates command signals which are proportional to the line of sight (LOS) angle so that the pursuing vehicle maintains its flight profile aligned with the LOS. In the other guidance technique, the vehicle speed is used as a means to formulate the guidance scheme. This is called a hybrid guidance system since it utilises various existing guidance laws during different phases of the mission as depicted in Figure 7.10.

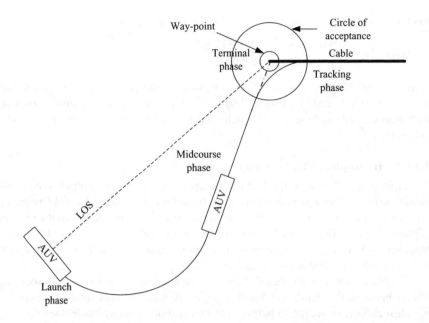

Figure 7.10 Planar view of the four phases of flight for cable tracking problem of an AUV

The idea is to gradually decrease the speed of the AUV as it approaches the cable/pipeline. Please see References 27 and 28 for a detailed description and evaluation of the pure pursuit and hybrid guidance strategies, respectively. From an implementation perspective, it is currently not possible to test the hybrid guidance law in *Hammerhead* since the AUV model is only available at one fixed speed. However, the concept has been verified by borrowing an AUV dynamic model from the literature which is represented in terms of vehicle velocity. In general, the proposed schemes require the vehicle's speed and orientation to evaluate the guidance signals which are available from an IMU onboard the *Hammerhead*.

7.5 *Hammerhead* autopilot design

Development of an autopilot for the *Hammerhead* AUV is of vital importance to the absolute design stage. This section presents controller design and results as applied to the *Hammerhead* AUV models identified in Section 7.3. It should be noted that there is a plethora of control systems available and a comprehensive review has been undertaken by Craven *et al.* [29]. However, the selection of a particular controller for an AUV is attributed to several factors. Some of them are

- Robustness to modelling errors (plant parameter variations).
- Disturbance handling characteristics.
- Set point tracking and trajectory following.

- Stability characteristics.
- Application to linear and nonlinear plants.

Two robust optimal control strategies and their variants have been selected as the candidate control schemes for the *Hammerhead* vehicle: a discrete-time LQG/LTR and the model-based predictive controller which has been modified to accommodate various AI techniques for improved performance. Simulation and experimental results of the application of the proposed controllers to the *Hammerhead* vehicle in the horizontal and dive planes are presented.

7.5.1 LQG/LTR controller design

LQG is an optimal controller whose name is derived from the fact that it assumes a linear system, quadratic cost function and Gaussian noise. Unlike the pole placement method, where the designer must know the exact pole locations, LQG places the poles at some arbitrary points within the unit circle so that the resulting system is optimal in some sense. A linear quadratic state feedback regulator (LQR) problem is solved which assumes that all states are available for feedback. However, this is not always true because either there is no available sensor to measure that state or the measurement is very noisy. A KF can be designed to estimate the unmeasured states. The LQR and KF can be designed independently and then combined to form an LQG controller, a fact known as the separation principle. Individually, the LQR and KF have strong robustness properties with gain margin up to infinity and over $60°$ phase margin [30]. Unfortunately, the LQG has relatively poor stability margins which can be circumvented by using LTR. The LTR works by adding fictitious noise to the process input which effectively cancels some of the plant zeros and possibly some of the stable poles, and inserts the estimator's zeros [31, 32]. Herein, a discrete time LQG/LTR design has been developed motivated by the work of Maciejowski [31]. A substantial amount of material is available on the state feedback LQG/LTR controller (see, e.g., References 30 and 33). Therefore, attention is focused only on the application of the autopilot to the *Hammerhead* and no controller design details are presented.

7.5.1.1 LOS following

The LQG/LTR controller has been tuned using the methodology proposed by Maciejowski [31]. In this technique, only the noise covariance matrices are required to be adjusted for the KF. The weighting matrices of the LQR are then selected according to an automatic procedure. See Reference 28 for more details regarding the LQG/LTR autopilot design for the *Hammerhead*. The algorithm is tested for a setpoint change in heading angle. The vehicle is assumed to be pointing in an arbitrary direction and is required to follow a certain heading angle closely without much control effort. A saturation block is inserted in series with the controller with cutoff limits of $\pm 20°$ and a desired heading angle of $100°$ is chosen with the vehicle initiating close to $0°$. The measurement and process noise covariance matrices are adjusted to achieve the desired closed-loop frequency response whereas weighting matrices for the objective function are selected according to Reference 31. The algorithm is simulated and the output response is depicted in Figure 7.11.

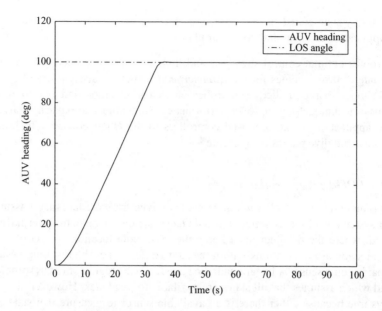

Figure 7.11   LQG/LTR control of the Hammerhead *showing LOS tracking*

The AUV heading bears a negligible overshoot and the settling time is less than 40 s.[1] However, the price to pay for this settling time is that the actuator saturation constraint becomes active for about 35 s in the beginning of the simulation run when the vehicle was making a turn.

Increasing the control input weighting matrix can help limit the rudder movement within the constrained boundaries and avoid this saturation but at the cost of large settling times and deviation from the desired stability margins.

7.5.1.2   Depth control

For depth controller design, the same procedure is adopted as before using the *Hammerhead* depth dynamic model. Simulations have been performed where the vehicle is assumed to be launched on the surface and the task is to follow a depth of 3 m below the sea surface. The simulation is run for 100 s and the *Hammerhead* response is depicted in Figure 7.12. With a small overshoot of less than 5 per cent and settling time approximately 15 s, the vehicle successfully follows the desired depth and stays on course throughout the rest of the mission duration.

7.5.2   Model predictive control

Model predictive control (MPC) refers to a class of algorithms that compute a sequence of manipulated variable adjustments in order to optimise the future

---

[1] Since $Ts = 1$, therefore 1 sample time corresponds to 1 s.

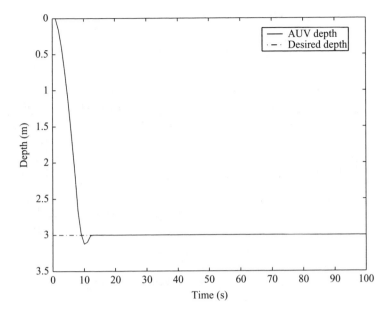

Figure 7.12   Depth control of the Hammerhead vehicle using the LQG/LTR controller showing a step change in depth

behaviour of a plant. Originally developed to meet the specialised control needs of power plants and petroleum refineries, MPC technology can now be found in a wide variety of application areas including chemicals, food processing, automotive, aerospace and metallurgy [34]. A good account of MPC technology from the past to the future has been reviewed by Morari and Lee [35], while a comparison between both theoretical and practical aspects of MPC has been undertaken by Carlos *et al.* [36]. For the interested reader, several other useful references on MPC can be found [37–41].

Herein, two modifications have been proposed to the standard predictive control problem. In the first technique, the conventional optimiser is replaced by a GA. One of the distinct advantages of using a GA is the possibility of employing various objective functions and the ability to deal with any type of process model and constraints, thus generalising a range of MPC technologies where each of them is defined on a fixed set of process model and objective function. In the other proposed strategy, a fuzzy performance index is used in place of the quadratic objective function and a GA is employed as an optimisation tool. Herein, only GA-based MPC with a standard cost function will be discussed and some real-time results are elaborated. The reader is referred to Naeem *et al.* [42] for the fuzzy based GA-MPC autopilot design. The genetic-based control algorithm is depicted in Figure 7.13. As shown, the GA replaces the optimiser block and the AUV model identified from SI on the trial data has been used. The GA-based controller uses the process model to search for the control moves, which satisfy the process constraints and optimises a cost function. A conventional

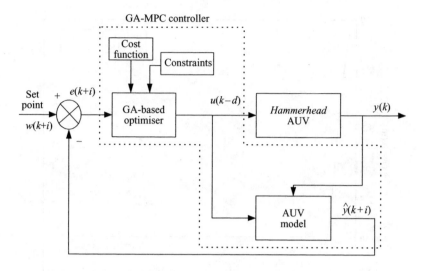

Figure 7.13   Genetic algorithm based model predictive controller

quadratic objective function is minimised to evaluate the control inputs necessary to track a reference trajectory and is given by Eq. (7.37).

$$J = \sum_{i=1}^{H_p} e(k+i)^T Q e(k+i) + \sum_{i=1}^{H_c} \Delta u(k+i)^T R \Delta u(k+i)$$
$$+ \sum_{i=1}^{H_p} u(k+i)^T S u(k+i) \qquad (7.37)$$

subject to input constraints

$$u^-_{\text{constraint}} \leq u(k) \leq u^+_{\text{constraint}}$$
$$\Delta u^-_{\text{constraint}} \leq \Delta u(k) \leq \Delta u^+_{\text{constraint}}$$

where the positive and negative signs represent the upper and lower constrained limits, respectively. Q is the weighting scalar on the prediction error given by

$$e(k) = \hat{y}(k) - w(k) \qquad (7.38)$$

where $w(k)$ is the reference or the desired setpoint. R and S are weights on the change in input, Δu, and magnitude of input, u, respectively. Please refer to Naeem *et al.* [27] for a comprehensive treatment on the GA-based MPC design.

The performance of the GA-MPC controller is next evaluated using the *Hammerhead* models identified in Section 7.3.1. As for the LQG/LTR controller explicated in Section 7.5.1, separate controllers are designed for the yaw and depth channels.

However, real-time results are shown here only for the horizontal plane whereas simulated data are illustrated for the depth output. It is important to point out the fact that the authors tested the controller on a 1960s-made vehicle hull with the bulk of the existing electronics, specifically the motors and their mechanical assembly, being retained. Due to this, the rudder movement could not be controlled precisely and the minimum deflection observed was 2°. For this reason, a minimum rate of change of input constraint was imposed on the rudder which could lead to chattering effects in the rudder movement.

7.5.2.1 Heading control

To simulate the controller, it is assumed that there is no model/plant mismatch. The constraints on the rudder were ±20° while the minimum allowed deflection is 2° as discussed in the previous section. The weighting matrices were adjusted heuristically and the prediction and control horizons (H_p and H_c) of MPC were set to minimise the control effort and to increase the speed of response.

The controller was simulated for a step change in heading and the result is depicted in Figure 7.14. Without taking any disturbance into account, the vehicle closely follows the desired course after initiating from an arbitrary direction. The response in Figure 7.14(b) bears a small overshoot which can be minimised by adjusting the weighting matrices but at the cost of slower response time. The rudder deflections

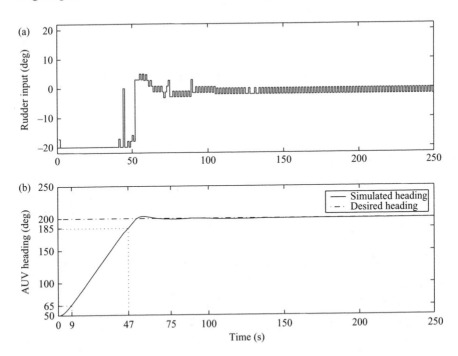

Figure 7.14 *GA-MPC simulation results: (a) rudder deflections generated by the controller and (b) AUV heading*

generated by the GA-based controller are also shown in Figure 7.14(a) requiring minimum control effort and staying within the specified bounds. There is a large movement in the rudder position around $t = 50$ s yet this does not affect the vehicle's motion because of its slow dynamics. The spike is due to the probabilistic nature of GA which produces such results unless accounted for in the code but has not been implemented owing to the extra computational burden that it imposes. The chattering phenomena can also be observed from this figure because of the 2° constraint. However, the effect of this is almost negligible on the vehicle's movement.

Next, the heading controller was tested in the *Hammerhead*. The parameters used in these trials were kept the same as in the simulation studies for a fair comparison. The test was carried out for a step change in heading where the initial and desired headings were the same as in the simulation. It is evident from Figure 7.15(b) that the GA-MPC was able to track the desired heading without any offset despite the presence of model uncertainty and external disturbances. There is an overshoot though, which could be blamed on the surface currents. Looking at the rudder deflections in Figure 7.15(a), again there is some expected chattering present. However, the rudder movement always remains within the specified constraints. There is a large spike followed by fluctuations in rudder movement at approximately 125 s in response to the change in vehicle's heading due to surface currents. However, the controller is robust enough to cope with it and attains the steady state input and output values in approximately

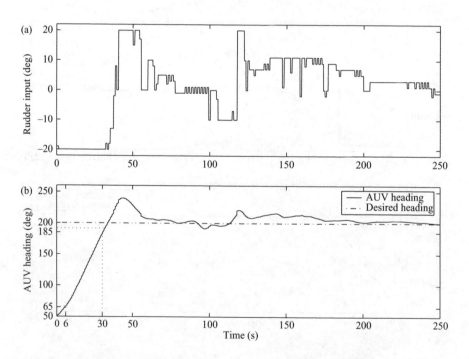

Figure 7.15 Controller trial results: (a) rudder deflections generated by the GA-MPC and (b) Hammerhead *heading obtained from IMU*

50 s. A statistical analysis reveals that the standard deviation of rudder deflections in experimental data is approximately 11°, while it is quite high (8°) in the simulation (even without any disturbance), possibly because of the chattering phenomena.

It is interesting to note that the rise time in the experimental data is much smaller (24 s) than in the simulation (38 s). One reason for this is the effect of surface currents pushing the vehicle unwantingly and causing an even a higher overshoot as compared to the simulated response. Model/plant mismatch could also be a potential source of this problem.

7.5.2.2 Depth control

A GA-based MPC depth autopilot was developed for *Hammerhead* in this section. The model used for controller design is given by Eq. (7.36) with a sampling time of 1 s and identified from trial data using SI techniques. The intent is to design a controller that is able to control the depth of the AUV as closely as possible despite the presence of external disturbances and modelling errors. The performance of the control strategy developed herein will eventually be assessed on the *Hammerhead* vehicle.

The front canards movement is restricted to ±25°, which was thought to be adequate to control the vehicle in the vertical plane and was obtained through a series of rigorous in water experiments. Please note that no minimum rate constraint is imposed on the hydroplane movement since these are newly installed in the vehicle. It is assumed that the vehicle is manoeuvring near the surface (zero depth) and is subjected to a depth command of 3 m. The *Hammerhead* response to a step change in depth when diving from the sea surface is shown in Figure 7.16.

It took less than 25 s for the vehicle to attain the desired depth of 3 m with little overshoot and no steady-state error. The diving rate stays uniform as the vehicle approaches the desired level with precision and maintains the specified depth throughout.

7.6 Concluding remarks

This chapter presents an overview of the navigation, guidance and control system design of the *Hammerhead* vehicle. Two navigation scenarios have been considered to validate the proposed approach: 2D/surface and 3D/surface–depth scenarios. In both scenarios, the data from the TCM2 electronic compass and IMU are fused with two other simulated sensors before being used in transforming data from the body to the NED coordinate frame, where integration between the INS and GPS data occurs. In the first scenario, as the vehicle operates on the surface only, the GPS data are available periodically and the proposed estimation process takes place between the GPS fixes. In the second scenario, the GPS fixes are available continuously when the vehicle operates on the surface, and the proposed estimation algorithm blends these data with the position solution produced by the dead reckoning method to find the best estimates of the vehicle's position. In this scenario, the vehicle uses only the dead reckoning method during an underwater mission and the accumulated errors

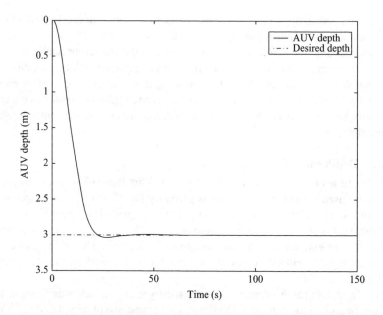

Figure 7.16 Hammerhead *step response for a change in depth obtained by employing a GA-MPC depth autopilot*

produced thereby are reset by GPS fixes the next time the vehicle gains access to their signals. It has been shown in both scenarios that the proposed algorithm has produced a significant improvement in accuracy and reliability of the navigation system of the vehicle.

The use of SI techniques on actual vehicle data for modelling is suggested where the models are subsequently used for autopilot development. Since the *Hammerhead* is a cruising type vehicle, guidance laws specifically for cable/pipeline inspection task are formulated. Finally, autopilots based on optimal control and AI techniques are conceived. Some real-time results based on GA-MPC controller are presented and a comparison is made with simulation examples. It is shown that the proposed autopilot is quite robust under a varying set of conditions including external disturbances and modelling uncertainty.

It is envisaged that the NGC techniques employed for *Hammerhead* which are presented within this chapter will be considered as an invaluable addition to the underwater research literature.

References

1 Størkersen, N., J. Kristensen, A. Indreeide, J. Seim, and T. Glancy (1998). Hugin – UUV for seabed surveying. *Sea Technology*, 39(2), pp. 99–104.

2. Wright, J., K. Scott, C. Tien-Hsin, B. Lau, J. Lathrop, and J. McCormick (1996). Multi-sensor data fusion for seafloor mapping and ordnance location. In Proceedings of the 1996 Symposium on Autonomous Underwater Vehicle Technology, Monterey, CA, USA, pp. 167–175.
3. Asakawa, K., J. Kojima, Y. Kato, S. Matsumoto, and N. Kato (2000). Autonomous underwater vehicle *Aqua Explorer 2* for inspection of underwater cables. In Proceedings of the 2000 International Symposium on Underwater Technology, Tokyo, Japan, pp. 242–247.
4. Naylies, I. (2000). The sensory requirement of a PC controlled AUV. Master's thesis, Offshore Technology Centre, Cranfield University.
5. Tetlow, S. and R.L. Allwood (1995). Development and applications of a novel underwater laser illumination system. *Underwater Technology*, 21(2), pp. 13–20.
6. Dalgleish, F.R. (2004). Applications of laser-assisted vision to autonomous underwater vehicle navigation. Ph.D. thesis, Cranfield University, Cranfield, UK.
7. Gade, K. and B. Jalving (1999). An aided navigation post processing filter for detailed seabed mapping UUVs. *Modeling, Identification and Control*, 20(3), pp. 165–176.
8. Grenon, G., P.E. An, S.M. Smith, and A.J. Healey (2001). Enhancement of the inertial navigation system for the morpheous autonomous underwater vehicles. *IEEE Journal of Oceanic Engineering*, 26(4), pp. 548–560.
9. Yun, X., R.E. Bachmann, R.B. McGhee, R.H. Whalen, R.L. Roberts, R.G. Knapp, A.J. Healey, and M.J. Zyda (1999). Testing and evaluation of an integrated GPS/INS system for small AUV navigation. *IEEE Journal of Oceanic Engineering*, 24(3), pp. 396–404.
10. Kinsey, J.C. and L.L. Whitcomb (2003). Preliminary field experience with the DVLNAV integrated navigation system for manned and unmanned submersibles. In Proceedings of the 1st IFAC Workshop on Guidance and Control of Underwater Vehicles, Newport, South Wales, UK, pp. 83–88.
11. Loebis, D., R. Sutton, and J. Chudley (2002). Review of multisensor data fusion techniques and their application to autonomous underwater vehicle navigation. *Journal of Marine Engineering and Technology*, AC-15(2), pp. 175–184.
12. Mehra, R.K. (1970). On the identification of variances and adaptive Kalman filtering. *IEEE Transactions on Automatic Control*, AC-15(2), pp. 175–184.
13. Fitzgerald, R.J. (1971). Divergence of the Kalman filter. *IEEE Transactions on Automatic Control*, AC-16(6), pp. 736–747.
14. Escamilla-Ambrosio, P.J. and N. Mort (2001). A hybrid Kalman filter-fuzzy logic multisensor data fusion architecture with fault tolerant characteristics. In Proceedings of the 2001 International Conference on Artificial Intelligence, Las Vegas, NV, USA, pp. 361–367.
15. Jetto, L., S. Longhi, and D. Vitali (1999). Localisation of a wheeled mobile robot by sensor data fusion based on a fuzzy logic adapted Kalman filter. *Control Engineering Practice*, 7, pp. 763–771.

16 Kobayashi, K., K.C. Cheok, K. Watanabe, and F. Munekata (1998). Accurate differential global positioning via fuzzy logic Kalman filter sensor fusion technique. *IEEE Transactions on Industrial Electronics*, 45(3), pp. 510–518.
17 Mohamed, A.H. and K.P. Schwarz (1999). Adaptive Kalman filtering for INS/GPS. *Journal of Geodesy*, 73, pp. 193–203.
18 Ljung, L. (1999). *System Identification, Theory for the User*, 2nd edition. PTR Prentice Hall, Englewood Cliffs, NJ.
19 Loebis, D., F.R. Dalgleish, R. Sutton, S. Tetlow, J. Chudley, and R. Allwood (2003). An integrated approach in the design of navigation system for an AUV. In Proceedings of MCMC 2003 Conference, Girona, Spain, pp. 329–334.
20 Dalgleish, F.R., S. Tetlow, and R.L. Allwood (2003). A preliminary experiments in the development of a laser based-imaging sensor for AUV navigation. In Proceedings of the 1st IFAC Workshop on Guidance and Control of Underwater Vehicles, Newport, South Wales, UK, pp. 239–244.
21 Naeem, W. (2004). Guidance and control of an autonomous underwater vehicle. Ph.D. thesis, The University of Plymouth, Plymouth, UK.
22 Loebis, D. (2004). An intelligent navigation system for an autonomous underwater vehicle. Ph.D. thesis, The University of Plymouth, Plymouth, UK.
23 Tinker, S.J., A.R. Bowman, and T.B. Booth (1979). Identifying submersible dynamics from free model experiments. In RINA Annual Report and Transactions, pp. 191–196.
24 Ippoliti, C.G., S. Radicioni, and A. Rossolini (2001). Multiple models control of a remotely operated vehicle: analysis of models structure and complexity. In Proceedings of the IFAC Conference on Control Applications in Marine Systems (CAMS'01), IFAC, Glasgow, Scotland, UK.
25 Goheen, K.R. and E.R. Jefferys (1990). The application of alternative modelling techniques to ROV dynamics. In Proceedings of the IEEE International Conference on Robotics and Automation, IEEE, Cincinnati, OH, USA, pp. 1302–1309.
26 Ahmad, S.M. and R. Sutton (2003). Dynamic modelling of a remotely operated vehicle. In Proceedings of the 1st IFAC Workshop on Guidance and Control of Underwater Vehicles GCUV 2003, IFAC, Newport, South Wales, UK, pp. 47–52.
27 Naeem, W., R. Sutton, and S.M. Ahmad (2004). Pure pursuit guidance and model predictive control of an autonomous underwater vehicle for cable/pipeline tracking. *IMarEST Journal of Marine Science and Environment*, C(1), pp. 25–35.
28 Naeem, W., R. Sutton, and J. Chudley (2003). LQG/LTR control of an autonomous underwater vehicle using a hybrid guidance law. In Proceedings of Guidance and Control of Underwater Vehicles 2003, Elsevier IFAC Publications, Newport, South Wales, UK, pp. 31–36.
29 Craven, P.J., R. Sutton, and R.S. Burns (1998). Control strategies for unmanned underwater vehicles. *The Journal of Navigation*, 51(1), pp. 79–105.
30 Burl, J.B. (1999). *Linear Optimal Control, H_2 and H_∞ Methods*. Addison-Wesley Longman, Reading, MA.
31 Maciejowski, J.M. (1985). Asymptotic recovery for discrete-time systems. *IEEE Transactions on Automatic Control*, AC-30(6), pp. 602–605.

32 Skogestad, S. and I. Postlethwaite (1996). *Multivariable Feedback Control: Analysis and Design Using Frequency-Domain Methods*. John Wiley and Sons, New York.
33 Franklin, G.F., J.D. Powell, and M. Workman (1998). *Digital Control of Dynamic Systems*, 3rd edition. Addison-Wesley Longman, Reading, MA.
34 Qin, S.J. and T.A. Badgewell (2000). An overview of nonlinear model predictive control applications. In *Nonlinear Model Predictive Control*, Switzerland.
35 Morari, M. and J.M. Lee (1999). Model predictive control: past, present and future. *Computers and Chemical Engineering*, 23, pp. 667–682.
36 Carlos, E.G., D.M. Prett, and M. Morari (1989). Model predictive control: theory and practice – a survey. *Automatica*, 25(3), pp. 335–348.
37 Maciejowski, J.M. (2002). *Predictive Control with Constraints*. Prentice Hall, Englewood Cliffs, NJ.
38 Clarke, D., Ed. (1994). *Advances in Model-Based Predictive Control*. Oxford Science Publications, Oxford, UK.
39 Soeterboek, R. (1992). *Predictive Control, A Unified Approach*. Prentice Hall, Englewood Cliffs, NJ.
40 Richalet, J. (1993). Industrial applications of model based predictive control. *Automatica*, 29(5), pp. 1251–1274.
41 Rawlings, J.B. (2000). Tutorial overview of model predictive control. *IEEE Control Systems Magazine*, pp. 38–52.
42 Naeem, W., R. Sutton, and J. Chudley (2004). Model predictive control of an autonomous underwater vehicle with a fuzzy objective function optimized using a GA. In Proceedings of Control Applications in Marine Systems (CAMS'04), IFAC, Ancona, Italy, pp. 433–438.

Chapter 8

Robust control of autonomous underwater vehicles and verification on a tethered flight vehicle

Z. Feng and R. Allen

8.1 Introduction

Autonomous underwater vehicles (AUVs) may be divided into two groups [1]: flight vehicles which have slender bodies, used for surveys, searches, object delivery, and object location; and the hovering vehicles, used for detailed inspection and physical work on and around fixed objects. While flight vehicles typically require three degrees of freedom to be controlled (longitudinally, and yaw and pitch), hovering vehicles require more than three degrees of freedom to be controlled, for example, dynamic positioning.

Automatic control of an AUV, either a flight vehicle or a hovering vehicle, presents several difficulties due to the fundamentally nonlinear dynamics of the vehicle, the model uncertainty resulting from, for example, inaccurate hydrodynamic coefficients and the external disturbances such as underwater currents. Depending on the degree of parameter uncertainty and disturbances, different levels of machine intelligence, from robust control to adaptive control are required. See, for example, Reference 2 for PID control, References 3 and 4 for adaptive control, References 5 and 6 for sliding mode control and References 7 and 8 for H_∞ control.

In the authors' opinion, robust control is adequate to deal with model uncertainty involved in vehicle dynamics if hydrodynamic coefficients can be realistically estimated by real tank tests or by theoretical estimation techniques.

Among robust control approaches, H_∞ design is becoming a standard method since it describes the robust control problem by a single linear fractional transformation (LFT) framework, the robust controller within which can be synthesised by the Riccati equation based approach [9] or the linear matrix inequality (LMI) based approach [10,11]. Both approaches can be implemented by existing software packages, for example, the Robust Control Toolbox and the LMI Toolbox in Matlab [12,13].

H_∞ control of flight vehicles is studied in this chapter. It is applied to devise robust autopilots for flight control in three degrees of freedom, that is, forward speed, heading and depth/altitude. Compared with the applications reported in the literature, our approach is non-standard and has the following new features:

(i) *Low order*: the standard H_∞ approach usually generates a robust controller of the same order as that of the generalised plant which is the augmented plant with weighting functions; our approach can produce reduced order controllers which maintain comparable performance to the control system with a full-order controller.
(ii) Instead of using quaternion representation concepts [3], representative singularity of Euler angles is considered in the design of depth control system.
(iii) Overshoot specification is also considered.

In order to verify the design via water trials in addition to nonlinear simulations, the Southampton University *Subzero III* vehicle was used as a test bed. It is self-powered and has the shape of a torpedo. Unfortunately, the thin cable, which is attached to *Subzero III* at the tail and transmits control commands and sensor data between the control computer on the shore (with a command joystick attached) and the underwater vehicle, means that *Subzero III* is effectively a tethered vehicle. Therefore, due to the dynamical coupling between the cable and the vehicle, *Subzero III* cannot be viewed as an untethered AUV.

Fortunately, as will be shown, it was found that cable effects on the dynamics of the vehicle can be compensated by feed-forward control. This suggests a composite control scheme for testing AUV autopilots in *Subzero III*: whereas AUV autopilots (feedback controllers) deal with the vehicle dynamics, feed-forward control counteracts the cable effects.

The remainder of this chapter is organised as follows. The design of robust autopilots for flight control of *Subzero III* is conducted in Section 8.2. The tether effects on *Subzero III* are then evaluated by a numerical scheme capable of dealing with cables of varying length in Section 8.3. Moreover, the feed-forward control is devised for the tether effects to be cancelled. Section 8.4 presents field tests of *Subzero III* in water tanks. Conclusions are drawn in Section 8.5.

8.2 Design of robust autopilots for torpedo-shaped AUVs

In this section, robust autopilots for flight control of torpedo-shaped AUVs will be designed using the reduced-order H_∞ approach [14]. Specifically, *Subzero III*, excluding its communication cable, is adopted as a prototype of a torpedo-shaped AUV to allow the design to be verified by nonlinear simulations.

8.2.1 Dynamics of Subzero III (excluding tether)

8.2.1.1 Motion equations

It is convenient to define two coordinate frames in order to describe the vehicle dynamics: the body-fixed frame $O\text{-}X_oY_oZ_o$ and the Earth-fixed frame $E\text{-}\xi\eta\zeta$. By defining the vehicle's translational and angular velocity along and around the axes of the body-fixed frame as $V = [u, v, w, p, q, r]^T$, and the vehicle's coordinates and attitude angles relative to the Earth-fixed frame as $\eta = [x, y, z, \phi, \theta, \psi]^T$, the following kinematical relationship can be obtained:

$$\dot{\eta} = J(\eta)V \tag{8.1}$$

where $J(\eta)$ can be found in Reference 15.

The motion equations of the vehicle can then be written in the compact form [15]:

$$(M_{RB} + M_A)\dot{V} + [C_{RB}(V) + C_A(V)]V + D(V) + g(\eta) = \tau \tag{8.2}$$

where M_{RB} and C_{RB} are the rigid body's inertia matrix and the Coriolis and centripetal matrix, respectively, M_A is the added mass and inertia matrix, C_A the added Coriolis and centripetal matrix, $D(V)$ the hydrodynamic damping force, $g(\eta)$ the restoring force and moments, and τ is the external force and moments exerted on the vehicle (from actuators such as thrusters and control surfaces, and environmental disturbances such as underwater currents). Note that the damping force is different from that in Reference 15 due to the geometric shape of torpedo AUVs.

Before the motion equations are specified for *Subzero III*, the following assumptions are made: (i) the vehicle is neutrally buoyant; (ii) the vehicle has two symmetric planes (top–bottom and port–starboard); (iii) the origin of the body-fixed frame is located at the geometric centre of the hull and (iv) the control surfaces consist only of a rudder (for heading control) and a stern-plane (for pitch/depth control). These assumptions maintain generality for dynamical modelling analysis of torpedo-shaped vehicles and lead to a simplified structure of model parameters as follows.

$$M_A = \begin{bmatrix} m_{11} & 0 & 0 & 0 & 0 & 0 \\ 0 & m_{22} & 0 & 0 & 0 & m_{26} \\ 0 & 0 & m_{33} & 0 & m_{35} & 0 \\ 0 & 0 & 0 & m_{44} & 0 & 0 \\ 0 & 0 & m_{35} & 0 & m_{55} & 0 \\ 0 & m_{26} & 0 & 0 & 0 & m_{66} \end{bmatrix} \tag{8.3}$$

$$C_A(\mathbf{v}) = \begin{bmatrix} 0 & C_{A1}(\mathbf{v}) \\ -C_{A1}^T(\mathbf{v}) & C_{A2}(\mathbf{v}) \end{bmatrix}$$

$$C_{A1}(\mathbf{v}) = \begin{bmatrix} 0 & m_{33}w + m_{35}q & -m_{22}v - m_{26}r \\ -m_{33}w - m_{35}q & 0 & m_{11}u \\ m_{22}v + m_{26}r & -m_{11}u & 0 \end{bmatrix} \quad (8.4)$$

$$C_{A2}(\mathbf{v}) = \begin{bmatrix} 0 & m_{26}v + m_{66}r & -m_{55}q \\ -m_{26}v - m_{66}r & 0 & m_{44}p \\ m_{55}q & -m_{44}p & 0 \end{bmatrix}$$

$$D(V) = \begin{bmatrix} -X_{u|u|}u|u| \\ \frac{1}{2}\rho C_d \int_{x_{\text{tail}}}^{x_{\text{nose}}} d(l)(v+lr)\sqrt{(v+lr)^2 + (w-lq)^2}\, dl \\ \frac{1}{2}\rho C_d \int_{x_{\text{tail}}}^{x_{\text{nose}}} d(l)(w-lq)\sqrt{(v+lr)^2 + (w-lq)^2}\, dl \\ 0 \\ -\frac{1}{2}\rho C_d \int_{x_{\text{tail}}}^{x_{\text{nose}}} ld(l)(w-lq)\sqrt{(v+lr)^2 + (w-lq)^2}\, dl \\ \frac{1}{2}\rho C_d \int_{x_{\text{tail}}}^{x_{\text{nose}}} ld(l)(v+lr)\sqrt{(v+lr)^2 + (w-lq)^2}\, dl \end{bmatrix} \quad (8.5)$$

$$\tau = \begin{bmatrix} T \\ Y_{\delta r} u^2 \delta r \\ Z_{\delta s} u^2 \delta s \\ 0 \\ M_{\delta s} u^2 \delta s \\ N_{\delta r} u^2 \delta r \end{bmatrix} \quad (8.6)$$

where T is the thrust produced by the propeller, and δr, δs are the deflections of the rudder and the stern-plane, respectively. The matrices M_{RB}, C_{RB}, $g(\eta)$ can be found in Reference 15.

The model parameters involved in (8.2) include rigid body data and hydrodynamic coefficients and can be found in References 16 and 17.

8.2.1.2 Dynamics of actuators

To model the dynamics of the vehicle from the manoeuvring commands to the vehicle's states, the dynamics of actuators need to be augmented with the motion equations (8.2). Typical actuators for torpedo-shaped AUVs include a thruster driven by a direct current motor, a rudder and a stern-plane which are driven by respective servo motors.

For *Subzero III*, the dynamics of the propulsion system can be written in the form [17] as follows:

$$\dot{n} = f(m_d, u) \quad (8.7)$$

$$T = \tfrac{1}{2}\pi \rho D^2 C_t \left[u^2 + (0.7\pi n D)^2\right] \quad (8.8)$$

where f is a nonlinear function over the motor command m_d (mark-space ratio of PWM drive), n is the propeller speed, D is the diameter of the propeller and C_t is the thrust coefficient.

The dynamics of the servo mechanisms of the control fins are simplified by linear systems subject to the saturation limits, that is,

$$\delta r(s) = \frac{0.9}{0.13s + 1} \delta r_d(s), \quad |\delta r| \leq \pi/9 \tag{8.9}$$

$$\delta s(s) = \frac{0.9}{0.087s + 1} \delta s_d(s), \quad |\delta r| \leq \pi/6 \tag{8.10}$$

where $\delta r_d, \delta s_d$ are the deflection commands.

To summarise, the vehicle dynamics are modelled by (8.1)–(8.10) with m_d, δr_d, δs_d being the manoeuvring commands.

8.2.2 Plant models for control design

Nonlinear simulations show that the interactions between the degrees of freedom to be controlled are slight and thus suggest that the degrees of freedom can be controlled independently to achieve an overall autopilot of decoupled structure [17,18]. Therefore, the overall autopilot consists of three autopilots (speed, heading and depth) each of which controls one degree of freedom and is designed independently. This is similar to the current practice in naval submarines [19].

To devise model-based linear controllers, the linearised models around the equilibrium points of the nonlinear model are required. For *Subzero III*, these design plant models are obtained by separating the nonlinear model into three subsystems (speed, heading and depth/altitude systems) and linearising them around the cruise condition (straight ahead motion with cruise speed of 1.3 m/s). These can be written as

$$\dot{x}_i = A_{p_i} x_i + B_{p_i} u_i, \quad y = C_{p_i} x_i + D_{p_i} u_i \quad (i = 1, 2, 3) \tag{8.11}$$

with $i = 1, 2, 3$ denoting the speed, the heading and the depth subsystems, respectively, and the states, the control inputs and the outputs are defined as

$$\mathbf{x}_1 := [u \quad n]^T, \quad \mathbf{u}_1 := m_d, \quad y_1 := u \tag{8.12}$$

$$\mathbf{x}_2 := [\psi \quad v \quad r \quad \delta r]^T, \quad \mathbf{u}_2 := \delta r_d, \quad y_2 := \psi \tag{8.13}$$

$$\mathbf{x}_3 := [z \quad \theta \quad q \quad \delta s]^T, \quad \mathbf{u}_3 := \delta s_d, \quad y_3 := z \tag{8.14}$$

The matrices involved in the linear models (8.11) can be found in Reference 20. Note for the heading subsystem $i = 2$, the roll modes (p, ϕ) have been removed due to their passive stability. The heave speed w has also been removed in the depth

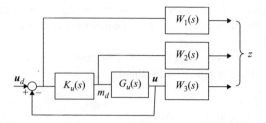

Figure 8.1   The speed control system

subsystem $i = 3$ due to its small magnitude for flight vehicles. This is consistent with the approach taken in Reference 2.

8.2.3   Design of reduced-order autopilots

8.2.3.1   Speed autopilot

As shown in Figure 8.1, the scheme of mixed sensitivity optimisation is adopted for speed control. In Figure 8.1, $G_u(s) := D_{p_1} + C_{p_1}(sI - A_{p_1})^{-1} B_{p_1}$ is the transfer function of the speed subsystem, $K_u(s)$ is the speed autopilot to be designed, $W_1(s)$, $W_2(s)$ and $W_3(s)$ are the weighting functions that penalise the tracking error $(u_d - u)$, control input m_d and the plant output u, respectively.

The generalised plant for speed control system is

$$P_u(s) = \begin{bmatrix} W_1(s) & -W_1(s)G_u(s) \\ 0 & W_2(s) \\ 0 & W_3(s)G_u(s) \\ 0 & -G_u(s) \end{bmatrix} \quad (8.15)$$

For the closed-loop system to achieve a steady state error less than 1%, $W_1(s)$ is chosen as

$$W_1(s) = \frac{10(s + 10)}{1000s + 1} \quad (8.16)$$

Since the control input for the motor is the mark-space ratio of the PWM drive which has a maximal value of 1, $W_2(s)$ is selected as

$$W_2(s) = 1 \quad (8.17)$$

The choice of $W_3(s)$ depends on the requirement on the attenuation rate in the high-frequency region. Here the restriction of the overshoot following step commands is also considered since too much overshoot could result in the vehicle's collision with underwater obstacles. The inclusion of the overshoot specification into $W_3(s)$ is achieved by specifying the characteristics of $W_3(s)$ in the low-frequency region [8]. Therefore $W_3(s)$ is chosen as

$$W_3(s) = \frac{(s + 14.33)^2}{215.34} \quad (8.18)$$

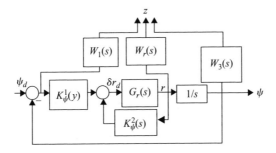

Figure 8.2 The heading control system

to achieve an attenuation rate of -40 dB/dec in the high-frequency region and an overshoot less than 10 per cent following step commands.

Inserting the weighting functions (8.16)–(8.18) into (8.15) yields the generalised plant of third order.

A reduced-order (second-order) speed autopilot with negligible performance degradation has been obtained [20].

8.2.3.2 Heading autopilot

The scheme for the design of the heading autopilot is similar to that for speed control. However, as depicted in Figure 8.2, in addition to the heading angle, the yaw rate is also fed back to improve the dynamic performance. This idea is also current practice in naval submarines [19]. Moreover, the weighting function $W_r(s)$ penalises the yaw rate instead of the rudder command. The reasons for this are: (1) confining the yaw rate is a way to restrict the rudder effort (to reduce energy consumption) since it reflects the rudder deflection and (2) confining the yaw rate can also limit the course overshoot which is important for the vehicle to fly in restricted waters. It is selected as

$$W_r(s) = 2 \tag{8.19}$$

to restrict the magnitude of the yaw rate less than $5°/s$ under a step command of $10°$.

The generalised plant for heading control is of fifth order and can be written as

$$P_\psi(s) = \begin{bmatrix} W_1(s) & -W_1(s)G_\psi(s) \\ 0 & W_r(s)G_r(s) \\ 0 & W_3(s)G_\psi(s) \\ I & -G_\psi(s) \\ 0 & G_r(s) \end{bmatrix} \tag{8.20}$$

and the multi-input, single-output (MISO) heading autopilot,

$$K_\psi(s) = [K_\psi^1(s) \quad K_\psi^2(s)] \tag{8.21}$$

where $G_\psi(s) = D_{p_2} + C_{p_2}(sI - A_{p_2})^{-1} B_{p_2}$ is the transfer function of the heading subsystem, $G_r(s)$ is the transfer function of the yaw rate system and $G_r(s) = sG_\psi(s)$ due to $\dot{\psi} = r$ under the cruise condition.

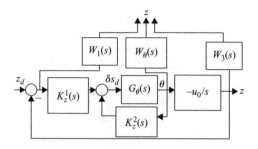

Figure 8.3 The depth control system

A second-order heading autopilot with slight performance degradation has been obtained [20].

8.2.3.3 Depth autopilot

The scheme for depth/altitude control is similar to that of heading control. As shown in Figure 8.3, the pitch angle is also fed back to improve the dynamic performance of the depth control since it reflects the rate of depth. Moreover, instead of the stern-plane command, the pitch angle is penalised for the following reasons: (1) it is a direct way to penalise the stern-plane effort and (2) it allows the representation singularity of the Euler angles to be avoided. Since the vehicle's attitude cannot be described by the Euler angles when the pitch angle is close to $\pm 90°$ (for *Subzero III*, the digital compass can only provide reliable attitude angles within a tilt of $50°$), the second advantage is more attractive. It is selected as

$$W_\theta(s) = \frac{6}{\pi} \tag{8.22}$$

to confine the pitch angle within $(-30°, 30°)$ under a step command of 1 m change in depth.

By defining $w = z_d$, $y = [z_d - z \quad \theta]^T$ and $u = \delta s_d$, the generalised plant for depth control is of fifth order and can be written as

$$P_z(s) = \begin{bmatrix} W_1(s) & -W_1(s)G_z(s) \\ 0 & W_\theta(s)G_\theta(s) \\ 0 & W_3(s)G_z(s) \\ I & -G_z(s) \\ 0 & G_\theta(s) \end{bmatrix} \tag{8.23}$$

where $G_z(s) = D_{p_3} + C_{p_3}(sI - A_{p_3})^{-1} B_{p_3}$ is the transfer function of the depth system, and $G_\theta(s)$ is the transfer function of the pitch system and $G_\theta(s) = -sG_z(s)/u_0$ due to $\dot{z} \approx -u_0\theta$ under the cruise condition. The depth autopilot is also a MISO controller,

$$K_z(s) = [K_z^1(s) \quad K_z^2(s)] \tag{8.24}$$

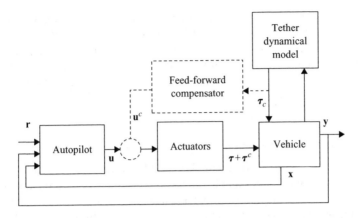

Figure 8.4   Composite control of the tethered vehicle

A second-order depth autopilot with slight performance degradation has been obtained [20].

From the nonlinear simulations [20], it has been demonstrated that the performance of all three autopilots is acceptable. Moreover, since they can deal with the time-varying dynamics of *Subzero III* from rest to its cruise state, they are robust to model uncertainty. Therefore it can be concluded that the robust autopilots exhibit good behaviour. However, this conclusion needs to be confirmed by water trials as the model uncertainty involved in the nonlinear model cannot be avoided.

8.3 Tether compensation for *Subzero III*

8.3.1 Composite control scheme

As stated in Section 8.1, although the communication cable allows *Subzero III* to be controlled from the shore, it effectively makes the vehicle tethered and thus influences the vehicle's motion. Since autopilots for AUV flight control do not have to counteract tether effects (an AUV is tether free), it is therefore desirable to remove the tether effects by external action rather than the autopilots in order to test AUV autopilots on *Subzero III*. This idea can be realised by introducing feed-forward control in addition to the feedback control if the tether effects can be predicted and can be fully counteracted by on-board actuators.

This leads to a composite control system, as depicted in Figure 8.4, which comprises a feedback controller with reference input **r**, plant output **y** and possible plant state **x** being its input, and a feed-forward controller with tether force/moment effects τ_c on the vehicle as its input. The actuating commands can be separated into two components, one from the feedback controller (autopilot) and the other from the feed-forward controller. In other words, the feed-forward controller counteracts the tether dynamics while the feedback autopilot deals with the dynamics of the vehicle.

To achieve the scheme of composite control as shown in Figure 8.4, evaluation and removal of tether effects are critical. The evaluation of tether effects is to predict disturbances (the force/moment exerted on the vehicle at the connecting point by the tether) and thus allows feed-forward control to be adopted. The removal of tether effects is based upon devising the compensation action to counteract the tether effects by the available actuators.

8.3.2 Evaluation of tether effects

Treated as a long, thin, flexible circular cylinder in arbitrary changes in gravity, inertial forces, driving forces and hydrodynamic loading which are taken to be the sum of independently operating normal and tangential drags, each given by a single coefficient, the dynamics of the tether can be described by the partial differential equations over time and arc length of the cable [21].

For a typical towed cable system, where the length of the cable is fixed, finite difference methods [21,22] can be applied to numerically solve the differential equations by discretising cable length and time. However, these numerical schemes cannot be applied to evaluate the tether effects directly due to non-fixed length of tether.

As shown in Figure 8.5, the communication cable attached to *Subzero III* is originally coiled around a cable drum on the shore and then pulled into the water by the moving vehicle. Therefore the length of wet cable increases as the vehicle moves forward. This is similar to the deployment of the umbilical cable attached to a remotely operated vehicle (ROV).

To evaluate the effects of tether of non-fixed length, the finite difference method can still be applied. However, unlike the numerical scheme in Reference 22 where the number of difference equations is fixed, the number of difference equations in our scheme is not fixed and increases as the cable is being deployed. Therefore this method is not suitable for online prediction of the effects of very long cables since increasing computational effort is involved in solving difference equations of increasing number. However, we argue that this scheme is still valid for those cases where the cable lengths are not very long and for computationally assessing tether effects.

8.3.2.1 Cable dynamics

To analyse the motion of the cable as well as its effect on the vehicle, it is convenient to define three coordinate systems, that is, the Earth-fixed frame and the local frames along the cable and the vehicle-fixed frame. As shown in Figure 8.5, the Earth-fixed frame ($\mathbf{i}, \mathbf{j}, \mathbf{k}$) is selected with $\mathbf{k}$ pointing vertically downwards. The vehicle-fixed frame ($\mathbf{i}_v, \mathbf{j}_v, \mathbf{k}_v$) is located at the centre of the hull, with $\mathbf{i}_v$ coinciding with the longitudinal axis and $\mathbf{j}_v$ pointing to starboard.

The relationship between the vehicle-fixed frame and the Earth-fixed frame can be expressed in terms of Euler angles [15], that is,

$$[\mathbf{i}_v \quad \mathbf{j}_v \quad \mathbf{k}_v] = [\mathbf{i} \quad \mathbf{j} \quad \mathbf{k}]R(\phi,\theta,\psi) \qquad (8.25)$$

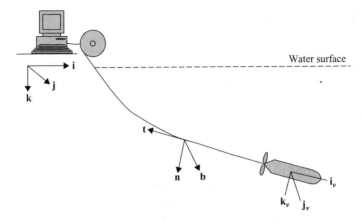

Figure 8.5 Definition of coordinate systems

with

$$R(\phi,\theta,\psi) = \begin{bmatrix} c\theta c\psi & -c\phi s\psi + s\phi s\theta c\psi & s\phi s\psi + c\phi s\theta c\psi \\ c\theta s\psi & c\phi c\psi + s\phi s\theta s\psi & -s\phi c\psi + c\phi s\theta s\psi \\ -s\theta & s\phi c\theta & c\phi c\theta \end{bmatrix} \quad (8.26)$$

where $c\cdot = \cos\cdot, s\cdot = \sin\cdot$ and ϕ, θ, ψ are the roll, pitch and heading angle of the vehicle, respectively.

The local frames (**t**, **n**, **b**) are located at points along the cable with **t** tangential to the cable in the direction of increasing arc length from the tow-point, and **b** in the plane of (**i**, **j**). They are obtained by three rotations of the Earth-fixed frame in the following order: (1) a counter-clockwise rotation through angle α about the **k** axis to bring the **i** axis into the plane of **t** and **n**; (2) a counter-clockwise rotation about the new the **i** axis through $\pi/2$ to bring the **k** axis into coincidence with $\mathbf{b}_i$ and (3) a clockwise rotation about **b** through β to bring **i** and **j** into coincidence with **t** and **n**. Thus the relationship between the local frames and the Earth-fixed frame can be expressed as follows [21,22], that is,

$$[\mathbf{t}\quad \mathbf{n}\quad \mathbf{b}] = [\mathbf{i}\quad \mathbf{j}\quad \mathbf{k}]W(\alpha,\beta) \quad (8.27)$$

with

$$W(\alpha,\beta) = \begin{bmatrix} c\alpha c\beta & -c\alpha s\beta & s\alpha \\ -s\alpha c\beta & s\alpha s\beta & c\alpha \\ -s\beta & -c\beta & 0 \end{bmatrix} \quad (8.28)$$

In terms of (8.25)–(8.28), the relationship between the local frames and the vehicle-fixed frame can be written as

$$[\mathbf{t}\quad \mathbf{n}\quad \mathbf{b}] = [\mathbf{i}_v\quad \mathbf{j}_v\quad \mathbf{k}_v]R^T(\phi,\theta,\psi)W(\alpha,\beta) \quad (8.29)$$

where the orthogonal property of R has been used.

To describe the configuration of the cable, it is necessary to define the vector

$$\mathbf{y}(s,t) := [T \quad V_t \quad V_n \quad V_b \quad \alpha \quad \beta]^T \tag{8.30}$$

where t and s denote the time and the arc length of the cable measured from the tow-point, respectively, T is the tension and $\mathbf{V}_c = [V_t \quad V_n \quad V_b]^T$ denotes the velocity vector in the local frames along the cable. The cable dynamics can be expressed in terms of the following partial differential equation [21]

$$\mathbf{M}\frac{\partial \mathbf{y}}{\partial s} = \mathbf{N}\frac{\partial \mathbf{y}}{\partial t} + \mathbf{q} \tag{8.31}$$

with $\mathbf{M}, \mathbf{N}, \mathbf{q}$ defined in References 21 and 22.

To solve the partial differential equation (8.31), the initial configuration of the cable and six boundary conditions are required.

While the initial configuration is given, the boundary conditions are determined by the constraints at both ends of the wet cable (vehicle end and drum end).

Since the cable has the same velocity as the vehicle at the tow-point (connecting point), three boundary conditions are obtained. Denote the position of the tow-point in the vehicle-fixed frame with $\mathbf{r}_c = [x_c \quad y_c \quad z_c]^T$, and the linear and angular velocity of the vehicle with $V = [u \quad v \quad w]^T$ and $\Omega = [p \quad q \quad r]^T$, where (u, v, w) denote surge, sway and heave speed, while (p, q, r) denote roll, pitch and yaw rates of the vehicle. The velocity of the tow-point in the vehicle-fixed frame is $(V + \Omega \times \mathbf{r}_c)$, where $\times$ denotes the cross product of two vectors. In terms of (8.29), the velocity of the tow-point can be expressed in the local frame of the cable, that is,

$$\mathbf{V}_c(0,t) = W^T(\alpha, \beta) R(\phi, \theta, \psi)(V + \Omega \times \mathbf{r}_c) \tag{8.32}$$

The remaining boundary conditions can be derived from the cable drum where the cable is deployed. Since the drum always deploys the cable in the tangential direction, the normal components of the cable velocity must be zero, that is,

$$V_n(S_t, t) = 0 \tag{8.33}$$
$$V_b(S_t, t) = 0 \tag{8.34}$$

where S_t is the total arc length of the immersed cable at time t.

Moreover, the last boundary condition can be obtained by considering the rolling motion of the cable drum. In terms of Newton's second law, one has

$$\frac{d}{dt}(I_d \Omega_d) = \Gamma - \Gamma_f$$

where I_d is the moment of inertia of the drum, Ω_d is the angular speed of the drum, Γ is the driving torque produced by the cable tension at the drum-end, and Γ_f is the resisting torque caused by the sliding friction between the drum and the pivot.

Since $\Omega_d = -V_t(S_t, t)/R_d$ and $\Gamma = T(S_t, t)R_d$, where R_d is the radius of the cable drum, the dynamic equation of the drum can be rewritten as

$$I_d \dot{V}_t(S_t, t) + T(S_t, t)R_d^2 = \Gamma_f R_d \tag{8.35}$$

To summarise, the boundary conditions have been determined by (8.32)–(8.35).

The tether effects on the vehicle are forces and moments at the tow-point. While the forces are actually the cable tension at the tow-point, that is, $\mathbf{T}(0,t)\mathbf{t}(0,t)$, the moments are caused by the cable tension due to non-coincidence of the tow-point with the origin of the vehicle-fixed frame. It is essential to describe these forces and moments in the vehicle-fixed frame since they need to be counteracted by the actuators on-board the vehicle.

According to the transformation of frames (8.27), the disturbing forces from the tether to the vehicle can be written as

$$\mathbf{F}_c(t) := \begin{bmatrix} F_{cX} \\ F_{cX} \\ F_{cX} \end{bmatrix} = R^T(\phi,\theta,\psi) W(\alpha(0,t),\beta(0,t)) \begin{bmatrix} T(0,t) \\ 0 \\ 0 \end{bmatrix} \quad (8.36)$$

Recall the position of the tow-point in the vehicle-fixed frame, the disturbing moments from the tether to the vehicle can be expressed by

$$\mathbf{M}_c(t) := \begin{bmatrix} M_{cX} \\ M_{cY} \\ M_{cZ} \end{bmatrix} = \mathbf{r}_c \times \mathbf{F}_c(t) = \begin{bmatrix} y_c F_{cZ} - z_c F_{cY} \\ z_c F_{cX} - x_c F_{cZ} \\ x_c F_{cY} - y_c F_{cX} \end{bmatrix} \quad (8.37)$$

8.3.2.2 Numerical scheme

To evaluate the tether effects on the vehicle as described by (8.35) and (8.36), it is necessary to solve the partial differential equation (8.31).

Since the length of the tether attached to the vehicle is not-fixed, a new numerical scheme based upon the finite difference method is applied [23]. Unlike the existing schemes where the number of nodes or segments (obtained by discretising the tether along its arc length) is fixed, our scheme allows the number of nodes or segments to be increased as the tether is being deployed as shown in Figure 8.6.

It can be seen from Figure 8.6 that during the time interval of $\Delta t = t_{k+1} - t_k$, a new segment (node) emerges during the deployment. Our scheme is to add the new segment (node) emerged at t_{k+1} to the old segments (nodes) which emerged at time instants $t_0, t_1, \ldots, t_k$. Therefore, although the length of each segment is constant over

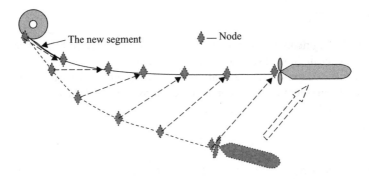

Figure 8.6   Evolution of the cable configuration from t_k to t_{k+1}

time, the number of the segments increases with time as long as the vehicle moves forward.

To specify the nodes along the tether, it is necessary to determine the arc length of the cable. The total length of the wet cable at time t is determined by the deployment speed $-V_t(S_\tau, \tau)$ with $0 \leq \tau < t$, that is,

$$S_t = -\int_0^t V_t(S_\tau, \tau) \, d\tau \tag{8.38}$$

where the initial length of the wet cable has been assumed to be zero.

Suppose the step for time discretisation is of a fixed value Δt, the cable length at time $k\Delta t$ can be written as

$$S_k = -\sum_{i=0}^{k-1} \int_{i\Delta t}^{(i+1)\Delta t} V_t(S_\tau, \tau) \, d\tau \tag{8.39}$$

and the cable length at time $(k+1)\Delta t$ can be written as

$$S_{k+1} = -\sum_{i=0}^{k} \int_{i\Delta t}^{(i+1)\Delta t} V_t(S_\tau, \tau) \, d\tau \tag{8.40}$$

Therefore, by subtracting S_k from S_{k+1}, the length of new segment is

$$\Delta S_k = -\int_{(k-1)\Delta t}^{k\Delta t} V_t(S_\tau, \tau) \, d\tau \tag{8.41}$$

Similarly, the lengths of the old segments are

$$\Delta S_j = -\int_{(j-1)\Delta t}^{j\Delta t} V_t(S_\tau, \tau) \, d\tau, \quad j = 1, 2, \ldots, k-1 \tag{8.42}$$

Provided the time step Δt is sufficiently small, the segment lengths can then be approximated by

$$\Delta S_j = -V_t(S_j, t_j)\Delta t, \quad j = 0, 1, \ldots, k \tag{8.43}$$

After the segments (nodes) along the immersed cable have been specified, the finite difference method of second-order approximation [22] can be applied. Suppose that the configuration of the cable at time t_k, that is, $\mathbf{y}(S_i, t_k)$ with $i = 0, 1, \ldots, k$, is known, the problem is then to predict the configuration of the cable at time t_{k+1}, that is, $\mathbf{y}(S_i, t_{k+1})$ with $i = 0, 1, \ldots, k+1$.

Define

$$\begin{aligned} \Delta S_j &:= S_j - S_{j-1} \\ S_{j-1/2} &:= \tfrac{1}{2}(S_j + S_{j-1}), \quad j = 1, 2, \ldots, k+1 \\ t_{k+1/2} &:= \left(k + \tfrac{1}{2}\right)\Delta t \end{aligned} \tag{8.44}$$

and

$$\mathbf{y}(S_{j-1/2}, t_k) := \tfrac{1}{2}[\mathbf{y}(S_j, t_k) + \mathbf{y}(S_{j-1}, t_k)]$$
$$\mathbf{y}(S_{j-1/2}, t_{k+1}) := \tfrac{1}{2}[\mathbf{y}(S_j, t_{k+1}) + \mathbf{y}(S_{j-1}, t_{k+1})]$$
$$\mathbf{y}(S_{j-1}, t_{k+1/2}) := \tfrac{1}{2}[\mathbf{y}(S_{j-1}, t_k) + \mathbf{y}(S_{j-1}, t_{k+1})] \quad j = 1, 2, \ldots, k+1$$
$$\mathbf{y}(S_j, t_{k+1/2}) := \tfrac{1}{2}[\mathbf{y}(S_j, t_k) + \mathbf{y}(S_j, t_{k+1})]$$

(8.45)

where $S_0 = 0$.

Note that since the new segment ΔS_{k+1} is not in the water at t_k, $\mathbf{y}(S_{k+1}, t_k)$ in (8.35) is prescribed according to the rolling motion of the drum, which is assumed to be uniform, that is, $\mathbf{y}(S_{k+1}, t_k) = [T(S_k, t_k) \; V_t(S_k, t_k) \; 0 \; 0 \; \alpha(S_k, t_k) \; \beta(S_k, t_k) - \Delta S_{k+1}/R_d]^\mathrm{T}$. Moreover, the governing equation (8.31) will still be applied to approximate the dynamics of the new segment although it is only partly true for the new segment which is gradually pulled into the water. Obviously, this may result in modelling error. However, as will be verified below, the modelling error can be negligible when the new segment is short.

Applying the governing equation (8.31) at the points $(S_{j-1/2}, t_{k+1/2})$, $j = 1, 2, \ldots, k+1$ yields the $6(k+1)$ difference equations as follows [22]:

$$M(\mathbf{y}(S_{j-1/2}, t_{k+1})) \frac{\mathbf{y}(S_j, t_{k+1}) - \mathbf{y}(S_{j-1}, t_{k+1})}{\Delta S_j}$$
$$+ M(\mathbf{y}(S_{j-1/2}, t_k)) \frac{\mathbf{y}(S_j, t_k) - \mathbf{y}(S_{j-1}, t_k)}{\Delta S_j}$$
$$= N(\mathbf{y}(S_j, t_{k+1/2})) \frac{\mathbf{y}(S_j, t_{k+1}) - \mathbf{y}(S_j, t_k)}{\Delta t} + N(\mathbf{y}(S_{j-1}, t_{k+1/2}))$$
$$\times \frac{\mathbf{y}(S_{j-1}, t_{k+1}) - \mathbf{y}(S_{j-1}, t_k)}{\Delta t} + \mathbf{q}(\mathbf{y}(S_{j-1/2}, t_{k+1})) + \mathbf{q}(\mathbf{y}(S_{j-1/2}, t_k))$$

(8.46)

for $j = 1, 2, \ldots, k+1$.

Note that the six equations under $j = k+1$ approximate the dynamics of the new segment.

The remaining six equations are determined by the boundary conditions (8.32)–(8.35), that is,

$$\mathbf{V}_c(S_0, t_{k+1}) = W^\mathrm{T}(\alpha(S_0, t_{k+1}), \beta(S_0, t_{k+1})) R(\phi, \theta, \psi)(\mathbf{V} + \mathbf{\Omega} \times \mathbf{r}_c) \quad (8.47)$$
$$V_n(S_{k+1}, t_{k+1}) = 0 \quad (8.48)$$
$$V_b(S_{k+1}, t_{k+1}) = 0 \quad (8.49)$$
$$I_d \frac{V_t(S_{k+1}, t_{k+1}) - V_t(S_k, t_k)}{\Delta t} + T(S_{k+1}, t_{k+1}) R_d^2 = \Gamma_f R_d \quad (8.50)$$

Combining (8.44)–(8.50), $6(k+2)$ nonlinear equations with $6(k+2)$ unknowns can be obtained from which the cable effects can be determined. Once the solutions to difference equations have been obtained, the cable effects at time t_{k+1} can be determined according to (8.36) and (8.37) where t is replaced by t_{k+1}.

To summarise, the numerical scheme to evaluate the effects of the communication cable comprises the following steps:

1. Set time step Δt say 1 s and time index $k = 0$ and determine the initial configuration of the cable. Since no cable is in water at $t_k = t_0 = 0$, the vehicle-end coincides with the drum-end. Therefore the initial configuration $\mathbf{y}(S_k, t_k)$ can be obtained by solving (8.32)–(8.35) with $t = 0$.
2. Given the vehicle states $(u, v, w, p, q, r, \phi, \theta, \psi)$ at time $t_{k+1} = (k+1)\Delta t$, the states can be measured by the on-board sensors or predicted by the dynamical model of the vehicle.
3. Solve the difference equations (8.46)–(8.50). This is an iterative process. Determine the cable effects in terms of (8.36) and (8.37) with t being replaced with t_{k+1}.
4. Update time index $k \leftarrow k + 1$.

8.3.2.3 Dynamics of the tethered vehicle

By coupling the tether effects with the dynamic model of the untethered vehicle as given by (8.1)–(8.10), the dynamic model of the tethered vehicle can be described by

$$\dot{\eta} = J(\eta)V$$
$$(M_{RB} + M_A)\dot{V} + [C_{RB}(V) + C_A(V)]V + D(V) + g(\eta) = \tau + \tau_c \qquad (8.51)$$

where τ_c denotes the tether effects, that is, $\tau_c = [F_c(t), M_c(t)]^T$ [cf. (8.36) and (8.37)].

The quality of the model is assessed by checking the reproduction error between the test data (measured by the sensors on-board the vehicle) and simulated data (predicted by the model). The test data of the vehicle accelerating and turning are shown in Figure 8.7. The test was to evaluate the ability of the model of the tethered flight vehicle in predicting tether effects on forward and turning motions. The vehicle was initially at rest. While the stern-plane remained in the central position, the propeller command in revolutions per second and the rudder deflection in degrees are shown in Figure 8.7(a) and (c), respectively. The forward speed and the heading of the vehicle are shown in Figure 8.7(b) and (d) (dotted lines), respectively. From Figure 8.7(b) it can be seen the vehicle accelerated in the early stage (0–10 s) and decelerated in the final stages (20–30 s). The acceleration in the early stage is due to the constant propeller thrust and negligible effects of the short cable. The deceleration at final stage is due to continuously increasing drag of the extending cable. From Figure 8.7(d), it can be seen that for a small turn (27–35°) the cable produces a slight influence on the heading in the final stage. Therefore, the cable effects on the heading of a flight vehicle can be neglected due to its small range of turn.

Simulation results of the models of the untethered vehicle are also shown in Figure 8.7(b) and (d) (dashed lines). It can be concluded that the model uncertainty involved in the tethered vehicle is tolerant.

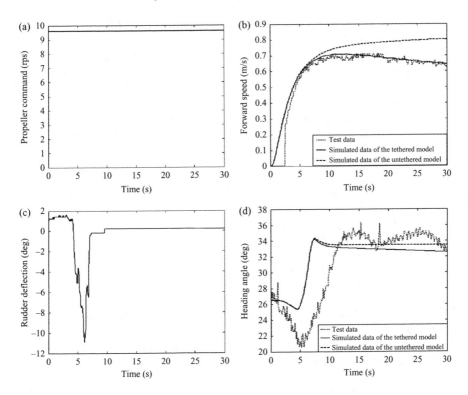

Figure 8.7 Comparison of simulated data against test data: (a) propeller revolutions, (b) forward speed, (c) rudder deflection command and (d) heading angle

8.3.3 Reduction of tether effects

After the tether effects on the motion of vehicle have been evaluated, it is possible to apply the feed-forward control to counteract it by the on-board actuators [24].

The tow-point is located at the tail of the vehicle and its position relative to the vehicle-fixed frame is

$$\mathbf{r}_c = [x_c, 0, 0]^T \tag{8.52}$$

with $x_c = -L/2$ and L is the length of the vehicle $L = 0.98$ m.

Substituting (8.52) into (8.36) and (8.37) yields the cable forces and moments on the vehicle as follows

$$\boldsymbol{\tau}_c = [F_X, F_Y, F_Z, 0, -x_c F_Z, x_c F_Y]^T \tag{8.53}$$

It can be seen from (8.53) that the tether attached to *Subzero III* affects three linear motions (surge, sway and heave) and two angular motions (pitch and yaw) of the vehicle. However, it does not affect the vehicle's roll motion. To counteract the five components of tether effects (8.53), the three on-board actuators (thruster, rudder and stern-plane) are used.

Generally speaking, it is impossible to achieve full compensation since it is an 'under-actuated problem' (the number of actuators is less than the number of degrees of freedom to be controlled). Fortunately, as will be presented, the tether effects can be cancelled in this way for Subzero III.

To counteract the tether effects, the counteracting action τ^c produced by the actuators needs to be the reverse of the tether effects (cf. Figure 8.4), that is,

$$\tau^c = -\tau_c \tag{8.54}$$

Or in componential equations [cf. (8.6) and (8.53)]

$$\begin{bmatrix} T^c \\ Y_{\delta r} u^2 \delta r^c \\ Z_{\delta s} u^2 \delta s^c \\ M_{\delta s} u^2 \delta s^c \\ N_{\delta r} u^2 \delta r^c \end{bmatrix} = - \begin{bmatrix} F_X \\ F_Y \\ F_Z \\ -x_c F_Z \\ x_c F_Y \end{bmatrix} \tag{8.55}$$

which implies that the three controls $(T^c, \delta r^c, \delta s^c)$ need to satisfy five constraints involved in (8.55).

Fortunately, this can be achieved due to the relationship between the manoeuvring coefficients listed as follows:

$$Y_{\delta r} = \tfrac{1}{2}\rho L^2 \times 2.16e - 2 \tag{8.56}$$

$$Z_{\delta s} = -\tfrac{1}{2}\rho L^2 \times 2.16e - 2 \tag{8.57}$$

$$M_{\delta s} = N_{\delta r} = -\tfrac{1}{2}\rho L^3 \times 1.02e - 2 \tag{8.58}$$

Since $x_c = -L/2$, it can be found from (8.56)–(8.58) that

$$N_{\delta r} \approx x_c Y_{\delta r} \tag{8.59}$$

$$M_{\delta s} \approx -x_c Z_{\delta s} \tag{8.60}$$

The relationship of manoeuvring coefficients (8.59) and (8.60) allows the counteraction constraints to be reduced and the compensation actions can be obtained, that is,

$$T^c = -F_X \tag{8.61}$$

$$\delta r^c = -F_Y/(Y_{\delta r} u^2) \tag{8.62}$$

$$\delta s^c = -F_Z/(Z_{\delta s} u^2) \tag{8.63}$$

where T^c, δr^c and δs^c are the thrust and the deflections of control surfaces required to cancel the tether effects.

To produce the compensation action (8.61)–(8.63), the manoeuvring commands for respective actuators need to be determined.

The dynamics of the control fins [cf. (8.9) and (8.10)] suggests that the control fin's commands be chosen as

$$\delta r_d^c = -F_Y/(0.9 Y_{\delta r} u^2) \tag{8.64}$$

$$\delta s_d^c = -F_z/(0.9 Z_{\delta s} u^2) \tag{8.65}$$

to generate deflections (8.62) and (8.63). Notice that the dynamics of the control surfaces are ignored to avoid improper transfer functions in (8.64) and (8.65).

To produce thrust (8.61), the dynamic model of the propulsion system [17] must be considered. By ignoring the dynamics of the DC motor and the nonlinearities, for example, the friction, it can be shown that

$$m_d^c = -g(u, n) F_x \tag{8.66}$$

with

$$g(u,n) = \frac{2500 D C_q(u,n)(R_a + R_f)}{C_t(u,n) K_\phi(n)(V_s - V_b - 2\pi n K_\phi(n))} \tag{8.67}$$

where $C_q(u, n)$ is the torque coefficient of the motor, R_a is the armature resistance of the DC motor, R_f is the effective resistance of the FET drive circuitry, $K_{\phi(n)}$ is the motor constant, V_s is the voltage supplied to the motor, V_b is the voltage drop across the brushes.

8.3.4 Verification of composite control by nonlinear simulations

To illustrate the effectiveness of the composite control scheme, nonlinear simulations are conducted for three motions of the tethered vehicle, that is, the straight ahead motion, turning and ascending.

Since our primary goal is to test AUV autopilots (i.e., feedback controllers) on the tethered flight vehicle (*Subzero III*), it is necessary for the tethered flight vehicle (controlled by the composite control) to behave the same as the untethered vehicle (controlled by a feedback controller only). Thus three cases are considered in the nonlinear simulations:

(i) feedback control of an untethered vehicle;
(ii) feedback control of a tethered vehicle;
(iii) composite control of a tethered vehicle.

Note that simulation results under case (i) reveals the true behaviour of AUV autopilots since an AUV is essentially untethered. This suggests that the responses under case (i) be regarded as standard responses in the evaluation of AUV autopilots on a tethered flight vehicle. In other words the comparison of the simulation results of cases (ii) and (iii) against case (i) provides a means to assess effectiveness of different control schemes (feedback and composite) for *Subzero III*.

The simulation results are shown in Figures 8.8–8.10. From comparisons, it can clearly be seen that the composite scheme allows the behaviour of AUV autopilots

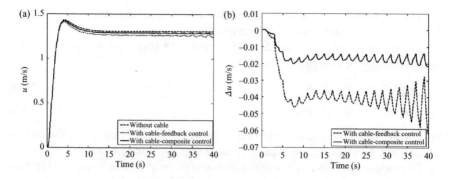

Figure 8.8 Comparison of composite and feedback control of forward speed: (a) step responses of forward speeds from 0 to 1.3 m/s and (b) tether-induced speed errors

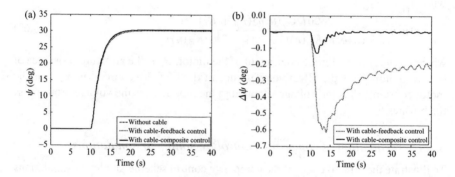

Figure 8.9 Comparison of composite and feedback control of heading angle: (a) step responses of heading from 0 to 30° and (b) tether-induced heading errors

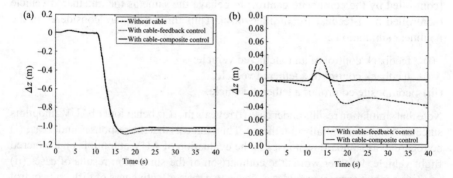

Figure 8.10 Comparison of composite and feedback control of depth: (a) step responses of depth with magnitude of −1 m and (b) tether-induced depth errors

(here reduced-order H_∞ autopilots) to be revealed in control of the tethered flight vehicle.

From Figures 8.8–8.10 it can also be seen that tether effects can be counteracted to a great extent by feedback autopilots. Thus the robustness of the reduced-order H_∞ autopilots to external disturbances appears very promising. This suggests that for short-term tests where the tether effects are not severe the feedback scheme is adequate. However, for long-term tests where the tether effects are significant, the composite scheme is necessary since feedback control cannot compensate for the increasing cable drag (e.g., Figure 8.10 where the depth error under feedback control increases with time).

8.4 Verification of robust autopilots via field tests

The reduced-order H_∞ autopilots developed in Section 8.3 are evaluated via water trials on the tethered flight vehicle – *Subzero III*. Since nonlinear simulation results in Section 8.3.4 show that for Subzero III water tests where the range of vehicle's motion is very limited, the feedback control scheme is adequate. The feed-forward control for tether compensation was, therefore, inactive during the water trials which were carried out in the Lamont Towing Tank of Southampton University and the Ocean Basin of QinetiQ Haslar.

A group of test data collected from the Ocean Basin are plotted in Figure 8.11 where the vehicle was at rest with zero pitch and roll angles before the test. The reference inputs for the degrees of freedom to be controlled simultaneously are 1 m/s, 100° and 2.5 m, respectively. The top two graphs show the transient response of the speed control system, the middle two graphs show the response of heading control system, and the bottom two show the depth response. It can clearly be seen from Figure 8.11 that the vehicle equipped with the autopilots exhibits good tracking performance when making large turns and deeper dives.

From Figure 8.11, it can be concluded that:

(i) The autopilots are robust to external disturbances, for example, short-length tether effects, since only feedback control has been applied.
(ii) The autopilots are robust to model uncertainty involved in the dynamics of Subzero III which contain several uncertain model parameters, for example, hydrodynamic coefficients.
(iii) The robust autopilots can deal with the time-varying dynamics of *Subzero III* since it was controlled by a fixed controller (without a gain schedule). Therefore gain-scheduling is not currently needed for flight control of *Subzero III*.

Finally, it should be noted here that for long-term water trials, for example, open water tests, the composite control scheme is necessary to deal with more pronounced tether effects over much longer range testing. It will also be necessary to test further the disturbance rejection of the robust autopilots to environmental disturbances, for example, underwater currents, via open water tests where the composite scheme can also be assessed due to much longer trial ranges.

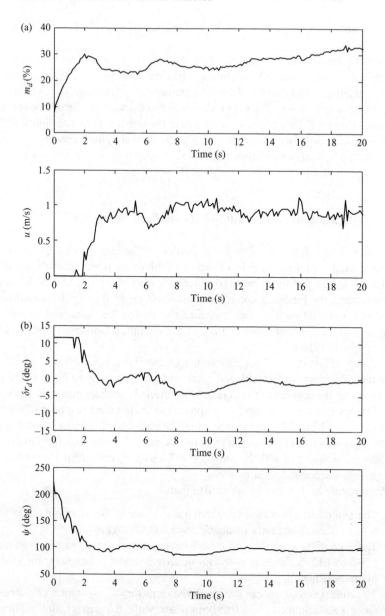

Figure 8.11　In-water behaviour of the robust autopilots (speed, heading and depth): (a) step response of forward speed from rest to 1 m/s (top: mark space ratio command for PWM drive of DC motor; bottom: forward speed); (b) step response of heading angle from 220 to 100° (top: deflection command for rudder; bottom: heading angle); and (c) step response of depth from 0.5 to 2.5 m (top: deflection command for sternplane; bottom: depth)

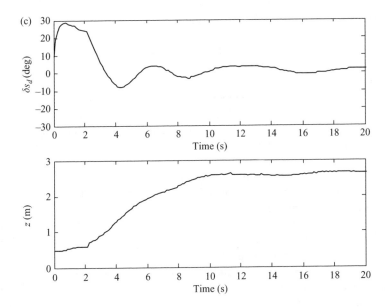

Figure 8.11   Continued

8.5   Conclusions

A tethered flight vehicle such as *Subzero III* can be a reliable test-bed for AUV control techniques provided that the tether effects be removed by feed-forward control. To achieve the composite control idea, a numerical scheme for prediction of tether effects has been proposed and assessed. We argue that the composite control scheme can also be applied to the control of an ROV during the deployment of the umbilical.

The H_∞ approach has been applied to the design of the autopilots for autonomous underwater vehicles. The autopilots developed have the following features: (1) they have simpler structures than conventional H_∞ design and the performance degradation caused by the controller order reduction is negligible; (2) rate feedback is applied for heading and depth control to improve tracking performance; (3) the overshoot specification is considered in the design by choosing suitable weighting functions.

The results of both nonlinear simulations and water trials in two different tanks show that the autopilots are robust to model uncertainty, external disturbances and varying dynamics of the vehicle and exhibit good tracking performance. Thus the effectiveness of robust control for AUVs appears promising.

Acknowledgements

The authors are very grateful to the EPSRC who supported the work reported in this chapter through the IMPROVES programme.

References

1. Valavanis, K.P., D. Gracanin, M. Matijasevic, R. Kolluru, and A. Demetriou (1997). Control architecture for autonomous underwater vehicles. *IEEE Control Systems*, pp. 48–64.
2. Jalving, B. (1994). The NDRE-AUV flight control system. *IEEE Journal of Oceanic Engineering*, 19, pp. 497–501.
3. Fjellstad, O. and T.I. Fossen (1994). Position and attitude tracking of AUVs: a quaternion feedback approach. *IEEE Journal of Oceanic Engineering*, 19(4), pp. 512–518.
4. Antonelli, G., S. Chiaverini, N. Sarkar, and M. West (2001). Adaptive control of an autonomous underwater vehicle: experimental results on ODIN. *IEEE Transactions on Control Systems Technology*, 9, pp. 756–765.
5. Healey, A.J. and D. Lienard (1993). Multivariable sliding mode control for autonomous diving and steering of unmanned underwater vehicles. *IEEE Journal of Oceanic Engineering*, 18, pp. 327–339.
6. Macro, D.B. and A.J. Healey (2001). Command, control and navigation experimental results with the NPS ARIES AUV. *IEEE Journal of Oceanic Engineering*, 26(4), pp. 466–476.
7. Fryxell, D., P. Oliveira, A. Pascoal, C. Silvetre, and I. Kaminer (1996). Navigation, guidance and control of AUVs: application to the MARIUS vehicle. *Control Engineering Practice*, 4(3), pp. 401–409.
8. Hu, J., C. Bohn, and H.R. Wu (2000). Systematic H_∞ weighting function selection and its application to the real-time control of a vertical take-off aircraft. *Control Engineering Practice*, 8(3), pp. 241–252.
9. Doyle, J.C., K. Glover, P.P. Khargonekar, and B.A. Francis (1989). State-space solutions to standard H_2 and H_∞ control problems. *IEEE Transactions and Automatic Control*, AC-34, pp. 831–847.
10. Gahinet, P. and P. Apkarian (1994). A linear matrix inequality approach to H_∞ control. *International Journal of Robust and Nonlinear Control*, 4, pp. 421–448.
11. Iwasaki, T. and R.E. Skelton (1994). All controllers for the general H_∞ control problem: LMI existence conditions and state space formulas. *Automatica*, 30(8), pp. 1307–1317.
12. The Mathworks, Robust Control Toolbox for use with MATLAB, 2001.
13. The Mathworks, LMI Toolbox for use with MATLAB.
14. Feng, Z. and R. Allen (2004). Reduced order H_∞ control of an autonomous underwater vehicle. *Control Engineering Practice*, 15(12), pp. 1511–1520.
15. Fossen, T.I. (1998). *Guidance and Control of Ocean Vehicles*. John Wiley and Sons, New York.
16. Lea, R.K. (1998). A comparative study by simulation and experimentation of control techniques for autonomous underwater flight vehicles. PhD thesis, University of Southampton.
17. Feng, Z. and R. Allen (2001). Modelling of Subzero II. Technical Memo 880, ISVR, University of Southampton.

18 Feng, Z. and R. Allen (2002). H_∞ autopilot design for an autonomous underwater vehicle. In Proceedings of the IEEE International Conference on Control Applications, Glasgow, UK, September, pp. 350–354.
19 Burcher, R. and L. Rydill (1998). *Concepts in Submarine Design*. Cambridge University Press, Cambridge, UK.
20 Feng, Z. and R. Allen (2003). Robust control of Subzero III in restricted water, Technical Memo No. 926, ISVR, University of Southampton, December.
21 Ablow, C.M. and S. Schechter (1983). Numerical simulation of undersea cable dynamics. *Ocean Engineering*, 10(6), pp. 443–457.
22 Milinazzo, F., M. Wilkie, and S.A. Latchman (1987). An efficient algorithm for the dynamics of towed cable systems. *Ocean Engineering*, 14, pp. 513–526.
23 Feng, Z. and R. Allen (2004). Evaluation of the effects of the communication cable on the dynamics of an underwater flight vehicle. *Ocean Engineering*, 31(8–9), pp. 1019–1035.
24 Feng, Z. and R. Allen (2004). Composite control of a tethered underwater flight vehicle. In Proceedings of the IFAC Conference on Control Applications in Marine Systems (CAMS04), Ancona, Italy, July, pp. 309–314.

Chapter 9

Low-cost high-precision motion control for ROVs

M. Caccia

9.1 Introduction

This chapter focuses on the problem of developing a low-cost station-keeping system for remotely operated vehicles (ROVs) that is able to handle external disturbances, as well as sea currents and tether forces, and uncertainty in system dynamics. This uncertainty includes poor knowledge of hydrodynamic derivatives, sensor measurements, that is, noise and low sampling rate, and actuator forces, which are generally affected by propeller–propeller and propeller–hull interactions.

If the sea current can generally be assumed irrotational and, at least locally, varying very slowly in time, its measurement may not be available, while the force exerted by the tether on the vehicle, especially in the case of small and mid-class ROVs working with a large amount of cable at sea, often constitutes the main disturbance on vehicle manoeuvring capabilities. Even if, in principle, it could be measured and some models of cable displacement and exerted force have been proposed in the literature [1], a real-time estimate of the action of the cable on the ROV is currently beyond realisation. Nevertheless, an exhaustive theoretical model of system dynamics could be given, representing the ROV as a floating rigid body, although the practical identification of hydrodynamic derivatives is quite difficult: conventional techniques of scaling the model parameter values from water-tank trials of scaled models of the open-frame vehicle could involve high parametric uncertainty [2], while on-board sensor-based identification techniques allow only the estimate of a subset of the model parameters [3]. On the other hand, the well-known effects of propeller–propeller and propeller–hull interactions are not negligible [4], and the dynamics of bladed thrusters is usually neglected [5].

In addition to the above-mentioned sources of uncertainty, as clearly discussed in Reference 7, only a few techniques are available for reliable and precise three-dimensional (3D) position and velocity sensing with an update rate compatible with fast and precise closed loop feedback control in all degrees of freedom and, thus experimental results for X–Y control of vehicles in the horizontal plane are rare.

Accurate, reliable and high sampling rate measurement of slow horizontal motion of unmanned underwater vehicles in the proximity of the seabed can be guaranteed by the employment of acoustic devices, such as high-frequency long base-line (LBL, 300 kHz) and Doppler velocimeter (1.2 MHz), combined with ring-laser gyro as in the case of archaeological applications of the Jason ROV [6]. High-frequency LBL, indeed, requires a very complex logistics in terms of careful placement of transponders and has a very limited maximum range [7].

Since in many applications in remote environments, ROVs are deployed by small support vessels and perform missions where a local positioning is sufficient, research is focused on the development of cheap, stand-alone optical vision devices for horizontal motion estimation. These efforts were supported by the exponential rise in computing and graphics capabilities and the availability of high-resolution digital cameras, which allowed us to extend the limits of underwater imaging up to direct estimation of motion from seafloor images [8] and mosaic-based visual navigation [9,10].

A brief overview of related research in the fields of modelling, identification, guidance, control and motion sensing technologies for ROVs is given in Section 9.2, while the following focuses on the research carried out with the Romeo prototype towards the development of a high-precision ROV for benthic applications. In this context, a practical approach to accurate ROV horizontal motion control consisting of the integration of a set of suitable subsystems in a harmonious blend of vehicle mechanical design, system modelling and identification, model-based motion estimation, guidance and control, and computer vision was followed by developing the Romeo ROV in order to satisfy scientific requirements in terms of manoeuvring capabilities. This involved, at first, the design of a fully controllable vehicle, with propeller allocation optimised for accurate horizontal motion control, as reported in Section 9.3. Then, the conventional ROV model was revised by including thruster installation coefficients, taking into account propeller–hull and propeller–propeller interactions, and on-board sensor-based identification techniques, defining a set of suitable manoeuvres for observing drag and inertia parameters, were developed. A dual-loop guidance and control architecture, which by decoupling the vehicle velocity control, handling system dynamics, from guidance laws managing the task kinematics, facilitates the development and tuning of a set of cinematic position controllers and was adopted as discussed in Section 9.4. The resulting system's performance proved satisfactory in vertical motion control, both depth and altitude, in many benthic scientific application, and in pool trials of horizontal manoeuvring where positioning was based on acoustic tracking of environmental features, demonstrating that the only remaining obstacle to good quality station-keeping at sea was the availability of suitable measurements of the horizontal motion. With this aim, a monocular video system, which includes laser beam triangulation measurements of image depth, was designed and developed, as discussed in Section 9.5. Experimental results, obtained by operating Romeo at sea, are reported and discussed in Section 9.6.

9.2 Related research

9.2.1 Modelling and identification

Basic elements of marine hydrodynamics and modelling of unmanned underwater vehicles can be found in References 11 and 12. The phenomenon of propeller–hull interactions, well-known in the case of surface vessels [11], was explicitly introduced in the UUV model in Reference 13 in the form of installation coefficients of a thruster, which, in their words "take into account the differences in force that the thruster provides when operating near to the ROV, as opposed to when it is tested in open water." Experimental identification of these coefficients for an operating ROV was carried out 10 years later [4].

Hydrodynamic derivatives are classically identified through towing tank trials of the vehicle itself or of a scaled model of the ROV as reported in References 14 and 15, respectively. Although these kind of trials allow a complete model identification, their cost in terms of time and money makes on-board sensor-based identification techniques preferable especially in the case of variable configuration ROVs, that is, changing their payload and shape according to the mission to be accomplished, where the identification procedure can be easily repeated when significant variations in the system structure occur. System identification techniques for UUVs, suggested in Reference 16, are limited by sensor constraints to the determination of hydrodynamic parameters for uncoupled one-dimensional (1D) models for the principal degrees of freedom, typically yaw, surge and heave. A first, interesting, practical example is reported in Reference 17, where a 1D system identification procedure was applied to the ROV *Hylas*. The requirement of defining a set of persistently exciting input signals by taking into account the model structure, actuator dynamics, and performance of available sensors was discussed in Reference 4, where the on-board sensor-based identification procedure of the *Romeo* ROV is presented. An on-line deterministic adaptive identification method, applied to the Johns Hopkins University remotely operated underwater robotic vehicle (JHUROV), is presented in Reference 3, while the aspect of considering the reliability of measurements in the identification process is discussed in Reference 18, where surge, pitch and yaw models of the *URIS* UUV are identified.

A conventional model of propeller-based actuators has been presented by Fossen [12], while for research on the dynamic modelling of bladed thrusters, in the perspective of designing and testing model-based thruster controllers that are able to reduce the oscillatory behaviour characteristics of dynamically positioned marine vehicles, the reader can refer to References 5 and 19–21. Complications in the practical application of dynamic thruster models are clearly discussed by Whitcomb and Yoerger [22] in terms of the instruments for measuring flow velocity in the duct and in the ambient.

9.2.2 Guidance and control

After the pioneering paper by Yoerger and Slotine [23], who proposed the use of robust sliding-mode control techniques for handling unpredictable disturbance and system nonlinearities of unmanned underwater vehicles, the first experimental validation of adaptive sliding control on a tethered underwater vehicle was performed on the RPV, a testbed vehicle for the development of the *Jason* ROV [24]. From then on the use of sliding-mode techniques for UUV control was quite common, see, for instance, the depth, altitude, heading and cross-track error controllers of the NPS ARIES AUV [25], and proved its capability in handling an uncertainty of the order of 50 per cent in the estimation of the vehicle's hydrodynamic parameters in the case of the heading and depth control of the autonomous underwater shuttle *SIRENE* [2]. Robustness was achieved by means of adaptive control schemes, as discussed by Fossen and Sagatun [26] and Yuh [27], and experimentally demonstrated, for instance, with the *ODIN* AUV by Antonelli *et al.* [28]. Adaptive variable structure control was demonstrated for automatic positioning of the passive-arm equipped Tatui and MKII ROV by Hsu *et al.* [29]. In the case of streamlined AUVs, the approach of scheduling on the basis of the vehicle forward speed, a set of linear finite static output feedback controllers was applied to the *Infante* AUV [30], while a PI gain scheduling controller that is able to reduce the robot dynamics to a nominal characteristic equation was demonstrated with the *Romeo* ROV by Caccia and Veruggio [31]. Latest results in underwater robotic vehicles dynamic positioning with an extended experimental comparison of trajectory-tracking controllers for low-speed manoeuvring (PD, exactly linearising and nonlinear controllers, and their adaptive extensions) can be found in Reference 32 together with a detailed review of useful research topics.

As far as UUV guidance is concerned, line-of-sight (LOS) algorithms generating the reference heading and, sometimes, surge velocity, are usually adopted (see, e.g., Reference 12 for algorithm description). In recent years, path-following techniques have been proposed for marine craft [33]. Among these techniques, a family of Lyapunov-based guidance laws for approaching a target area with different constraints on the desired orientation was derived by the field of wheeled robotics [34], and integrated, first with the conventional autopilot of the *Roby2* ROV performing auto-heading, auto-depth and auto-speed [35], and then in the dual-loop guidance and control architecture of the *Romeo* ROV [31]. Exploiting the potential of a Lyapunov-based design of the guidance system, which naturally generates reference velocities for the control level, this could also be used to implement sonar-based guidance tasks [36].

9.2.3 Sensing technologies

9.2.3.1 Acoustic devices

Conventional navigation systems for unmanned underwater vehicles used in the 1990s are presented in Reference 37, where performance of dead reckoning combining compass with Doppler log, inertial navigation systems, long baseline and ultrashort baseline acoustic positioning systems are discussed and compared. State-of-the-art acoustic positioning systems, both LBL and USBL, are discussed by Kinsey and Whitcomb [7], where advanced Doppler-based navigation is presented and compared with LBL navigation. For an application of high-precision combined with high-frequency LBL and Doppler log, the reader is referred to the study by Whitcomb *et al.* [6].

9.2.3.2 Optical devices

Basic optical properties of underwater images, that is, non-uniform lighting, suspended particles in water, and limited range, are clearly introduced in Reference 38, while the presence of optical flow induced in a stationary scene by the motion of a light source, that is, in the typical operating condition in underwater vision, where the light sources are mounted on the ROV and move together with the camera(s), motivated a revised definition of optical flow as 'the perceived transformation of brightness patterns in an image sequence', which is discussed by Negahdaripour [39].

A station-keeping technique based on obtaining the robot position by tracking texture features using image filtering and correlation was proposed in Reference 40, and then transferred to the MBARI *Ventana* ROV [41], demonstrating a precision of the order of 10 cm when operating at sea at an altitude from the seabed of 1 m in the hypothesis of no yaw rotations. The sign of the Laplacian of Gaussian was used to perform bandpass filtering and to highlight the zero-crossing points of the spatial intensity gradient.

A visual servoing approach to station-keeping of underwater vehicles was proposed in Reference 42, where tests were performed in a tank with a Cartesian robot emulating the surge and sway dynamics of the *Angus* 003 ROV.

Direct estimation of linear and yaw motion from seafloor images through optical flow computation is presented in Reference 8, where accurate station-keeping is demonstrated in experiments with a three-thruster floating vehicle in a water tank, and then tested at sea with a *Phantom* XTL ROV, pointing out the strong coupling between the constraints on robust motion sensing from images and the vehicle control [43]. These techniques were improved and integrated in a mosaic-based concurrent mapping and localisation scheme in the work by Negahdaripour and Xu [9], where, in spite of a high-degree of robustness of the gradient-based motion estimation and mosaicking methods, the inability of the control system of the *Phantom* XTL ROV to execute corrective actions promptly for maintaining station was confirmed. Quite interesting results in combined vision-based motion estimation and mosaicking were demonstrated with a *Phantom* 500SP ROV by Gracias *et al.* [10]. In video mosaicking research it is worth remembering the pioneer work of Marks *et al.* [44], where the observability of rotations about axes in the image plane and translations parallel to the image plane are discussed, and recent results published by Pizarro and Singh [45], who demonstrated a purely image-based approach on real data obtained using the *Jason* ROV at an archaeological site.

Stereovision systems are conventionally used for extracting 3D information on the operating environment such as size of objects, slope of the bottom and relative position and orientation of objects [46].

Quantitative information from standard undersea video can be extracted using photogrammetric laser/video systems based on optical triangulation of observed laser blobs [47]. The basic idea of directly measuring the image depth by locally structuring the environment with a set of laser spots of known orientation with respect to the camera axis is already well known in the literature [48,49] and has been proposed for underwater applications by Chen and Lee [50] and Caccia [51].

9.3 Romeo ROV mechanical design

The *Romeo* ROV was designed for robotics research and scientific applications, in particular oriented to the study of benthos in harsh and remote environments, in the mid-1990s, when the underwater robotics community was discussing the need for designing homogenous mechanical and control architectures for unmanned underwater vehicles in order to foster cooperation between research groups around the world [52]. The result was a ROV characterised by a networked architecture and an interchangeable toolsled for carrying payloads of different nature, as well as *Victor* 6000 [53], developed by IFREMER, and MBARI *Tiburon* [54].

Particular attention was dedicated to the geometry of the propulsion system in order to design a vehicle that is able, at least in principle, and if suitable position and speed measurements are available, to manoeuvre with high precision both in the vertical and horizontal planes thereby minimising the interactions with the sea floor. Thus, as shown in Figure 9.1, the four vertical propellers were positioned on the top vehicle corners, and the four horizontal thrusters were aligned to the horizontal diagonals in the mid corners of the frame. The symmetry in thruster location allowed a smooth distribution of the control actions over the actuators, and the redundancy

Figure 9.1 The Romeo ROV

of the actuation system allowed the vehicle to handle faults in the propulsion system without altering its motion control performances.

The core vehicle is composed of a frame ($130 \times 90 \times 66$ cm, length × width × height or $l \times w \times h$), equipped with a number of titanium cylindrical canisters for electronics (100×32 cm, ld), batteries (80×15 cm ld), DC/DC converters (80×15 cm ld), and compass, gyro and inclinometers (60×15 cm ld). The standard toolsled, which measures $130 \times 90 \times 30$ cm (lwh), brings additional batteries.

Further details on the *Romeo* system design and applications can be found in Reference 55–57.

9.4 Guidance and control

The design of a three-level hierarchical architecture (see Figure 9.2), with the ability to uncouple the management of the actuation system, the vehicle's dynamics and its kinematics, increased the system modularity, thus facilitating the development and testing of different control algorithms and easily tuneable guidance task functions.

The guidance level executes the user-defined task functions dealing with the cinematic interactions between the robot and the operating environment, that is, position control of the vehicle with respect to an environment-fixed reference frame or detected environmental features. The control level is task independent and tracks the required velocities that handle the vehicle's dynamics by means of a set of model-based uncoupled controllers. The required control actions, force and torque, generated by the speed regulators are mapped onto the vehicle actuator thrusts by the actuation system module, which takes into account the vehicle's thruster configuration. It is worth noting that conventional ROV autopilots can be implemented in this framework by a suitable combination of guidance position task functions and velocity controllers. The mapping of the control force and torque onto the vehicle thrusters is performed by applying the classical Moore–Penrose pseudoinverse method [12] to a lumped parameter actuator model including propeller–hull interactions. A set of gain scheduling proportional–integral (PI) velocity controllers constitutes the control level, where unmodelled dynamics, errors in the estimate of the hydrodynamic parameters and external disturbances are compensated by the integrator. The guidance level consists of a set of Lyapunov-based controllers executing motion task functions: a relevant

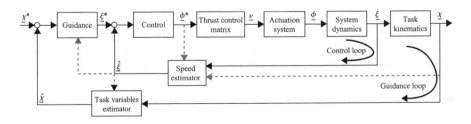

Figure 9.2 Guidance and control architecture

enhancement in precision performances is provided by the introduction of PI-type task functions, which enable the system to handle any bias in the speed estimate and unmodelled, that is, not measured or estimated, kinematic interactions between the robot and the operating environment.

The limited sensing capabilities of generic ROVs can induce some adaptations (grey lines in Figure 9.2) of the ideal dual-loop architecture. First, the position measurements are also used to estimate the robot velocities (unbroken line). Second, the estimates of the system velocities and task variables are respectively enhanced by the knowledge from the corresponding inner loop of the applied control action (force and/or torque) and estimated speed (broken lines), which allow the prediction of the system behaviour on the basis of the physical relations embodied in the models of the system dynamics and task kinematics. Third, the knowledge of the estimated speed (dotted lines) enhances the performance of the guidance module which can predict the task kinematics.

9.4.1 Velocity control (dynamics)

The design and tuning of task kinematics controllers is simplified by reducing the vehicle's dynamics to a second-order system with fixed time constants. This can be done, at least at the level of linear approximation, by means of a PI controller with gains scheduled as continuous functions of the desired velocity. The controller compensates for unmodelled dynamics, errors in the estimation of the hydrodynamic parameters and external disturbances with the integral action.

In particular, the control action ϕ is the result of a feed-forward component ϕ^*, leading the system to the desired operating point $\xi = \xi^*$ and $\phi = \phi^*(\xi^*): \dot{\xi}(\xi^*, \phi^*) = 0$, plus a feedback component ϕ_δ yielding to a closed-loop linearised system with the desired characteristic equation

$$s^2 + 2\sigma s + \sigma^2 + \omega_n^2 = 0 \tag{9.1}$$

The conventional six degrees freedom (DOF) ROV model, presented, for instance, in Reference 12 is usually reduced, in practical applications, to a set of uncoupled single DOF equations by neglecting the off-diagonal elements of the added mass matrix, the Coriolis and centripetal kinematics terms and drag coupling terms:

$$m_\xi \dot{\xi} = -k_\xi \xi - k_{\xi|\xi|}\xi|\xi| + \phi_\xi + \nu_\xi \tag{9.2}$$

where m_ξ is the inertia (including added inertia) relative to the considered degree of freedom, ξ is the 1D velocity (surge, sway, heave, yaw, pitch or roll rate), k_ξ and $k_{\xi|\xi|}$ are the linear and quadratic drag coefficients, ϕ_ξ is the applied force or torque and ν_ξ is a disturbance modelling otherwise unmodelled phenomena such as cable effects or the weight/buoyancy force.

Linearisation of (9.2) about each constant operating velocity ξ^* assumes the form

$$\dot{\xi}_\delta = -\frac{k_\xi \xi + 2k_{\xi|\xi|}|\xi^*|}{m_\xi} \xi_\delta + \frac{1}{m_\xi}\phi_{\xi_\delta} \tag{9.3}$$

with $\xi_\delta = \xi - \xi^*$ and $\phi_{\xi_\delta} = \phi_\xi - \phi_\xi^*$, and, according to the gain-scheduling technique presented by Khalil [58], the PI controller is designed as

$$\phi_\delta = k_P e + k_I \gamma, \dot{\gamma} = e = \xi - \xi^* = \xi_\delta \qquad (9.4)$$

yielding a closed-loop linearised system with the characteristic equation

$$s^2 + \left(k_\xi + 2k_{\xi|\xi|}|\xi^*| - \frac{k_P}{m_\xi}\right)s - \frac{k_I}{m_\xi} = 0 \qquad (9.5)$$

In order to obtain the desired characteristic Equation (9.1), the gains k_P and k_I are scheduled as functions of ξ^* as follows:

$$k_P = k_\xi + 2k_{\xi|\xi|}|\xi^*| - 2m_\xi \sigma \qquad (9.6)$$

and

$$k_I = -m_\xi(\sigma^2 + \omega_n^2) \qquad (9.7)$$

In operating conditions an anti-windup mechanism is implemented such that $|\gamma| \leq \gamma_{max}$.

9.4.2 Guidance (task kinematics)

In the proposed approach, the synthesis of guidance controllers is based on the definition of a positive definite function V of a task function $\mathbf{e}$, $V(\mathbf{e}) = \frac{1}{2}\mathbf{e}^T\mathbf{e}$. If the reference velocities are chosen so that $\dot{\mathbf{e}} = -\lambda \mathbf{e}$, then $\dot{V}$ is a negative definite function and V is a Lyapunov function with a stable equilibrium point $\mathbf{e} = \mathbf{0}$.

Assuming the form $\mathbf{e} = \boldsymbol{\chi} - \boldsymbol{\chi}^*$, the task function $\mathbf{e}$ can represent the position/orientation errors, and, if a linear invertible mapping $\dot{\boldsymbol{\chi}} = L\boldsymbol{\xi}$ exists, a suitable choice of the desired velocities is $\boldsymbol{\xi}^* = -\lambda^{-1}(\boldsymbol{\chi} - \boldsymbol{\chi}^*)$.

In underwater robotic applications, the presence of a bias in the speed measurements and estimates is quite common (e.g., Doppler velocimeter bias or model-based nonlinear biased estimators such as an extended Kalman filter). This fact determines a non-zero mean error in the velocity tracking, which acts as a disturbance in the guidance loop and could be counteracted by a high gain λ, which can cause instability or limit-cycling problems due to system nonlinearities and bandwidth limitations. To avoid this problem, an integrator can be included in the guidance regulator, defining a task function of PI-type, which in the case of hovering control assumes the form

$$\mathbf{e} = (\mathbf{x} - \mathbf{x}^*) + \mu \int_0^t (\mathbf{x} - \mathbf{x}^*) d\tau \qquad (9.8)$$

where $\mathbf{x}$ denotes the horizontal position of the vehicle. In this case, the kinematics controller assumes the form

$$\boldsymbol{\xi}^* = -g_P \mathbf{L}^{-1}(\mathbf{x} - \mathbf{x}^*) - g_I \mathbf{L}^{-1} \int_0^t (\mathbf{x} - \mathbf{x}^*) d\tau \qquad (9.9)$$

where ξ^* represents the desired surge and sway, **L** the rotation between a vehicle-fixed and an Earth-fixed frame, $g_\mathrm{P} = \lambda + \mu$ and $g_\mathrm{I} = \lambda\mu$, $\lambda > 0$ and $\mu \geq 0$.

In order to minimise wind-up effects, the integrator is enabled/disabled with an hysteresis mechanism when the range from the target $r = \sqrt{(\mathbf{x} - \mathbf{x}^*)^\mathrm{T}(\mathbf{x} - \mathbf{x}^*)}$ gets lower/higher than $I_e^{\mathrm{ON}}/I_e^{\mathrm{OFF}}$, respectively. Steady-state properties of the proportional regulator guarantee the system reaches the switching condition.

9.5 Vision-based motion estimation

Research and experimental results, examined in Section 9.2.3.2, pointed out the theoretically expected difficulties, in an unstructured environment, related to feature extraction and matching, and, in the case of monocular systems, to the estimate of the ratio between the motion components orthogonal and parallel to the image plane. In particular, in station-keeping applications, it has been shown that system performance is often limited by interactions with the control system of ROVs characterised by under-actuated strongly coupled propulsion systems.

In this framework, a mono-camera vision system for measuring the linear motion of an ROV executing basic operational tasks at low speed, such as station-keeping and near bottom traverses, in order to provide satisfactory performances, should:

(i) increase the reliability and stability of the motion estimate through a direct measurement of the image depth;
(ii) determine unambiguous areas of interest to be tracked in order to compute the camera motion;
(iii) compensate the lighting variations induced by the light motion;
(iv) compensate or filter the effects of pitch and roll oscillations on apparent linear motion;
(v) be properly integrated within the vehicle guidance and control system.

9.5.1 Vision system design

Following the above-mentioned guidelines, a mono-camera vision system for measuring the horizontal motion of a ROV executing basic operational tasks at low speed was developed. At first, the possibility of obtaining scene depth information from dedicated sensors was investigated. In order to reduce costs and circumvent the difficulties related to the calibration of sensors of different nature, for example, optical and acoustic devices, the basic idea of directly measuring the image depth by locally structuring the environment with a set of laser spots parallel to the camera axis was explored on the basis of previous satisfactory results. This was done in terms of reliability in a large variety of operating conditions, in the employment on the *Romeo* ROV of a small matrix of parallel red laser beams for immediate sizing of the observed objects by the human operator, typically a marine scientist. The result was the design of a video system comprising a video camera and four parallel red laser diodes for measuring distance and orientation from surfaces, as shown in Figure 9.3.

Figure 9.3   Optical laser triangulation–correlation sensor

The video camera is mounted inside a suitable steel canister, while the four red laser diodes are rigidly connected in the corners of a 13 cm side square, with their rays perpendicular to the image plane. The selected camera is the high sensitivity (0.8 lux F1.2, 50 IRE; 0.4 lux F1.2, 30 IRE) Sony SSC-DC330 1/3″ High Resolution Exwave HAD CCD Color Camera which features 480 TV lines of horizontal resolution with 768H × 594V picture elements. After calibration, a focal length f of about 1063.9 [pixel] was computed. A camera-fixed reference frame $\langle c \rangle$, with origin in the image centre and z-axis directed towards the scene, is defined. The resulting optical device was mounted downward-looking below the Romeo ROV (see Figure 9.4).

In order to minimise the ambiguity in the conventionally defined optical flow originated by the motion of the light source together with the vehicle, that is, with the camera [39], a special illumination system with diffuse light was built.[1] Two 50 W halogen lamps covered by suitable diffusers illuminate the camera scene almost uniformly. The illumination of the camera scene, since the vehicle works in the proximity of the seabed, is not affected by the lamps mounted in front of the ROV for pilot/scientist video and photo cameras.

As far as the problem of horizontal motion estimation from video images is concerned, given the assumption of slow motion characterised by small rotations and changes in the scene depth, the approach of computing motion from the displacement

---

[1] The system was planned and built by Giorgio Bruzzone.

Figure 9.4 Optical laser triangulation–correlation sensor mounted below the ROV

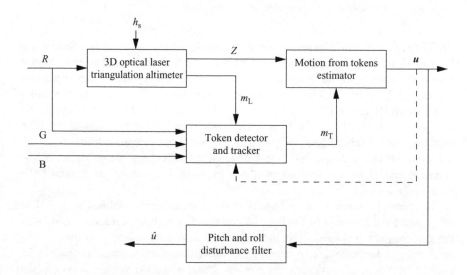

Figure 9.5 Optical triangulation-correlation sensor architecture

of image templates tracked through correlation techniques was followed. Thus, from a computing and image processing point of view, the system, as shown in Figure 9.5, comprises a 3D optical laser triangulation altimeter, a token detector and tracker, a motion from tokens estimator and a pitch and roll disturbance filter, which will be discussed in the next section.

9.5.2 Three-dimensional optical laser triangulation sensor

After in-lab system calibration (for details the reader can refer to Reference 59), the processing of the red component of the image to measure range and orientation from the seabed surface is performed in three steps:

1. *Detection and tracking of the laser spots in the image coordinates.* Each laser blob is detected by searching the pixel with maximum intensity in a suitable sub-image, and then computing the intensity centre of the light spot in its neighbourhood [50]. Reliability is increased by masking the image with stripes centred on the epipolar lines where the laser spot images lie. During algorithm iteration, each laser spot is searched in a sub-image of suitable size, related to the sensitivity of its image coordinates to image depth variations, centred in its predicted location in the image plane.
2. *Estimation of the spatial coordinates $[X \; Y \; Z]^T$ of the laser spots in the camera(vehicle)-fixed frame.* This is done by combining the laser ray equations, whose parameters are known from device calibration, with the camera perspective model

$$\begin{bmatrix} m \\ n \end{bmatrix} = \frac{f}{Z} \begin{bmatrix} X \\ Y \end{bmatrix} \qquad (9.10)$$

where f is the focal length, and $[m \; n]^T$ are the image point coordinates in the image plane.
3. *Estimation of the bottom range and orientation.* Here, it is assumed that the seabed profile is not vertical, and is locally linearly approximated by a plane

$$Z = -a_\pi X - b_\pi Y + c_\pi \qquad (9.11)$$

where $c_\pi = h$ is the vehicle altitude, and a_π and b_π represent the signed slope of the seabed in the X- and Y-directions, in the surge and sway directions in the case of parallel vehicle- and Earth-fixed reference frames. Given the coordinates in the camera-fixed frame of four laser spots, the coefficients of the plane (9.11) can be estimated using an LS algorithm.

Then, the vision-based estimate of the image depth and seabed orientation are integrated with altitude measurements supplied by acoustic altimeters mounted on the vehicle to increase system reliability and field of work. A rigorous sensor fusion module would require an accurate inter-calibration of optical and acoustic devices, and should consist of a multi-rate filter able to integrate the optical-based signal with the acoustic-based one characterised by a lower sampling rate and a relatively high noise. In practice, the image depth is assumed as the true altitude when available, and acoustic measurements are used only when the optical altimeter cannot provide any measurement, that is, laser spots are not tracked because of seabed conditions or out-of-range altitude.

9.5.3 Template detection and tracking

The choice of interest areas in underwater images, where the typical tokens considered in computer-vision literature such as corners, circular symmetric features and logarithmic spirals [60] are not widespread, is usually based on the detection of areas with locally varying intensity. In particular, an approach based on averaging, Laplacian filtering and local variance, originally proposed by Misu *et al.* [61] to automatically choose and track image templates for autonomous landing of spacecraft, was adopted and is summarised in the following. In order to guarantee robust and accurate tracking, templates should be characterised by shading pattern of wavelength comparable with pre-selected block size and be distinctive in the sense of contrast.

Their extraction procedure consists of three steps:

1. *Two-dimensional bandpass filtering to enhance specific spatial wavelength.* In order to reduce computation bandpass filtering is performed executing averaging as low-pass filtering and sub-sampling and Laplacian filtering as high-pass filtering. Averaging and sub-sampling are performed simultaneously:

$$I_S(\mu, \nu) = \frac{1}{S_m S_n} \sum_{i=1}^{S_m} \sum_{j=1}^{S_n} I \left[S_m(\mu - 1) + i, S_n(\nu - 1) + j \right] \quad (9.12)$$

while the eight-neighbour Laplacian is computed as

$$I_L(\mu, \nu) = \sum_{i=-1}^{1} \sum_{j=-1}^{1} (I_S(\mu + i, \nu + j) - I_S(\mu, \nu)) \quad (9.13)$$

where I, I_S and I_L represent the intensity of the original, averaged (and sub-sampled), and Laplacian-filtered image respectively, and S_m and S_n indicate the sub-sampling interval.

2. *Local variances computation to evaluate contrast.* The roughness of the bandpass filtered image I_L is evaluated by computing the statistical variance within a sliding window of size W by W, equal to the size of the token that is to be extracted, to cover the entire image.

3. *Extraction of high local variance areas as templates.* Tokens are extracted according to their local variance (starting from the highest). In order to avoid clustered templates, which are more sensitive to observation noise, all the neighbours of a selected token, i.e. token at a range lower than $3W$ by the selected one, are eliminated. Since laser spots usually denote image areas with high local variance, their neighbourhoods are excluded by template detection and tracking.

In the assumed operating conditions of an ROV working at constant heading in the proximity of the seabed, disturbances induced to correlation of two consecutive frames by variations in the scene depth, that is, vehicle's altitude, and rotations, namely, small oscillations of the auto-heading system, can be considered negligible. Thus, tokens are tracked by looking for the highest correlation displacement in a suitable neighbourhood of their predicted image position. Token tracking fails when the correlation gets lower than a suitable threshold.

9.5.4 Motion from tokens

Considering a vehicle-fixed downward-looking camera moving at linear and angular speed $[u \ v \ w]^T$ and $[p \ q \ r]^T$, respectively, and neglecting pitch and roll, the motion field of a generic 3D point in the camera frame is

$$\begin{bmatrix} \dot{m} \\ \dot{n} \end{bmatrix} = -\frac{f}{Z} \begin{bmatrix} u \\ v \end{bmatrix} + \frac{w}{Z} \begin{bmatrix} m \\ n \end{bmatrix} + r \begin{bmatrix} n \\ -m \end{bmatrix} \tag{9.14}$$

In the case when the image depth is assumed to be constant (this hypothesis is reasonable given the small area covered by the image), defining the normalised speed $\tilde{u} = u/Z, \tilde{v} = v/Z$ and $\tilde{w} = w/Z$, the following over-constrained system can be obtained given N tracked image templates and solved with an LS algorithm:

$$\begin{bmatrix} -f & 0 & m_1 & n_1 \\ 0 & -f & n_1 & -m_1 \\ \vdots & \vdots & \vdots & \vdots \\ -f & 0 & m_N & n_N \\ 0 & -f & n_N & -m_N \end{bmatrix} \begin{bmatrix} \tilde{u} \\ \tilde{v} \\ \tilde{w} \\ r \end{bmatrix} = \begin{bmatrix} \dot{m}_1 \\ \dot{n}_1 \\ \vdots \\ \dot{m}_N \\ \dot{n}_N \end{bmatrix} \tag{9.15}$$

The vehicle linear speed is then computed by multiplying the estimated vector $[\tilde{u} \ \tilde{v} \ \tilde{w}]^T$ by the image depth Z measured by the optical laser altimeter.

9.5.5 Pitch and roll disturbance rejection

When pitch and roll rotations are considered, neglecting the terms of order higher than 1, the motion field of a generic 3D point in the camera frame is

$$\begin{bmatrix} \dot{m} \\ \dot{n} \end{bmatrix} \approx -\frac{f}{Z} \begin{bmatrix} u \\ v \end{bmatrix} + \frac{w}{Z} \begin{bmatrix} m \\ n \end{bmatrix} + r \begin{bmatrix} n \\ -m \end{bmatrix} + f \begin{bmatrix} -q \\ p \end{bmatrix} \tag{9.16}$$

Equation (9.16) shows that, for small variations, surge and sway displacements and pitch and roll rotations are undistinguishable. It is worth noting that for typical ROV benthic operations at an altitude of about 1 m, an angular rate of 1°/s corresponds to a disturbance of about 1.75 cm/s on the estimated linear speed in the case one pixel corresponds to 1 mm at a range of 1 m. This suggests that the effects on surge and sway estimates of roll and pitch rates are not negligible at low speed. In particular, experiments carried out with the *Romeo* ROV showed that small oscillations in uncontrolled pitch and roll induce quasi-sinusoidal disturbance on the measured surge and sway [62]. For the case when fast response-time pitch and roll rate sensors are available on-board the vehicle, their measurements could be used to compensate these effects, otherwise, this disturbance can be rejected by suitable bandstop filtering introducing some delay. In particular, Butterworth filters of the form

$$a_1 y(n) = b_1 x(n) + b_2 x(n-1) + b_3 x(n-2) + b_4 x(n-3) + b_5 x(n-4)$$
$$- a_2 y(n-1) - a_3 y(n-2) - a_4 y(n-3) - a_5 y(n-4) \tag{9.17}$$

were designed and implemented for surge and sway.

9.6 Experimental results

System trials were performed with the *Romeo* ROV equipped with the optical laser triangulation-correlation sensor in the Ligurian Sea, Portofino Park area.

The optical device was mounted downward-looking in the toolsled below the ROV as described in Section 9.3. Images were acquired and processed in real time at five frames per second at the resolution of 360×272 RGB pixels by a PC equipped with a Leutron PicPort-Color frame-grabber and a Pentium III running at 800 MHz. The application, written in C++, using Intel Integrated Performance Primitives v2.0 for image processing and signal filtering, received the ROV telemetry, including acoustic altimeter data, via datagram sockets from the vehicle control system.

During the trials the ROV operated in auto-altitude mode at a working depth of about 45 m over a mixed rocky and posydonia-populated seabed. Altitude data were provided by a Tritech PA500-6 acoustic echo-sounder (500 kHz, operating range between 0.2 and 50 m, beam-width 6° conical), and the ROV pitch and roll were measured by two inclinometers. In order to evaluate the effectiveness of the specially designed ROV illumination system, first trials were performed at night without any influence from sunlight. The vehicle motion was computed from token displacements and image depth according to (9.15) assuming a horizontal profile of the seabed. Results are shown in Figure 9.6.

Zooming-in on the estimated surge, as in the top graph of Figure 9.7(a), it is possible to note that the speed estimate is affected by an approximately sinusoidal noise of a few centimeters per second of amplitude. By comparing the estimated surge and pitch rate, plotted in the bottom graph of the same figure, speed estimate oscillations seem to be correlated to attitude rates as discussed in Section 9.5.5.

This hypothesis is confirmed by the analysis of the power spectral density of the vision-based 5 Hz estimated surge and 10 Hz inclinometer-measured pitch, which, as shown in Figure 9.7(b), presents a local peak in correspondence of the same frequencies. However, the low resolution of angular rates computed as first derivative of the corresponding angle measurements provided by inclinometers and their delay with respect to the fast-response vision-based speed sensor make impossible a direct compensation according to (9.16).

As discussed in Reference 62, pitch and roll oscillations are concentrated around approximately the same frequencies during different Romeo missions, that is, in the proximity of the resonance frequencies of the system with respect to the corresponding degrees of freedom. This allowed the estimate of these frequency intervals on the basis of inclinometer data and the design and implementation of surge and sway Butterworth bandstop filters of the type given in (9.17). The passband and stopband edge frequencies are [0.15 0.55] and [0.25 0.45] Hz for the surge/pitch filter, and [0.20 0.60] and [0.30 0.50] Hz for the sway/roll filter; the desired loss is lower than 1.0 dB in the passband and the attenuation is at least 5.0 dB in the stopband. Using the Matlab *buttord* function, Butterworth filters of fourth order were designed for surge and sway. Results obtained by applying the designed filters to the measured surge are plotted in Figure 9.8, which shows how the strong reduction of the oscillating

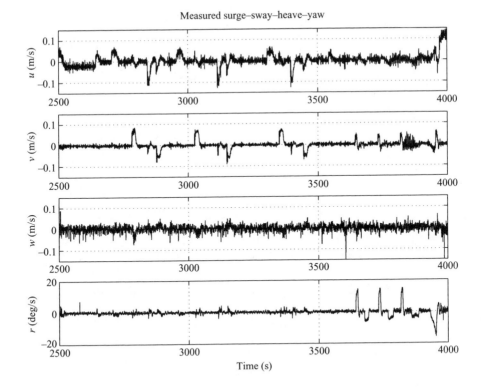

Figure 9.6 Estimated surge, sway, heave and yaw rates

disturbance is achieved but at the cost of an introduction of a delay of eight samples, that is, about 1.6 s.

In order to evaluate the performance of the adopted vision-based motion estimation algorithm, at least in terms of zero mean noise, that is, capability of enabling the pilot to guide the ROV back to a desired location, a sub-sequence of images in which *Romeo* flew over the same area at different times, navigating at constant heading through the corners of a square, was considered. The vehicle trajectory in the 10 min time interval between 2700 and 3300 s, plotted in Figure 9.9(a), was computed by integrating the vision-measured surge and sway (before stopband filtering) with the compass heading measurements according to the equations

$$\dot{x} = u \cos(\psi) - v \sin(\psi)$$
$$\dot{y} = v \cos(\psi) + u \sin(\psi)$$
(9.18)

The difference between the estimated displacements, denoted by *hat*, obtained by integrating the vision-based surge and sway measurements and the compass heading, and the camera motion, denoted by Δx, Δy and $\Delta \psi$, directly computed from the displacements of tokens tracked in reference images according to a discrete version of (9.15), was considered. As shown in Figure 9.9(b)–(f), the camera images at times

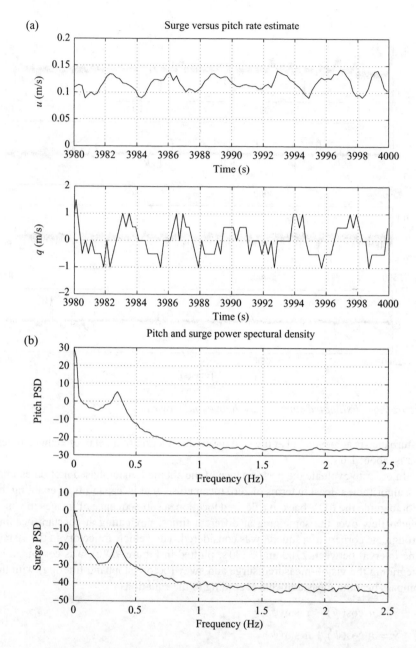

Figure 9.7 Relationship between the Romeo ROV vision-based estimated surge and inclinometer-measured pitch. (a) Estimated surge and pitch rates. (b) Power spectral density of estimated surge and pitch

Low-cost high-precision motion control for ROVs 205

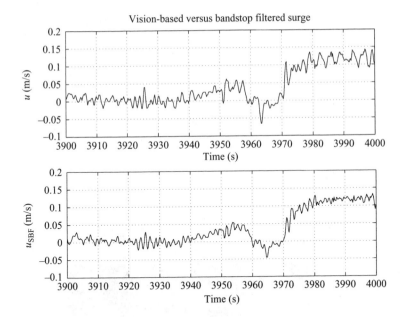

Figure 9.8   Estimated and bandstop filtered surge

t_1 and t_2 were considered as reference, and displacements were computed for images at times t_3 and t_5, and t_4.

The results are reported in Table 9.1, together with the estimated image depth. It is worth noting that here, as elsewhere in the paper, time is expressed in seconds, linear position in metres and angles in degrees.

The maximum error between the integral of surge and sway and the absolute inter-frame displacement is less than 5 cm over a time interval of about 10 min, demonstrating that the vision-based speed sensor is characterised by an approximately zero mean noise and could be reasonably employed to measure the path covered by the vehicle.

The satisfactory performance of the vision-based motion estimator was established and guidance and control trials were performed. In the experiments discussed in the following, *Romeo* worked at an altitude, that is, image depth, of about 80 cm, which corresponds to a field of view of about 21 × 28 cm in the images shown below.

In the first test the ROV travelled along a square twice. The vehicle path is plotted in Figure 9.10(a), where the red and green lines indicate the first and second lap path, while way-points are denoted by up-triangle, diamond, down-triangle and asterisk, respectively. As shown in Figure 9.10(b) the horizontal motion controller was quite precise with small overshoot. Coupling effects, resulting in strong disturbance in surge, are visible during fast sway motion.

Ground-truth verification of the system performance was performed, as discussed above, by directly computing the vehicle displacements at time t_B and t_C with respect

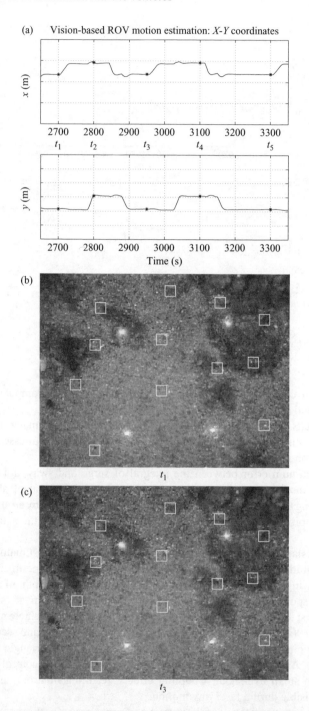

Figure 9.9 Vision-based ROV estimated position during sea trials by night. Verification of the behaviour of the diffuse light illumination system

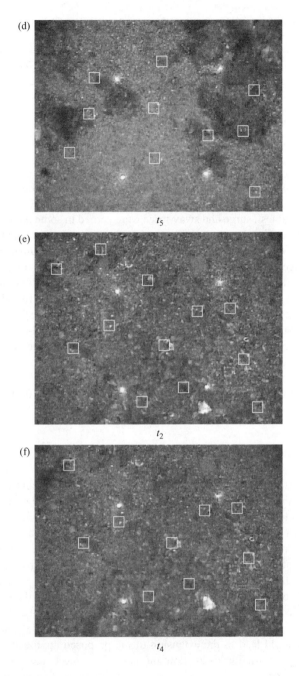

Figure 9.9 Continued

Table 9.1 Speed integral versus camera displacement from couple of images

Δt	Z	Δx	Δy	$\Delta \psi$	$\dot{\xi}_\delta = -(k_\xi \xi + 2k_{\xi|\xi|}|\xi^*|)/(m_\xi)\xi_\delta$		$+(1/m_\xi)\phi_{\xi_\delta}$
$t_3 - t_1$	1.03	0.01	−0.00	0.43	0.05	0.00	0.13
$t_5 - t_1$	1.02	0.03	0.00	−0.18	0.07	0.01	0.34
$t_4 - t_2$	0.99	−0.02	−0.01	−1.56	−0.00	−0.00	−2.13

to a reference position at time t_A. The results are reported in Table 9.2, while the processed images are shown in Figure 9.10(c)–(e).

In the second test, surge and sway force were zeroed in some intervals in order to show the effects of environmental disturbance, that is, sea current and tether tension, on the vehicle. Spike-like variations in the reference x and y coordinates are forced by an automatic reconfiguration system of the execution level in order to maintain system consistency and stability [63]. As shown in Figure 9.11(a) the ROV drifted, but the controller was able to drive it again over the operating point as revealed by images acquired at time t_D, t_E, t_F and t_G (see Figure 9.11(c)–(f), and Table 9.3 for comparison between direct displacement computation and vision-based dead-reckoning position estimate). The control force exerted to compensate current and tether disturbance is shown in Figure 9.11(b).

It is worth noting that during trials the integral-based motion predictor had a drift such that in the second test the reference point drifted 10 cm in the x- and y-directions.

9.7 Conclusions

A low-cost system for high-precision ROV horizontal motion control based on the integration of a vision-based motion estimator with a dual-loop hierarchical guidance and control architecture has been presented in this chapter, showing fully satisfying results in extensive preliminary at-sea trials carried out with the Romeo ROV. For a direct comparison of the proposed approach with results obtained by controlling the Romeo motion in a pool, where position was estimated from acoustic range measurements from environmental features [31], the system was preliminarily tested with the ROV operating at constant heading and altitude, using the same controller structure, neglecting more advanced gain-scheduling and integral control techniques presented in Reference 58. The promising results in vision-based estimate of yaw motion [51] should lead to the extension of the proposed approach to full, that is, linear and angular, motion estimation and control on the horizontal plane. In this direction, current research is focusing on the development of simultaneous mapping and localisation system based on image templates, and on the test and comparison of different guidance and control algorithms in the framework both of trajectory and path following.

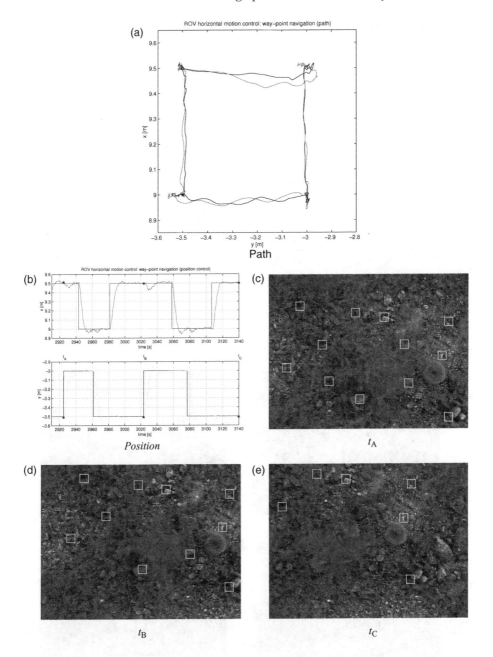

Figure 9.10 Vision-based ROV way-point navigation through the corners of a square

Table 9.2 Speed integral versus camera displacement from a couple of images: execute square test

Δt	Z	Δx	Δy	$\dot{\xi}_\delta = -(k_\xi \xi + 2k_{\xi|\xi|}|\xi^*|)/(m_\xi)\xi_\delta$	$+(1/m_\xi)\phi_{\xi_\delta}$
$t_B - t_A$	0.81	−0.03	−0.01	−0.01	0.00
$t_C - t_A$	0.79	−0.04	−0.05	−0.01	0.00

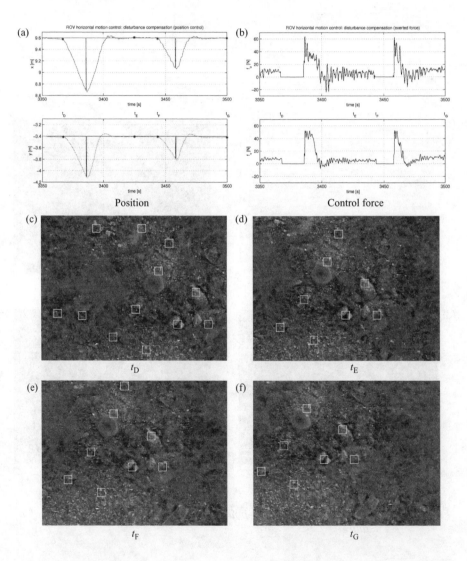

Figure 9.11 Vision-based ROV station-keeping counteracting external disturbance

Table 9.3 Speed integral versus camera displacement from couple of images: external disturbance compensation test

Δt	Z	Δx	Δy	$\dot{\xi}_\delta = -(k_\xi \xi + 2k_{\xi|\xi|}|\xi^*|)/(m_\xi)\xi_\delta$	$+(1/m_\xi)\phi_{\xi_\delta}$
$t_E - t_D$	0.80	−0.01	0.06	0.03	0.00
$t_F - t_D$	0.80	−0.03	0.07	0.01	0.00
$t_G - t_D$	0.81	−0.04	0.10	0.02	0.00

Acknowledgements

The research presented in this chapter is the result of the work carried out by the Robotics Department of CNR-IAN in Genova, Italy. Thus, a special mention is made of the members of CNR-IAN Robotlab starting with its leader, Gianmarco Veruggio, who introduced the author to underwater robotics and, with his competence and enthusiasm, played a key role in robot design, development and exploitation. Thanks also to my colleagues Riccardo Bono, Gabriele Bruzzone, Giorgio Bruzzone and Edoardo Spirandelli, who gave fundamental contributions to the robot hardware and software design and development, maintenance and operation at sea. Particular thanks are for Prof. Giuseppe Casalino for introducing me to the topics of mobile robotics and dual-loop architectures.

References

1. Buckham, B., M. Nahon, and G. Cote (2000). Validation of a finite element model for slack ROV tethers. In Proceedings of the OCEANS 2000, vol. 2, pp. 1129–1136.
2. Silvestre, C., A. Aguiar, P. Oliveira, and A. Pascoal (1998). Control of the SIRENE underwater shuttle: system design and tests at sea. In Proceedings of the 17th International Conference on Offshore Mechanics and Arctic Engineering, Lisbon, Portugal.
3. Smallwood, D.A. and L.L. Whitcomb (2003). Adaptive identification of dynamically positioned underwater robotic vehicles. *IEEE Transactions on Control Systems Technology*, 11(4), pp. 505–515.
4. Caccia, M., G. Indiveri, and G. Veruggio (2000). Modelling and identification of open-frame variable configuration unmanned underwater vehicles. *IEEE Journal of Oceanic Engineering*, 25(2), pp. 227–240.
5. Bachmayer, R., L.L. Whitcomb, and M. Grosenbaugh (2000). An accurate four-quadrant nonlinear dynamical model for marine thrusters: theory and experimental validation. *IEEE Journal of Oceanic Engineering*, 25(1), 146–159.
6. Whitcomb, L., D. Yoerger, H. Singh, and J. Howland (1999). Advances in underwater robot vehicles for deep ocean exploration: navigation, control, and

survey operations. In Proceedings of the 9th International Symposium of Robotics Research, Snowbird, USA.
7 Kinsey, J.C. and L.L. Whitcomb (2004). Preliminary field experience with the DVLNAV integrated navigation system for oceanographic submersibles. *Control Engineering Practice*, 12(12), pp. 1541–1550.
8 Negahdaripour, S., X. Xu, and L. Jin (1999). Direct estimation of motion from seafloor images for automatic station keeping of submersible platforms. *IEEE Journal of Oceanic Engineering*, 24(3), pp. 370–382.
9 Negahdaripour, S. and X. Xu (2002). Mosaic-based positioning and improved motion-estimation methods for automatic navigation of submersible vehicles. *IEEE Journal of Oceanic Engineering*, 27(1), pp. 79–99.
10 Gracias, N.R., S. Van der Zwaan, A. Bernardino, and J. Santos-Victor (2003). Mosaic-based navigation for autonomous underwater vehicles. *IEEE Journal of Oceanic Engineering*, 28(4), pp. 609–624.
11 Newman, J.N. (1977). *Marine Hydrodynamics*. MIT Press, Cambridge, MA, USA.
12 Fossen, T.I. (1994). *Guidance and Control of Ocean Vehicles*. John Wiley and Sons, England.
13 Goheen, K.R. and E.R. Jefferys (1990). Multivariable self-tuning autopilots for autonomous and remotely operated underwater vehicles. *IEEE Journal of Oceanic Engineering*, 15(3), pp. 144–150.
14 Goheen, K.R. (1986). The modelling and control of remotely operated underwater vehicles. Ph.D. thesis. University of London.
15 Nomoto, M. and M. Hattori (1986). A deep ROV Dolphin 3K: design and performance analysis. *IEEE Journal of Oceanic Engineering*, 11(3), pp. 373–391.
16 Goheen, K.R. and E.R. Jefferys (1990). The application of alternative modelling techniques to ROV dynamics. In Proceedings of the IEEE ICRA '90. Cincinnati, OH, USA, pp. 1302–1309.
17 Morrison, A.T. and D.R. Yoerger (1993). Determination of the hydrodynamic parameters of an underwater vehicle during small scale, nonuniform, 1-dimensional translation. In Proceedings of the OCEANS '93, vol. 2, pp. 277–282.
18 Carreras, M., A. Tiano, A. El-Fakdi, A. Zirilli, and P. Ridao (2004). On the identification of nonlinear models of unmanned underwater vehicles. *Control Engineering Practice*, 12(12), pp. 1483–1500.
19 Yoerger, D.R., J.G. Cooke, and J.E. Slotine (1990). The influence of thruster dynamics on underwater vehicle behavior and their incorporation into control system design. *IEEE Journal of Oceanic Engineering*, 15(3), pp. 167–178.
20 Healey, A.J., S.M. Rock, S. Cody, D. Miles, and J.P. Brown (1995). Toward an improved understanding of thruster dynamics for underwater vehicles. *IEEE Journal of Oceanic Engineering*, 20(4), pp. 354–361.
21 Whitcomb, L.L. and D.R. Yoerger (1999). Development, comparison, and preliminary experimental validation of non-linear dynamic thruster models. *IEEE Journal of Oceanic Engineering*, 24(4), pp. 481–494.

22 Whitcomb, L.L. and D.R. Yoerger (1999). Preliminary experiments in model based thruster control for underwater vehicle positioning. *IEEE Journal of Oceanic Engineering*, 24(4), pp. 495–506.
23 Yoerger, D.R. and J.E. Slotine (1985). Robust trajectory control of underwater vehicles. *IEEE Journal of Oceanic Engineering*, 10(4), pp. 462–470.
24 Yoerger, D.R. and J.E. Slotine (1991). Adaptive sliding control of an experimental underwater vehicle. In Proceedings of the IEEE International Conference on Robotics and Automation, pp. 2746–2751.
25 Marco, D.B. and A.J. Healey (2001). Command, control, and navigation experimental results with the NPS ARIES AUV. *IEEE Journal of Oceanic Engineering*, 26(4), pp. 466–476.
26 Fossen, T.I. and S.I. Sagatun (1991). Adaptive control of nonlinear underwater robotic systems. In Proceedings of the IEEE Conference on Robotics and Automation, pp. 1687–1695.
27 Yuh, J. (1990). Modelling and control of underwater robotic vehicles. *IEEE Transactions on Systems, Man and Cybernetics*, 20(6), pp. 1475–1483.
28 Antonelli, G., S. Chiaverini, N. Sarkar, and M. West (2001). Adaptive control of an autonomous underwater vehicle: experimental results on ODIN. *IEEE Transactions on Control Systems Technology*, 9(5), pp. 756–765.
29 Hsu, L., R.R. Costa, F. Lizarralde, and J.P.V.S. Da Cunha (2000). Dynamic positioning of remotely operated underwater vehicles. *IEEE Robotics and Automation Magazine*, 7(3), pp. 21–31.
30 Silvestre, C. and A. Pascoal (2003). Control of the INFANTE AUV using gain scheduled static output feedback. *Control Engineering Practice*, 12(12), pp. 1501–1510.
31 Caccia, M. and G. Veruggio (2000). Guidance and control of a reconfigurable unmanned underwater vehicle. *Control Engineering Practice*, 8(1), pp. 21–37.
32 Smallwood, D.A. and L.L. Whitcomb (2004). Model-based dynamic positioning of underwater robotic vehicles: theory and experiment. *IEEE Journal of Oceanic Engineering*, 29(1), pp. 169-186.
33 Encarnaçao, P. and A. Pascoal (2001). Combined trajectory tracking and path following: an application to the coordinated control of autonomous marine craft. In Proceedings of the 40th IEEE Conference on Decision and Control, vol. 1, pp. 964–969.
34 Aicardi, M., G. Casalino, A. Bicchi, and A. Balestrino (1995). Closed loop steering of unicycle like vehicles via Lyapunov techniques. *IEEE Robotics and Automation Magazine*, 2(1), pp. 27–35.
35 Caccia, M., G. Casalino, R. Cristi, and G. Veruggio (1998). Acoustic motion estimation and control for an unmanned underwater vehicle in a structured environment. *Control Engineering Practice*, 6(5), pp. 661–670.
36 Caccia, M., G. Bruzzone, and G. Veruggio (2001). Sonar-based guidance of unmanned underwater vehicles. *Advanced Robotics*, 15(5), pp. 551–574.
37 Ageev, M.D., B.A. Kasatkin, and A. Scherbatyuk (1995). Positioning of an autonomous underwater vehicle. In Proceedings of the International Program

Development in Undersea Robotics and Intelligent Control, Lisbon, Portugal, pp. 15–18.
38 Balasuriya, B.A.A.P., M. Takai, W.C. Lam, T. Ura, and Y. Kuroda (1997). Vision based autonomous underwater vehicle navigation: underwater cable tracking. In Proceedings of the OCEANS '97, vol. 2, pp. 1418–1424.
39 Negahdaripour, S. (1998). Revised definition of optical flow: integration of radiometric and geometric cues for dynamic scene analysis. *IEEE Transactions on Pattern Analysis and Machine Intelligence*, 20(9), pp. 961–979.
40 Marks, R.L., H.H. Wang, M.J. Lee, and S.M. Rock (1994). Automatic visual station keeping of an underwater robot. In Proceedings of the OCEANS '94, vol. 2, pp. 137–142.
41 Leabourne, K.N., S.M. Rock, S.D. Fleischer, and R. Burton (1997). Station keeping of an ROV using vision technology. In Proceedings of the OCEANS '97, vol. 1, pp. 634–640.
42 Lots, J.-F., D.M. Lane, E. Trucco, and F. Chaumette (2001). A 2D visual servoing for underwater vehicle station keeping. In Proceedings of the IEEE International Conference on Robotics and Automation, vol. 3, pp. 2767–2772.
43 Xu, X. and S. Negahdaripour (1999). Automatic optical station keeping and navigation of an ROV; sea trial experiments. In Proceedings of the OCEANS '99, vol. 1, pp. 71–76.
44 Marks, R.L., S.M. Rock, and M.J. Lee (1995). Real-time video mosaicking of the ocean. *IEEE Journal of Oceanic Engineering*, 20(3), pp. 229–241.
45 Pizarro, O. and H. Singh (2003). Toward large area mosaicing for underwater scientific applications. *IEEE Journal of Oceanic Engineering*, 28(4), pp. 651–672.
46 Singh, H., F. Weyer, J. Howland, A. Duester, D. Yoerger, and A. Bradley (1999). Quantitative stereo imaging from the autonomous benthic explorer (ABE). In Proceedings of the OCEANS '99, vol. 1, pp. 52–57.
47 Kocak, D.M., F.M. Caimi, T.H. Jagielo, and J. Kloske (2002). Laser projection photogrammetry and video system for quantification and mensuration. In Proceedings of the OCEANS '02, vol. 3, pp. 1569–1574.
48 Clark, J., A.K. Wallace, and G.L. Pronzato (1998). Measuring range using a triangulation sensor with variable geometry. *IEEE Transactions on Robotics and Automation*, 14(1), pp. 60–68.
49 Marques, L., U. Nunes, and A.T. de Almeida (1998). A new 3D optical triangulation sensor for robotics. In Proceedings of the 5th International Workshop on Advanced Motion Control, pp. 512–517.
50 Chen, H.H. and C.J. Lee (2000). A simple underwater video system for laser tracking. In Proceedings of the OCEANS 2000, vol. 3, pp. 1543–1548.
51 Caccia, M. (2005). Laser-triangulation optical-correlation sensor for ROV slow motion estimation. *IEEE Journal of Oceanic Engineering*, in press.
52 Wang, H.H. and E. Coste-Manière (1994). A look at the vocabulary of various underwater robotic control architectures. In Proceedings of the 2nd IARP Workshop on Mobile Robots for Subsea Environments. Monterey, CA, USA.

53 Nokin, M. (1996). ROV 6000 – a deep teleoperated system for scientific use. In Proceedings of the 6th IARP Workshop on Underwater Robotics, Toulon, France.
54 Newman, J.B. and D. Stokes (1994). Tiburon: development of an ROV for ocean science research. In Proceedings of the OCEANS '94, vol. 2, pp. 483–488.
55 Caccia, M., R. Bono, G. Bruzzone, and G. Veruggio (2000). Unmanned underwater vehicles for scientific applications and robotics research: the ROMEO project. *Marine Technology Society Journal*, 24(2), pp. 3–17.
56 Caccia, M., R. Bono, Ga. Bruzzone, Gi. Bruzzone, E. Spirandelli, and G. Veruggio (2002). Romeo–ARAMIS integration and sea trials. *Marine Technology Society Journal*, 2, pp. 3–12.
57 Bruzzone, G., R. Bono, M. Caccia, and G. Veruggio (2003). Internet-based teleoperation of a ROV in Antarctica. *Sea Technology*, 44(10), pp. 47–56.
58 Khalil, H.K. (1996). *Nonlinear Systems*. Prentice Hall, Englewood Cliffs, NJ.
59 Caccia, M. (2002). Optical triangulation-correlation sensor for ROV slow motion estimation: experimental results (July 2002 at-sea trials). Rob-02. CNR-IAN.
60 Haralick, R.M. and L.G. Shapiro (1992). *Computer and Robot Vision*. Addison-Wesley, Reading, MA.
61 Misu, T., T. Hashimoto, and K. Ninomiya (1999). Optical guidance for autonomous landing of spacecraft. *IEEE Transactions on Aerospace and Electronic Systems*, 35(2), pp. 459–473.
62 Caccia, M. (2003). Pitch and roll disturbance rejection in vision-based linear speed estimation for UUVs. In Proceedings of the MCMC 2003, pp. 313–318.
63 Bruzzone, Ga., M. Caccia, P. Coletta, and G. Veruggio (2003). Execution control and reconfiguration of navigation, guidance and control tasks for UUVs. In Proceedings of the MCMC 2003, Girona, Spain, pp. 137–142.

Chapter 10

Autonomous manipulation for an intervention AUV

G. Marani, J. Yuh and S.K. Choi

10.1 Introduction

An intervention autonomous underwater vehicle (AUV) is a significant step forward from a conventional AUV, mainly used in survey tasks. Most underwater manipulation tasks have been performed by manned submersibles or remotely operated vehicles (ROVs) in tele-operation mode. Today there are a few AUVs equipped with manipulators, for example, SAUVIM (Semi Autonomous Underwater Vehicle for Intervention Mission, University of Hawaii) [1], and ALIVE (AUV for light interventions on deepwater subsea fields, Cybernétix, France). For intervention AUVs, the human intervention during manipulation tasks is limited mainly due to the low bandwidth and significant time delay inherent in acoustic subsea communication. This is the main issue that the manipulator control system must address, along with some hardware constraints due to the limited on-board power source.

In this chapter, the evolution of the hardware of underwater manipulators will be described by introducing an electromechanical arm of SAUVIM, and some theoretical issues with the arm control system will be discussed, addressing the required robustness in different situations that the manipulator may face during intervention missions. An advanced user interface will then be briefly discussed, which helps to cope with the communication limits and provides a remote programming environment where the interaction with the manipulator is limited only to a very high level. An application example with SAUVIM will be presented before conclusions.

10.2 Underwater manipulators

Most intervention ROVs use electrohydraulic arms. Table 10.1 lists a few commercial ROV manipulators and Figure 10.1 shows some of their photographs. However, energy-efficient electrically actuated arms are preferred for AUVs that have limited on-board power and use noise-sensitive sensors. The *MARIS 7080* (Figure 10.2) is an example of an AUV manipulators graph.

The *MARIS 7080* is an electromechanical 7 + 1 DOF (degrees of freedom) redundant manipulator manufactured for SAUVIM by *ANSALDO* in Italy. Seven DOF are used to control the arm positioning while the eighth degree is for controlling the gripper. Each joint is actuated by a brushless motor keyed into a reduction unit. The power consumption during a generic operation is less than 200 W. The accuracy of each joint angular measurement is assured by two resolvers that are mounted before and after the gear reduction unit. The combined use of both resolvers allows calibrating the joint offsets at any initial position of the arm, thus guaranteeing high precision and repeatability. The manipulator was designed for underwater applications at 6000 m depth. It is oil-filled and the difference between the internal and external pressure is maintained slightly positive by a compensating system, in order to prevent internal leakages. A force/torque sensor installed between the wrist and the gripper measures forces and torques acting on the gripper. The manipulator is also equipped with a camera installed on the gripper and is ready for the installation of other sensors or tools for different intervention missions.

10.3 Control system

The primary purpose of an intervention AUV is to perform underwater intervention missions with limited or no human assistance. The main issue in designing and implementing a control system for its manipulator is autonomous manipulation, ensuring a reliable behaviour within the workspace, and avoiding collisions, system instabilities and unwanted motions while performing the required task that is theoretically executable. Autonomous manipulation capability is crucial especially for intervention operations in a deep ocean where the low bandwidth and significant time delay are inherent in acoustic underwater communication.

The control system must also address other general manipulation issues, such as task-space oriented, task priority assignments, and dynamic priority changes. The system should be capable of following high-level input commands and providing feedback when the task cannot be executed. The control software for the SAUVIM manipulator, *MARIS 7080* has been developed to satisfy the above objectives and to be modular and flexible for future expansion.

10.3.1 Kinematic control

Consider the schematic representation of a 7 DOF arm workcell as in Figure 10.3.

In Figure 10.3, T_e is the transformation matrix of the end-effector frame $\langle e \rangle$ with respect to the base frame $\langle o \rangle$, T is the (constant) transformation matrix of the tool

Autonomous manipulation for an intervention AUV 219

Table 10.1 Commercial underwater manipulators

	Tecnomare, Italy	Western Space and Marine, Inc.	Kraft Telerobotics, Inc.	Schilling Robotics Systems, Inc.	International Submarine Engr. Ltd, Canada
Model	Telemanipulator	The Arm-66	Predator	Titan III S	Magnum 7F
DOF	6 plus gripper	6 plus gripper	6 plus gripper	6 plus gripper	6 plus gripper
Master/slave	Master/slave	Master/slave	Master/slave	Master/slave	Master/slave
Power source	220 V-50 Hz/ 110 V-60 Hz optional	110–240 V AC, 50/60 Hz, hydraulic power – 7.6 LPM at 207 bar, 5–25 μm absolute	47–63 Hz, 105–250 V AC and hydraulic power at 138 bar 19 LPM 25 μm absolute	50/60 Hz, 90–260 V AC and hydraulic power at 207 bar, 5–19 LPM, 10–200 cSt	Hydraulic power at 86 bar LPM, 25 μm filter
Material	Aluminium alloy type 6000	Aluminium, stainless steel composites, corrosion isolation system	Aluminium with teflon coating	6-4 Titanium and 316 stainless	6061-T6 Aluminium
Joint sensors	Resolver at each joint and torque sensor at the output shaft	Position, velocity and torque	Position and force feedback	Resolver	Potentiometers
Force/torque	Jaw closure force: 700 N	Jaw force controls and sets grip 0–1556 N (162 Nm)	Jaw closure force: 1334 N, wrist torque: 135 Nm	Gripping force: 4448 N, wrist torque: 169 Nm	Gripping force: 1468 N, wrist torque: 190 Nm
Actuator	DC motor – Brushless	Hydraulic cylinders	Hydraulic cylinders	Hydraulic cylinders	Hydraulic cylinders
Max. reach	2.07 m	1.68 m	2.032 m	1.915 m	1.5 m
Payload	40 kg	65.8 kg	90.7 kg	113.4 kg	295 kg at 1.4 m

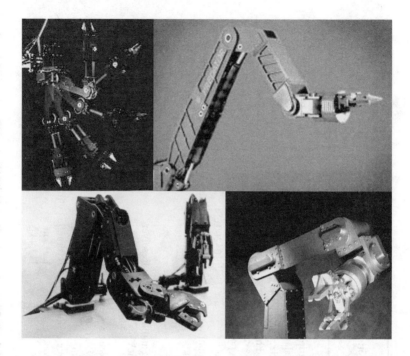

Figure 10.1 From top left: The Arm (Western Space and Marine), Predator (Kraft Telerobotics), Titan 3 (Shilling Robotics Systems), Magnum 7F (International Submarine Engineering)

centre frame $\langle t \rangle$ with respect to $\langle e \rangle$, while T^*, generally time varying, is the transformation matrix of the reference frame $\langle g \rangle$ with respect to the base frame $\langle o \rangle$. The reference frame $\langle g \rangle$ is usually solidal with the target, while the base frame $\langle o \rangle$ is solidal with the vehicle. In the general case, both the above frames are time dependent and moving with respect to the Earth-fixed frame (not shown in Figure 10.3).

The general goal is to track the reference frame $\langle g \rangle$ by the tool frame $\langle t \rangle$. With this aim, the global error e is automatically defined by a vector

$$e \doteq [r_{gt}, \rho_{gt}]^{\mathrm{T}} \tag{10.1}$$

where vectors r_{gt} and ρ_{gt} (both projected on the base frame $\langle o \rangle$) represent the distance and the misalignment (equivalent rotation vector) of the reference frame $\langle g \rangle$ with respect to $\langle t \rangle$. The objective of the control scheme is to make the global error e asymptotically converging towards zero or, alternatively, asymptotically confined within acceptable norm bounds. This goal could be achieved with the closed-loop scheme shown in Figure 10.4. In this scheme, the block named 'Robot + VLLC' (very low level control) represents the physical arm equipped with its seven joint drives, each one implementing a closed-loop velocity control at the corresponding joint. The overall block can be seen as a compact one, receiving the vector $\dot{\bar{q}}$ of the seven reference joint velocities as input, and giving the vector q of the corresponding

Autonomous manipulation for an intervention AUV 221

Figure 10.2 MARIS 7080 *underwater manipulator*

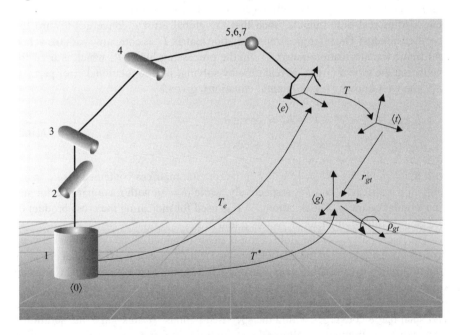

Figure 10.3 *Schematic representation of the arm workcell*

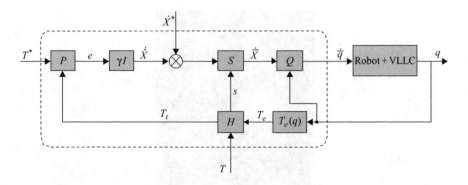

Figure 10.4   Fundamental closed-loop control scheme

seven joint positions as output. The last substantially coincides with the time integral of $\dot{q}$, provided that sufficiently high bandwidth loops are guaranteed by the VLLC system itself.

It is noted that, with the presented approach, the hardware PID-type control loop would be effective enough for the VLLC system in the application field. The joints are decoupled from the motors by a high gear ratio that helps in reducing the joint velocity error.

The remaining part of the control system represents the medium level control (MLC) loop of the arm. The joint velocity reference signals $\dot{q}$ are appropriately generated as real-time outputs, such that the global error e converges toward the specified bounds. The reference transformation matrix T^* is compared with the actual tool frame transformation matrix T_t via the processing block P, which is used for evaluating the global error e in real time by solving, for the rotational error part ρ_{gt} only, the well-known 'versor lemma' equations, given by

$$i_t \wedge i^* + j_t \wedge j^* + k_t \wedge k^* = \tfrac{1}{2} z \sin \theta$$
$$i_t^T \cdot i^* + j_t^T \cdot j^* + k_t^T \cdot k^* = 1 + \cos \theta \tag{10.2}$$

with $R_t \doteq [i_t, j_t, k_t]$, $R^* \doteq [i^*, j^*, k^*]$ the rotation matrices contained inside transformation matrices T_t and T^*, respectively, while $\rho \doteq z\theta$ with z a unitary vector and θ an angular quantity. The notation $a \wedge b$ is used for indicating the cross product of two generic three-dimensional vectors a and b.

The linear part r_{gt} of the global error is easily obtained as the difference between the first three elements of the last columns of T^* and those of T_t. The global error e is then multiplied by a suitable gain matrix γI. The result is the generalised Cartesian velocity $\hat{X} \doteq [\hat{\omega}, \hat{v}]^T \in \Re^6$ (projected on $\langle o \rangle$), where $\hat{\omega} \in \Re^3$ and $\hat{v} \in \Re^3$ are the angular and linear velocity, respectively, which are assigned to the tool frame $\langle T_t \rangle$ such that e converges within the specified bounds. At this stage, the additional Cartesian velocity input $\dot{X}^*$ allows direct control of the end-effector velocity, which is needed for force-feedback control.

The generalised velocity control input $\dot{X}$ is then translated into the one-to-one related velocity $\dot{X} \doteq [\bar{\omega}, \bar{v}]$ to be assigned to the end-effector frame $\langle e \rangle$. The velocity translation is performed by the block S, using the well-known rigid body velocity relationships while the block H provides the vector distance s (projected on $\langle o \rangle$) of the frame $\langle t \rangle$ with respect to $\langle e \rangle$. The end-effector Cartesian velocity control signal $\dot{X}$ is transformed into a corresponding set of seven joint velocity reference input vector $\dot{q}$ by the functional block Q. The interface block Q is critically important for the overall system and will be described in the following section.

10.3.2 Kinematics, inverse kinematics and redundancy resolution

The interface between the assigned task velocity and the joint velocity concerns several important aspects of a critical nature. The control system for autonomous robotic manipulators must ensure a reliable behaviour within the workspace, avoiding collision, system instabilities and undesired motions while the human intervention is usually limited to very few high-level input commands, or sometimes absent.

A robust and efficient singularity avoidance approach is described in this section. The presented algorithm has the property that the task priority is dynamically assigned: when the distance from a singular configuration is approaching to zero, the highest priority is given to the task for maintaining the distance over a predefined threshold. The approach is suitable for avoiding kinematic and algorithmic singularities while providing satisfactory performance near the singular configurations. The conventional approach to achieve a singularity-free motion often uses off-line path planning that requires a preliminary knowledge of all the singular configurations of the manipulator. However, it is not always possible for large DOF systems. The presented approach is based on a real-time evaluation of the measure of manipulability, overcoming the above drawback. Changing the measure of manipulability with a different index function (such as the minimum distance between the arm and the obstacles), means that the same approach could be used for collision avoidance as well as for the mechanical limits of the joints.

10.3.3 Resolved motion rate control

The kinematic output of a generic robotic manipulator is usually represented by a m-dimensional manipulation variable $r \in \Re^m$. The manipulation variables may be, but not limited to, the position and orientation of the end-effector. The relationship between r and the joint angles q is represented by the following equation:

$$r = f(q) \tag{10.3}$$

Invoking small variation, the relationship between δr and δq is given by

$$\delta r = \frac{\partial f(q)}{\partial q} \delta q = J(q) \delta q \tag{10.4}$$

where $J(q) \in \Re^{m \times n}$ is the Jacobian matrix of the manipulation variable r. In resolved motion rate control [2], δq is computed for a given δr and q by solving the linear system (10.4). In general, this is done by using the pseudoinverse of the Jacobian matrix as in Reference 3:

$$\delta q = J^+(q)\delta r + [I_n - J^+(q)J(q)]y \quad (10.5)$$

where $J^+(q) \in \Re^{n \times m}$ is the pseudoinverse of $J(q)$, $y \in \Re^n$ is an arbitrary vector and $I_n \in \Re^{n \times n}$ indicates an identity matrix.

A singular point is defined by the joint configuration vector value $\bar{q}$ where $J(\bar{q})$ is not full rank. Its pseudoinverse $J^+(\bar{q})$ is not defined at such a configuration. Moreover, in the neighbourhood of singular points, even a small change in δr requires an enormous change in δq, which is not practical in the manipulator operation and dangerous to the structure.

The damped least-squares method [9] is a classical and simple way to overcome this drawback. It introduces a regularisation term acting only in the neighbourhood of the singularities

$$J^+ = J^T(JJ^T + \lambda I)^{-1} \quad (10.6)$$

The main disadvantages for the above approach are a loss of performance and an increased tracking error [4]. The choice of damping constant must balance between the required performance and the error allowed. To overcome these defects, Nakamura [3,5] introduced a variable damping factor. Chiaverini [4] also proposed a modified inverse by adding only the damping parameter to the lowest singular values. Their results are shown to be better than those of the damped least-squares method, but they still present a tuning problem of the damping coefficient.

However, all the above solutions do not guarantee that the manipulator does not fall into a singularity configuration, from where the successive departure may require a set of all complex manoeuvrings in the wide range of the arm motion. Especially for autonomous systems, where the human intervention may be very limited or absent, the possible presence of the above drawback would naturally suggest avoidance of all the singular configurations. The main idea of the presented approach is to reconstruct the task to avoid the occurrence of any singularity. This solution allows a direct control of the overall performance by limiting the minimum value of the distance from the singular configuration.

10.3.4 Measure of manipulability

The first step in avoiding singularities is to locate them in the joint space. Yoshikawa [6] proposed a continuous measure that evaluates the kinematic quality of the robot mechanism:

$$\text{mom} = \sqrt{\det\{JJ^T\}} \quad (10.7)$$

The measure of manipulability (mom) takes a continuous non-negative scalar value and becomes equal to zero only when the Jacobian matrix is not full rank. In fact, mom is exactly the product of the singular values of J [7] and may be regarded as a distance from singularity.

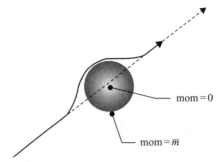

Figure 10.5   Conceptual diagram of the task reconstruction method

One of the advantages of this choice is that its derivative can be found with respect to the joint configuration vector q, which is needed for the controller. It is simply calculated as follows [7,8]:

$$\frac{\partial \text{mom}(q)}{\partial q_k} = \text{mom}(q) \cdot \text{trace}\left\{\frac{\partial J}{\partial q_k} J^+\right\} \qquad (10.8)$$

Equation (10.8) shows that the derivative of mom can be expressed by some known quantities: mom and the pseudoinverse of the Jacobian matrix J^+. The derivative of each element of J with respect to q_k can be easily computed in a symbolic form and is computationally less expensive than the Jacobian itself.

10.3.5   Singularity avoidance for a single task

For a given manipulation variable, usually a singularity-free motion path may be achieved with off-line path planning. However, this approach requires an *a priori* knowledge of all the singular configurations of the manipulator but this is not always possible for large DOF systems.

The proposed method, based on a real-time evaluation of mom, allows movement along a singularity-free path for a generic manipulator whose singular configurations are not known in advance. The basic idea is to circumscribe singularities by moving, when approaching them, on a hyper-surface where mom is constant. Figure 10.5 shows this concept.

For now, let us consider $y = 0$ in Eq. (10.5) (null motion absent):

$$\delta q = J^+(q)\delta r \qquad (10.9)$$

The differential of mom is given by:

$$\partial \text{mom}(q) = \frac{d\text{mom}(q)}{dq}\partial q = \frac{d\text{mom}(q)}{dq}J^+\partial r \qquad (10.10)$$

In order to have $\partial \text{mom}(q) = 0$, Eq. (10.10) implies that the given task must be orthogonal to the vector

$$\frac{d\text{mom}(q)}{dq} J^+ \tag{10.11}$$

or, equivalently, that $\dot{r}$ must be lie on the surface defined by:

$$\left\{ x \in \Re^m : \left(\frac{d\text{mom}(q)}{dq} J^+ \right) \cdot x = 0 \right\} \tag{10.12}$$

Let n_m be the unitary vector orthogonal to the surface (10.12):

$$n_m = \frac{\left((\partial \text{mom}/\partial q) J^+ \right)^T}{\| (\partial \text{mom}/\partial q) J^+ \|} \tag{10.13}$$

Consequently, the projection of the given task is:

$$\delta r_p = \delta r - (\delta r \cdot n_m) n_m \tag{10.14}$$

As stated above, such a projection must be done only when approaching singularities. Eq. (10.14) is modified with a weight as follows:

$$\delta r_p = \delta r - (\delta r \cdot n_m) n_m k(\text{mom}, \bar{m}) \tag{10.15}$$

where $k(\text{mom}, \bar{m})$ is a positive, well-shaped function of mom, to be equal to zero for values of mom greater than a predefined threshold $\bar{m}$ and equal to 1 for values of mom smaller than $\bar{m}/2$:

$$k(m, \bar{m}) = \begin{cases} 0 & \bar{m} < m \\ 4\left(\dfrac{4m^3 - 9m^2\bar{m} + 6\bar{m}^2 m - \bar{m}^3}{\bar{m}^3} \right) & \dfrac{\bar{m}}{2} < m < \bar{m} \\ 1 & m < \dfrac{\bar{m}}{2} \end{cases} \tag{10.16}$$

Figure 10.6 shows an example of this function for $\bar{m} = 0.04$.

It is noted that the first derivative is equal to zero in correspondence of the $\bar{m}$ and $\bar{m}/2$. This makes it possible to progressively lay down the task solution δr on the surface where mom is constant, without introducing instabilities on the controller when closing the loop. However, when mom is already smaller than the value on the surface, Eq. (10.15) does not guarantee escape from the enclosed volume. Furthermore, numerical errors may introduce a small drift term driving the task below the surface. To avoid the above drawback, a third term is introduced to Eq. (10.15)

$$\delta r_p = \delta r - (\delta r \cdot n_m) n_m k(\text{mom}, \bar{m}) k(\text{mom}, \bar{m}/2) n_m \tag{10.17}$$

This produces a recalling action towards the surface, starting when δr_p has no more components along the gradient (i.e., when mom $< \bar{m}/2$). Finally, it must be ensured to leave from the surface by initiating the task correction in (10.17) only when the scalar

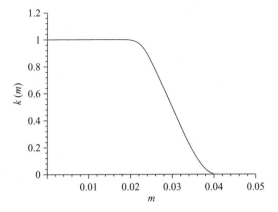

Figure 10.6   Weight function

product $\delta r \cdot n_m$ is positive, that is, when mom is decreasing. Therefore, the final form becomes

$$\delta r_p = \delta r - \delta r_{\text{corr}} \qquad (10.18)$$

where

$$\delta_{\text{corr}} = \frac{1 - \text{sign}(\delta r \cdot n_m)}{2}(\delta r \cdot n_m)n_m k(\text{mom}, \bar{m}) - k\left(\text{mom}, \frac{\bar{m}}{2}\right)n_m \qquad (10.19)$$

The above reconstructed task is used in Eq. (10.9) in place of the original task, guaranteeing a total singularity-free motion within the workspace.

10.3.6   Extension to inverse kinematics with task priority

For autonomous robotic systems, the subtask decomposition between the position and orientation would be advantageous since it could enlarge the reachable workspace of the first-priority manipulation variable (usually position) by allowing the second priority subtask to be incomplete. The concept of task priority in the inverse kinematics of manipulators was introduced by Nakamura [3]. In this approach, the occurrence of algorithmic singularities arises from conflicts between two subtasks when the corresponding non-prioritised task is not feasible. The performance and error of the secondary task depend on the method used for solving the inverse kinematics of the second manipulation variable. This is usually done by using classical methods like the singularity-robust inverse [5] or the damped least-squares method [9]. The main disadvantages of these approaches are a loss of performance and an increased tracking error [4]. The choice of damping constant must balance between the required performance and the error allowed. To eliminate the occurrence of algorithmic singularities, Chiaverini [4] proposed to solve the secondary task separately and then project the solution onto the null space of the first manipulation variable.

The presented approach uses and modifies the lower priority task in order to avoid the occurrence of both kinematic and algorithmic singularities whenever possible.

This solution enhances the performance of the first priority task and allows a direct control of the overall performance by limiting the minimum value of mom. When this is not feasible, a similar correction to the primary task ensures a totally singularity-free path of the manipulator within its workspace.

For simplicity, let us consider only two manipulation variables:

$$r_1 = f_1(q) \tag{10.20}$$

$$r_2 = f_2(q) \tag{10.21}$$

Invoking again small variation, the relationship between δr and δq is given by [10]:

$$\delta q = J_1^+ \delta r_1 + \hat{J}_2^+ \cdot (\delta r_2 - J_2 J_1^+ \delta r_1) \tag{10.22}$$

$$\hat{J}_2 = J_2(I_n - J_1^+ J_1) \tag{10.23}$$

which is the inverse kinematics taking account of the priority of the subtasks [3]. Algorithmic singularities are configurations at which the matrix $\hat{J}_2$ loses rank with J_1 and J_2 full rank. The goal of the presented approach is to avoid both J_1 and $\hat{J}_2$ becoming singular. For this goal, let us consider a new matrix, J_a obtained by stacking J_1 and J_2:

$$J_q = \begin{bmatrix} J_1 \\ J_2 \end{bmatrix} \tag{10.24}$$

It is possible to show that [10]

$$\sqrt{\det[J_a J_a^T]} = \sqrt{\det[J_1 J_1^T]} \cdot \sqrt{\det[\hat{J}_2 \hat{J}_2^T]} \tag{10.25}$$

Therefore, the problem becomes to limit the minimum value of mom of the augmented Jacobian J_a in Eq. (3.24):

$$\text{mom}(q) = \sqrt{\det[J_a J_a^T]} \tag{10.26}$$

The differential of mom is given by:

$$\partial \text{mom}(q) = \frac{d\text{mom}(q)}{dq} \partial q = \frac{d\text{mom}(q)}{dq}((J_1^+ - \hat{J}_2^+ J_2 J_1^+) \partial r_1 + \hat{J}_2^+ \partial r_2) \tag{10.27}$$

If $\text{mom}(q)$ is smaller than a predefined threshold $\bar{m}$, the goal is to have $\partial \text{mom}(q) = 0$ (see also Reference 7), or equivalently from Eq. (10.27)

$$\frac{d\text{mom}(q)}{dq}(J_1^+ - \hat{J}_2^+ J_2 J_1^+) \partial r_1 + \frac{d\text{mom}(q)}{dq} \hat{J}_2^+ \partial r_2 = 0 \tag{10.28}$$

In order to realise Eq. (10.28) for any arbitrary ∂r_1, first let us consider only a secondary (lowest priority) task correction by adding a correction term ∂r_{2c} to ∂r_2 in order to compensate for the decreasing of $\text{mom}(q)$

$$\frac{d\text{mom}(q)}{dq}(J_1^+ - \hat{J}_2^+ J_2 J_1^+) \partial r_1 + \frac{d\text{mom}(q)}{dq} \hat{J}_2^+ (\partial r_2 + \partial r_{2c}) = 0 \tag{10.29}$$

The exact solution for ∂r_{2c} of the above equation does not generally exist. However, it is possible to obtain ∂r_{2c} that minimises (10.29) in the least-squares sense by using again the pseudoinverse [10]:

$$\partial r_{2c} = J_{2c}^+ \cdot \partial \text{mom}(q) \tag{10.30}$$

where $J_{2c} = [d\text{mom}(q)/dq]\hat{J}_2^+$.

Similarly to the single task case, the secondary task correction must be done only if the variation $\partial \text{mom}(q)$ is negative and smaller than a predefined threshold $\bar{m}$.

Finally, because the solution of Eq. (10.28) does not always exist, after the secondary task correction, the task projection procedure must be applied to ensure a complete singularity avoidance, similarly to the single task case [7]. Therefore, noting the modified secondary task with

$$\partial r_{2m} = \partial r_2 + \partial r_{2c} \tag{10.31}$$

the final form of the inverse kinematics, taking account of the priority of the subtasks Eq. (10.22), becomes:

$$\delta q = J_1^+ \delta r_{1p} + \hat{J}_2^+ \cdot (\delta r_{2p} - J_2 J_1^+ \delta r_{1p}) \tag{10.32}$$

where:

$$\partial r_{1p} = \partial r_1 - \frac{1 - \text{sign}(\partial r_1 \cdot n_1)}{2}(\partial r_1 \cdot n_1)n_1 k(\text{mom}, \bar{m}/2) + k(\text{mom}, \bar{m}/4)n_1 \tag{10.33}$$

$$\partial r_{2p} = \partial r_{2m} - \frac{1 - \text{sign}(\partial r_{2m} \cdot n_2)}{2}(\partial r_{2m} \cdot n_2)n_2 k\left(\text{mom}, \frac{\bar{m}}{2}\right) + k\left(\text{mom}, \frac{\bar{m}}{4}\right)n_2 \tag{10.34}$$

$$n_1 = \frac{((\partial \text{mom}(q)/\partial q)(J_1^+ - \hat{J}_2^+ J_2 J_1^+))^T}{\|(\partial \text{mom}(q)/\partial q)(J_1^+ - \hat{J}_2^+ J_2 J_1^+)\|} \tag{10.35}$$

$$n_2 = \frac{((\partial \text{mom}(q)/\partial q)\hat{J}_2^+)^T}{\|(\partial \text{mom}(q)/\partial q)\hat{J}_2^+\|} \tag{10.36}$$

This solution allows avoidance of both kinematic and algorithmic singularities in task-priority based kinematic controllers. The proposed approach uses a secondary task correction and a successive task projection in order to maintain mom over a predefined threshold. The main advantage is a better tracking error in the proximity of singular configurations, with respect to the order of priority of the tasks. Because mom is never zero, the presented algorithm uses the exact pseudo-inversion and the resulting task errors depend only on the choice of the lower limit of mom. This approach is suitable for autonomous systems because it ensures avoidance of every kind of singularity regardless of the input task. If, for example, the task planner requires the arm to reach a particular configuration close to or at a singular point, the manipulator will execute the task with an error as small as possible, avoiding

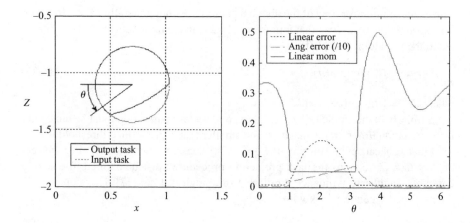

Figure 10.7   Task reconstruction method: linear task in the x–z plane (left); task errors and measure of manipulability, with respect to the angle θ of the circle (right)

the singular point. The path planner could be informed of this error and then would make an appropriate action.

10.3.7   Example

In this example, the given task is a circle partially enclosed in the workspace: the manipulator tries to follow the path, giving the highest priority to the position when far from a singular configuration. When approaching the boundary of the workspace, the first manipulation variable task is performed with an error necessary to maintain mom over the predefined limit.

Figure 10.7 shows a result with the MARIS manipulator. A comparison with Nakamura's singular-robust pseudoinversion [5] in Figure 10.8 shows that the presented task reconstruction method has a faster error recovery. In fact, since mom will never be zero, the exact pseudoinversion can be used minimising the (projected) task error. It differs from the original task only when this is necessary for limiting mom.

It is noted that it is possible to regard the proposed method as a dynamic priority-changing algorithm. As shown in Figure 10.7, when mom of the given task approaches zero, the highest priority is given to the distance from the singularity.

10.3.8   Collision and joint limits avoidance

With a different choice of the index function, the approach described in previous sections could be used to avoid obstacle collisions. The new index function is the minimum distance between the arm and the obstacle represented by simplified solid boxes (Figure 10.9). It makes the procedure for computing the minimum distance simple and efficient (see chapter Fast Overlap Test for OBBs in Reference 11). However, the measure of object distance does not have a closed symbolic form of its

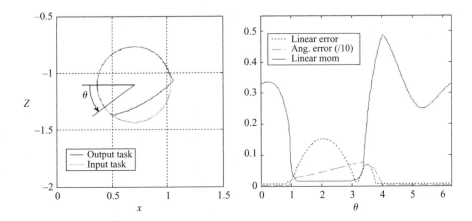

Figure 10.8   Nakamura singular-robust: linear task in the x–z plane (left); task errors and measure of manipulability, with respect to the angle θ of the circle (right)

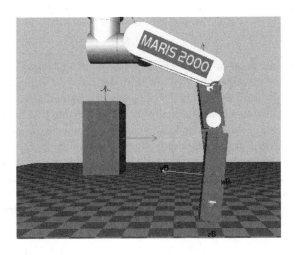

Figure 10.9   Bounding boxes for computing the minimum object distance

derivative with respect to the joint configuration vector q. Therefore, the derivative has been numerically computed for this experiment.

Another index suitable to use with the precedent method is a potential function of the joint position, taking large values when the joints are far from their limits. Indicating $\theta_{i\text{Min}}$ and $\theta_{i\text{Max}}$ as the mechanical limits of the ith joint, its potential function is

$$Jpf_i = -4 \frac{\theta_i^2 - \theta_1 \theta_{i\text{Min}} - \theta_i \theta_{i\text{Max}} + \theta_{i\text{Min}} \theta_{i\text{Max}}}{(\theta_{i\text{Max}} - \theta_{i\text{Min}})^2} \tag{10.37}$$

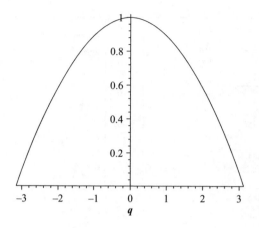

Figure 10.10 The joint potential function with $\theta_{i\,\text{Min}} = -\pi$ and $\theta_{i\,\text{Max}} = \pi$

The global index is:

$$Jpf = \prod_{i=1}^{\text{dof}} Jpf_i \qquad (10.38)$$

Figure 10.10 shows a plot of Eq. (3.37) for $\theta_{i\,\text{Min}} = -\pi$ and $\theta_{i\,\text{Max}} = \pi$.

Its derivative with respect to the joint configuration vector q can be easily computed in a symbolic form. Similar to the singularity avoidance approach, this solution may also be regarded as a dynamic change of the task priority. Maintaining the minimum distance between the arm and the environment objects over a minimum limit can be considered as a primary task only when running too close to the obstacle.

As an immediate effect, the final path between two generic points within the workspace will be automatically adjusted by the controller to avoid collision, singular points and joint limits, whenever possible. This is one of the main requirements in autonomous manipulation. With this advanced feature in the arm controller, the task planner can focus on various high-level aspects of the task.

10.4 Vehicle communication and user interface

The most common communication link for AUVs is acoustic. However, the low bandwidth and a significant time delay inherent in acoustic subsea communication make it very difficult to remotely operate the manipulator.

Robot teleprogramming was proposed as an intermediate solution between supervised control systems and direct tele-operations when a significant delay appears in the communication [12–14]. The main idea is to make the operator feel the system as a common tele-operation without the communication delay. It decouples the local and remote zones by limiting the data exchange to a few symbolic information entities, as opposed to the common robot tele-operation system that uses low level information (e.g., the joint position, motor velocity reference, etc.). Usually, in teleprogramming systems, a partial copy of the remote information is used to create a virtual reality model of the remote environment. In this way, the main component of the virtual model is a predictive simulator. The user–system interaction is performed over the simulated environment and the stability of the local zone is not affected by the delay of the communication channel. The remote system, if not able to cope with an error, can transmit this information to the local system that may assume a safety state. This is to have the remote system avoiding these critical errors as much as possible in tele-operation.

Another useful concept is semi-autonomous operation when significant communication delays exist. The user may provide, instead of directly operating the manipulator, higher level commands during a particular mission, such as 'unplug the connector'. SAUVIM uses this concept in order to perform its intervention tasks. It is an interesting concept between the fully autonomous mode and the classic tele-operation mode. In this approach, the function of the operator is to decide, after an analysis of the data, which particular task the vehicle is ready to execute, and then to send the decision command. The low-level control commands are provided by a pre-programmed on-board subsystem, while the virtual reality model in the local zone uses only the few symbolic information entities received through the low bandwidth channel in order to reproduce the actual behaviour of the system. The robot must be capable of acting and reacting to the environment in an autonomous way, with the extensive use of sensor data processing.

The SAUVIM system consists of the communication interface (client–server architecture); task execution (programming environment); simulation server; and graphic interface and development environment. Details of each component will be described in future articles.

10.5 Application example

The overall autonomous manipulation system described in this chapter was successfully tested during the first intervention mission of SAUVIM where major features of the system were evaluated, such as the task space controller, the communication link, and the programming environment. The experiment was conducted in the ocean, executing the following sequence of autonomous tasks while keeping the vehicle stationary:

1. extract the arm from its parking position;
2. locate, grab and pull-out a connector from its socket;
3. insert the plug in a different socket;
4. park the arm and turn off the power.

When not used, the arm must be docked in a safe position to avoid any kind of physical stress or damage. The parking solution (Figure 10.11) allows the arm to fold over its tray and lock itself as a result of the sequence of several different joint movements. The extraction uses the reverse of this sequence.

It is important to have an exact calibration of the joint offsets of the arm in order to ensure the repeatability of the operation. The final steps of the parking sequence include locking two joints into a fixed device, which is done by an open-loop controller. A small error would result in physical damage of the mechanical system. After powering on the system, the calibration is carried out automatically with the combined use of the joint resolver providing an absolute position (even if less accurate) and the motor resolver providing a relative position (but more accurate for the effect of the gear reduction unit). Figure 10.11 shows underwater scenes of the above undock sequence. Figure 10.12 shows scenes of the plug–unplug operation with the following sequence of movements:

(a) move the gripper in front of the plug;
(b) open the gripper;
(c) move the gripper ahead over the plug;
(d) close the gripper;
(e) move the gripper backward (extract the plug).

After moving the end-effector in front of the destination socket, the above sequence can be reversed in order to insert the plug in a different socket.

All the above operations were coded in a subset of procedures, using a dedicated robot programming language developed for SAUVIM. The above procedures are organised in different levels of hierarchy such that the main execution task would result in the simple sequence of statements:

```
// ExecuteTask()
```

```
//
// Description: main execution task.
// Variables: None.
ExecuteTask := proc()

        // Extract the arm
        ExtractArm();

        // Unplug
        GrabFromSlot();

        // Plug in the first slot
        PutInSlot(1);

        // Done
        BayPark();

end proc:
```

Using this approach, the task can be started by the following command from the operator at the ground station:

```
ExecuteTask();
```

After this command is issued, the autonomous manipulation operates. However, the operator can continue monitoring the task execution and may intervene in case of unexpected behaviours, as long as the internet link provides a reliable connection.

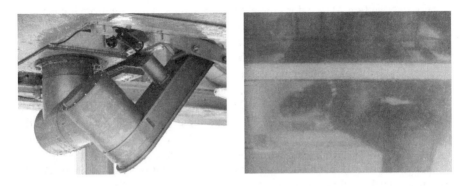

Figure 10.11   The arm in its parking configuration (left) and its underwater extraction (right)

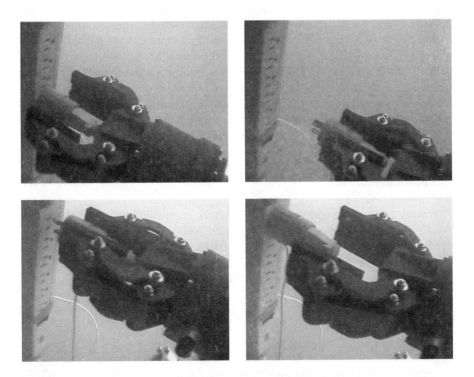

Figure 10.12   Some sequences of the plug/unplug underwater test

10.6 Conclusions

This chapter described the evolution of the hardware of underwater manipulators, addressed some theoretical issues with the arm control system for autonomous manipulation, and briefly discussed vehicle communication and user interface. Preliminary experimental results with SAUVIM were also presented.

For intervention AUVs, it is crucial to make a gradual passage from human tele-operated robotic systems to autonomous manipulation. Within this passage, the most noticeable aspect is the increase of the level of information exchanged between the system and the human supervisor. In tele-operation with ROVs, the user sends and receives low level information in order to directly set the position of the manipulator with the aid of a visual feedback. As the system becomes more autonomous, the user may provide only a few higher level, decision commands while the management of lower level functions (i.e., driving the motors to achieve a particular task) is left to the on-board system. The level of autonomy is related to the level of information needed by the system in performing the particular intervention.

Autonomous manipulation requires the on-board control system to manage various tasks, such as singularity, collision and joint limits avoidance as discussed in this chapter. At the task execution level, the system must be capable of acting and reacting to the environment with the extensive use of sensor data processing. Another important aspect is the target localisation since most AUV position sensors are not accurate enough for precise manipulation. Additional sensors are required to precisely measure an AUV's relative position with respect to the target. Among them are video processing, laser or ultrasonic 3D scanners, motion trackers (ultrasonic, magnetic or inertial) or shape tape. SAUVIM uses a new sensor package for target localisation, including an optical homing device, a passive arm, an acoustic/laser ranger and an ultrasonic motion tracker. An ultrasonic motion tracker as a real-time target position sensor has been successfully tested in generic manipulation tasks and details on target localisation will be presented in future articles.

Acknowledgements

This work was sponsored in part by NSF under grant BES97-01614, ONR under grant N00014-97-1-0961, N00014-00-1-0629, N00014-02-1-0840, N00014-03-1-0969, and N00014-04-1-0751 and KRISO/KORDI via MASE. Any opinions, findings and conclusions or recommendations expressed in this material are those of the authors and do not necessarily reflect the views of the funding agencies.

References

1 Yuh, J. and S.K. Choi (1999). Semi-autonomous underwater vehicle for intervention mission: an AUV that does more than just swim. *Sea Technology*, 40(10), pp. 37–42.

2. Whitney, D.E. (1969). Resolved motion rate control of manipulators and human prostheses. In *IEEE Transactions on Man–Machine Systems* vol. MMS-10, no 2, pp. 47–53.
3. Nakamura, Y. (1991). *Advanced Robotics: Redundancy and Optimization*. Addison Wesley, Reading, MA.
4. Chiaverini, S. (1997). Singularity-robust task-priority redundancy resolution for real-time kinematic control of robot manipulators. *IEEE Transactions, on Robotics and Automation*, 13, pp. 398–410.
5. Nakamura, Y. and H. Hanafusa (1986). Inverse kinematic solutions with singularity robustness for robot manipulator control. *Journal of Dynamic Systems, Measurement and Control*, 108, pp. 163–171.
6. Yoshikawa, T. (1984). Analysis and control of robot manipulators with redundancy. *Robotics Research*, M. Brady and R. Paul, Eds., MIT Press, Cambridge, MA.
7. Marani, G., J. Kim, and J. Yuh (2002). A real-time approach for singularity avoidance in resolved motion rate control of robotic manipulators. In 2002 IEEE International Conference on Robotics and Automation, Washington, DC, 11–15 May.
8. Park, J. (1999). Analysis and control of kinematically redundant manipulators: an approach based on kinematically decoupled joint space decomposition. Ph.D. thesis, Pohang University of Science and Technology (POSTECH).
9. Wampler, C.W. (1986). Manipulator inverse kinematic solutions based on vector formulations and damped least-squares methods. *IEEE Transactions on Systems, Man, and Cybernetics*, SMC-16(1), pp. 93–101.
10. Marani, G., J. Kim, J. Yuh, and W.K. Chung (2003). Algorithmic singularities avoidance in task-priority based controller for redundant manipulators. In 2003 IEEE/RSJ International Conference on Intelligent Robots and Systems, Las Vegas, 27–31 October.
11. Gottschalk, S., M.C. Lin, and D. Manocha (1996). OBBTree: a hierarchical structure for rapid interference detection. In Computer Graphics (SIGGRAPH'96 Proceedings), pp. 171–180.
12. Funda, J., T.S. Lindsay, and R.P. Paul (1992). Teleprogramming: toward delay invariant remote manipulation. *Presence*, 1, pp. 29–44.
13. Paul, R.P., C.P. Sayers, and M.R. Stein (1993). The theory of teleprogramming. *Journal of the Robotics Society of Japan*, 11(6), pp. 14–19.
14. Sayers, C.P., R.P. Paul, J. Catipovic, L.L. Whitcomb, and D. Yoerger (1998). Teleprogramming for subsea teleoperation using acoustic communication. *IEEE Journal of Oceanic Engineering*, 23(1), pp. 60–71.

Chapter 11
AUV 'r2D4', its operation, and road map for AUV development
T. Ura

11.1 Introduction

Starting by simply lowering sensing equipment into the ocean from boats, the development of new technology and methods to gain access to the ocean has led to the birth of submarines, remotely operated vehicles (ROVs) and underwater observation stations connected to land via cables. These advancements have made possible observations that were unimaginable 100 years ago. Now, autonomous underwater vehicles (AUVs) are being introduced as a new observation platform.

Up until 2004, numerous AUVs of various size and configuration, ranging from 10 kg to 10 tonnes in weight are under operation at sea all over the world. The hostile environment of the ocean means that the experience gained through their experiments at sea is crucial for further AUV development. The AUV 'r2D4' [1] (cf. Figure 11.1) built in 2003, is an example of a full-scale ocean going AUV and is now used for the observation of knolls and fault lines. Having successfully observed the NW Rota 1 Underwater Volcano in May 2004, it is planned that the vehicle will be operated for exploration of hydrothermal ore deposit in the Izu-Ogasawara back-arc basin of Japan. In addition, the small AUV 'Tri-Dog 1' (cf. Figure 11.9) successfully observed the breakwater caissons and their base rock mound at Kamaishi Port in Japan in October 2004. Here we detail the practical considerations for r2D4's dives at Rota Underwater Volcano and furthermore discuss what types of AUV are required for underwater observation and create a road map for future AUV development.

Figure 11.1   r2D4 under deployment work from R/V Hakuho-maru

11.2   AUV 'r2D4' and its no. 16 dive at Rota Underwater Volcano

11.2.1   R-Two project

The University of Tokyo started research and development of AUVs in 1984, and since then has constructed more than ten AUVs to investigate potential application of AUVs for various tasks [2]. The R-One and R-Two projects are a series of large scale projects that have culminated in the development of two AUVs, 'R-One Robot' and 'r2D4'. The R-One Robot, initially planned in 1990, was equipped with a closed cycle diesel engine (CCDE) system and succeeded in 12-h continuous operation in 1998. In 2000, the robot went on to take high-resolution side scan images of the Teisi Knoll in full autonomy [3]. The second stage of this program, the R-Two project, was initiated in 2001 based on the experience gained in the construction and operation of the R-One Robot. The major application of these AUVs is in the scientific survey of hydrothermal vent areas. Along mid-ocean ridge systems and back-arc basins, it is thought that there are many active volcanoes and hydrothermal vents, though these had never been investigated or even found. They are thought to be distributed over extremely wide areas, making cruising type AUVs almost ideal for this application since unlike ROVs they can cover wide areas without the restriction of cables.

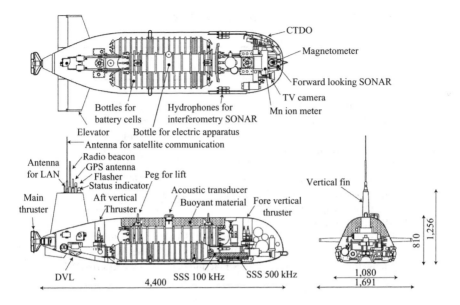

Figure 11.2   General arrangement of r2D4 and fitted equipment

The AUV r2D4 was designed based on the R-One Robot where the primary issues were downsizing and a 4000 m maximum depth rating. This was established by introducing lithium ion secondary battery packages and a compact inertial navigation system (INS). Since the configuration of effectors for controlling the dynamics of r2D4 is the same as that of R-One Robot, the software for R-One Robot, which had been debugged through long term operation, could be directly introduced by changing only the control gains. Therefore, it was possible to carry out the first sea trial in July 2003 soon after completion of its hardware only 2 years from conceptualisation. Next were a series of deployments along a fault line in Ryotsu Bay in Sea of Japan, Kurosima Knoll (E124°10′, N24°7′, depth 620 m) south of Okinawa archipelagos in December 2003 [4], where spouts of cold water containing a high concentration of carbon dioxide were reported, and NW Rota 1 Underwater Volcano in the Mariana Trough, all of which were successfully carried out.

11.2.2   AUV 'r2D4'

11.2.2.1   Development of hardware

Figures 11.1 and 11.2, and Table 11.1 show an image and the specifications of r2D4 and its general arrangement, which consists of a main body, a pair of elevators, a vertical fin with antennas and a main thruster. The following are the special features of r2D4:

1. Compact
   - a small ship can deploy r2D4;
   - full autonomous mission execution does not involve any real-time operator.

Table 11.1   Principal particulars of r2D4

AUV	r2D4	R-One Robot
Length OA (m)	4.4	8.27
Breadth of body (m)	1.08	1.15
Height of body (m)	0.81	1.15
Dry weight (ton)	1.63	4.74
Without payload (ton)	1.51	4.55
Depth rating (m)	4000	400
Cruising range (km)	60	100
Energy source	Lithium ion 2nd bat.	CCDE system
Energy capacity (km/h)	15	60
Max. cruising speed (knots)	3	3
Main CPU	Power PC 233 MHz	MC68040*2
OS	VxWorks	VxWorks
Navigation	INS(FOG)+DVL	INS(RLG)+DVL

2. Self-contained
   - high accuracy positioning is possible without the need for transponders in the sea;
   - high reliability of mission execution based on sufficient countermeasures for possible accidents, making the operator comfortable during the operation.
3. Control and data acquisition from the payload equipment to the main computer
   - complicated environmental changes can be recognised by r2D4.
4. Plan of tracklines can be changed during the mission
   - when an anomaly is encountered along the trackline, r2D4 tries to find out its source.

A 1.5 kW main thruster propels the vehicle with 3 knots forward speed. The main thruster can be rotated horizontally by up to 20° to control yaw motion. The minimum turning circle at 3 knots has a radius of about 25 m.

Roll and pitch motion can be controlled by trimming the elevators. Although the maximum physical angles of descent and ascent depend on the vehicle's righting moment, they are limited to 25° at 3 knots in order to avoid stalling. Two vertical thrusters are used for ascent and descent for close approach to the seabed and for emergency ascent.

Side scan and interferometry SONARs [5] are equipped as standard apparatus for observation of the sea floor. The forward region of the robot is reserved for payload so it is possible to equip the robot with various sensors.

11.2.2.2   Behaviour control and mission planning

The behaviour of the robot is controlled by a hardware level computer, named 'M-CPU', which is connected to the 'P-CPU' responsible for the management of

Table 11.2   Datasets for operation which can be specified by the users

---

Group 1: Mission definition
   1.1 Course data
      Sequence of action patterns
   1.2 Emergency return course data
      Route to go back to the recovery point if
      the mission is terminated during diving
   1.3 Map of safety
      Roughly drawn map of depth up to which r2D4 can
      reach without encountering obstacles
Group 2: Control and limits
   2.1 Criteria of unusual situation and countermeasures
   2.2 Control gains and miscellaneous

---

payload sensors in the payload bay. The user of r2D4 specifies the behaviour of the robot by loading five datasets, divided into two parts, to M-CPU which are shown in Table 11.2. The user, in practice, needs only to prepare three datasets of group (1) since the default settings for datasets of group (2) are adequate for standard operation.

The sequence behavioural patterns set by the user in the course data 1.1 is crucial to the performance of the vehicle and must be specified in consideration of the mission, bathymetry of the seafloor and the manoeuvrability of the robot. The course data consist of a series of way-points defined by their longitude, latitude and depth or height off the sea floor, between which one of the behavioural patterns shown in Table 11.3, is selected to specify the AUV motion control. These patterns are defined by a number of parameters. For example,

- the time frame before carrying out the next manoeuvre,
- the obstacle avoidance algorithm,
- the surfacing procedure in the case that the mission is ended prematurely.

11.2.2.3   Emergency countermeasures

In the event that M-CPU detects a problem with the vehicle hardware or its behaviour during the mission, one of the following emergency procedures is carried out depending on the level of emergency:

1. Keep a log and continue the mission.
2. Use the vertical thrusters to reach a safe altitude and continue the mission.
3. Follow the emergency return course.
4. Use the vertical thrusters to surface and end the mission.
5. Use the vertical thrusters to reach a safe depth, then use the main thruster and elevators to surface and end the mission.
6. Release the emergency ballast to surface and end the mission.

Table 11.3   Performance patterns which identify the actions

Pattern number	Action	Note
0	No action	1st line of table
1	Vertical descent and ascent	By vertical thrusters
2	To a specified depth	Keeping a bearing
3	Turn to a direction	Keeping a depth
4	Go to next way-point	Keeping a depth
5	Go to next way-point	Repetitive descent and ascent
6	Go to next way-point	Keeping an altitude
7	Direct to current	Keeping 0 ground speed
8	Stop by astern rotation of main thruster	–
9	–	–
10	No control	Stop trimming of elevators
11	Descend from rough sea surface	Try to descend in four different directions
12	TV camera on/off	–
13	Payload on/off	–
14	Hover at the present depth	–
15	Descend to a specified altitude	After hitting the altitude, move to the next sequence
16	–	–
17	Discharge a ballast weight	–
18	Discharge a recovery buoy	–
100	No action	End of table

These emergency countermeasures taken for various types of problem are defined by the datasets of group (2) except the route of emergency return course which should be defined in the dataset 1.2. The user does not have to make any special considerations to emergency countermeasures when programming the mission. The robustness of these emergency countermeasures determines the level of confidence in the vehicle's reliability.

11.2.3   Dive to Rota Underwater Volcano

11.2.3.1   North West Rota 1 Underwater Volcano

In May 2004, the vehicle was deployed at the North West Rota 1 Underwater Volcano, which is located about 100 km northwest of Rota island in Mariana archipelagos as shown in Figure 11.3 and was reported to be very active by the NOAA Group [6]. Although we enriched our experiences during the exploration of the Kurosima Knoll, which is relatively flat, investigating an active underwater volcano is notably more

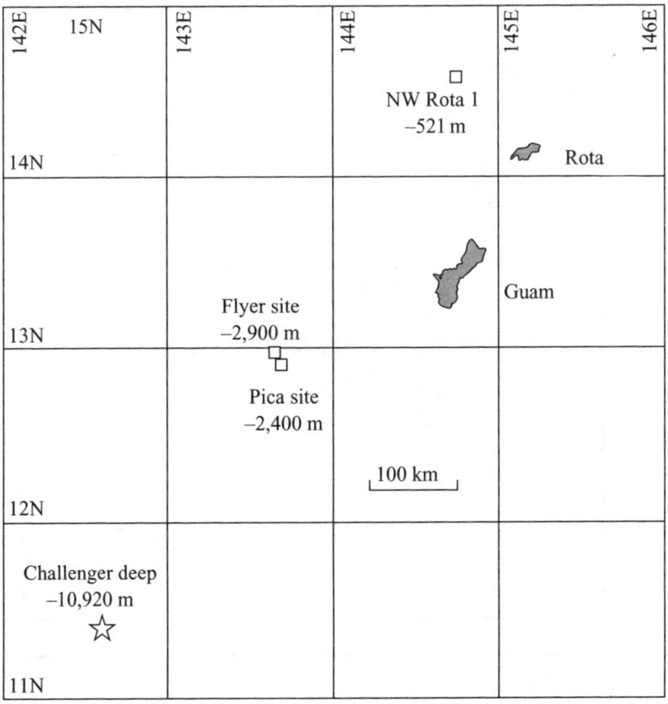

Figure 11.3 Location of NW Rota 1 Underwater Volcano in Mariana Trough

adventurous. The Rota Underwater Volcano is extremely attractive for us because of its shape being similar to Mount Fuji, which is a symbol of Japan.

The Robot r2D4 was deployed a total of seven times off the mother vessel *Hakohu-maru* of JAMSTEC, and used its side scan sonar and various other scientific instruments to record data in the vicinity of the peak at depths from 514 to 1400 m. In this chapter, we limit our discussion to the behaviour of the vehicle, the data from the environmental sensors will be discussed in separate articles.

11.2.3.2 Number 16 dive

The mission plan for the No. 16 dive on the 31 May, 2004 is shown in Figure 11.4(a) and (b). The abstract of motion between the way-points are shown in Table 11.3, which includes the time when r2D4 hits the way-point. The time history of depth and altitude of the robot is shown in Figure 11.5. Figure 11.6 shows the three-dimensional view of trajectory of the robot.

The mission was planned by considering the following points:

1. The Doppler SONAR which assists the INS, can measure the ground speed only when the robot's altitude is less than 200 m off the sea floor. To minimise the period in which the ground speed is not available, the robot should first descend to the highest peak.

246 *Advances in unmanned marine vehicles*

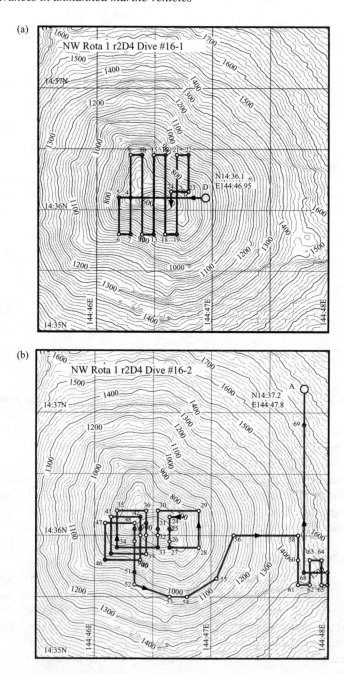

Figure 11.4 (a) Way-points and tracklines in the (a) first and second stages of No. 16 dive, (b) third, fourth and fifth stages of No. 16 dive

AUV 'r2D4', its operation, and road map for AUV development   247

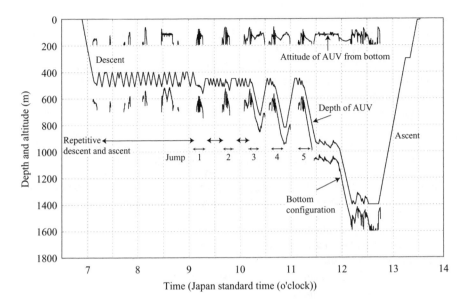

Figure 11.5   Vertical profile of trajectory of r2D4 during No. 16 dive

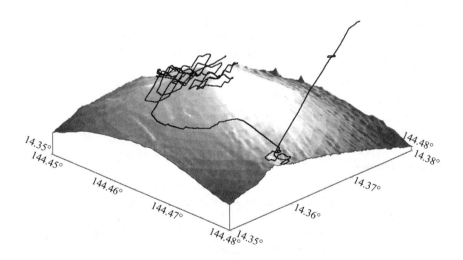

Figure 11.6   Three-dimensional view of the trajectory of r2D4 during No. 16 dive

2. As the vehicle approaches the peak it maintains an adequate altitude to avoid any unexpected problems caused by the sea floor.
3. The manoeuvre of diving over the peak and jumping off to the bottom of the cliff (called the 'Bungee Jump') is crucial since there is a high probability of making observations of the hydrothermal plumes during this period.

4. Upon confirmation of hydrothermal activity of the peak, the robot should search for hydrothermal activity in areas other than south of the peak, which had been investigated by the NOAA Group.
5. First the robot repeatedly changes its depth to observe a large volume of water surrounding the peak, and later cruises closely following the sea floor.
6. It investigates the small mound southeast of the main peak.
7. In the middle of surfacing, the vehicle draws a figure of eight to calibrate its magnetometer.
8. Minimise the period during which the ground speed cannot be measured.
9. The maximum angles of ascent and descent are limited to 25°.

The No. 16 dive consisted of five stages as shown in Table 11.4. The major events described in the pre-programmed course data are divided into five stages as follows:

Stage 1: Descend to the peak
  To way-point 4: approach the vicinity of the peak.
Stage 2: Mid-layer survey
  To way-point 23: change the depth between about 500 and 400 m to cover a large volume of water.
Stage 3: Bungee jumping
  To way-point 52: perform a sequence of five bungee jump manoeuvres over the peak to reach an altitude of 120 m.
Stage 4: Survey of the mound at 1400 m depth
  To way-point 56: maintain an altitude of 120 m off the sea floor of depth 1050 m.
  To way-point 61: approach the small mound southeast of the main peak.
  To way-point 68: observe the mound from an altitude of 120 m.
Stage 5: Ascent and surfacing
  From way-point 68: begin surfacing routine.
  At way-point 69: draw a figure of eight at a depth of 300 m to calibrate the magnetometer. Then surface.

11.2.3.3 Plumes

The data obtained from the onboard sensors suggest that the robot encountered a number of hydrothermal plumes as it traversed the peak. The video cameras were turned off during No. 16 dive in order to conserve power. However, later in No. 19 dive, video images of the hydrothermal plume were recorded. Figure 11.7 shows an image of a dense hydrothermal plume taken by a TV camera fitted on the bottom of r2D4.

11.3 Future view of AUV research and development

The exploration of Rota Underwater Volcano by r2D4 has been presented as one example of underwater survey, where in fact underwater observation is carried out in many forms, ranging from obtaining information and mapping underwater bathymetry to

Table 11.4 Mission plan for No. 16 dive at NW Rota 1 Underwater Volcano

Way-point and stage	Event (start)	Time (JST)	Elapsed Time
Stage 1	Descend to the peak		
D	Descend to 150 m altitude	6:53:44	–
Stage 2	Mid-layer survey		
4	Go to 400 m depth	7:11:08	0:17:24
5	Rep. desc. and asc. 510–400 m	7:13:53	0:20:09
6	Descend to 400 m depth	7:20:57	0:27:13
7	Rep. desc. and asc. 510–400 m	7:22:57	0:29:13
9	Go to 400 m depth	7:37:57	0:44:13
10	Rep. desc. and asc. 510–400 m	7:39:55	0:46:11
12	Go to 400 m depth	7:58:38	1:04:54
13	Rep. desc. and asc. 500–400 m	8:01:01	1:07:17
15	Go to 400 m depth	8:16:19	1:22:35
16	Rep. desc. and asc. 510–400 m	8:19:15	1:25:31
18	Go to 400 m depth	8:34:52	1:41:08
19	Rep. desc. and asc. 510–400 m	8:37:42	1:43:58
21	Go to 400 m depth	8:55:25	2:01:41
22	Rep. desc. and asc. 510–400 m	8:58:42	2:04:58
23	Go to 500 m depth	9:07:43	2:13:59
Stage 3	Jump off to the bottom of cliff (bungee jumping: BJ)		
24	Go to 120 m altitude (BJ1)	9:10:45	2:17:01
27	Go to 450 m depth	9:16:40	2:22:56
28	Rep. desc. and asc. 500–450 m	9:22:42	2:28:58
29	Rep. desc. and asc. 510–400 m	9:30:01	2:36:17
30	Go to 120 m altitude (BJ2)	9:40:19	2:46:35
33	Go to 450 m depth	9:47:53	2:54:09
34	Rep. desc. and asc. 510–450 m	9:55:13	3:01:29
36	Go to 120 m altitude (BJ3)	10:09:07	3:15:23
40	Go to 450 m depth	10:25:20	3:31:36
41	Rep. desc. and asc. 500–450 m	10:32:31	3:38:47
42	Go to 120 m altitude (BJ4)	10:38:35	3:44:51
46	Go to 450 m depth	10:54:20	4:00:36
47	Rep. desc. and asc. 500–450 m	11:06:57	4:13:13
48	Go to 120 m altitude (BJ5)	11:13:09	4:19:25
Stage 4	Survey of the mound at 1400 m depth		
Stage 5	Asending and surfacing		
68	Go to 300 m depth	12:44:20	5:50:36
69	Turn for calibr. of magnetometer	13:14:57	6:21:13
A	Surface	13:32:39	6:38:55

Figure 11.7   Picture of the plume taken by r2D4 during No. 19 dive

sampling materials from the sea floor, by using a big/small support ship in calm/strong current. Since the nature of these missions vary widely, it follows that the design and operation of the AUVs must also vary to meet the requirements of their missions. This can be explained by analogy with the design of aeronautic vehicles, for example, fighter planes, passenger carriers and even helicopters, all vary tremendously to meet the specific requirements of their missions. r2D4 is one example of a specific type of AUV. Consequently, we would like to emphasise that the future of AUV development lies in the variety of vehicle types designed to perform a variety of missions.

11.3.1   AUV diversity

The future of AUV design is determined by opinions on what underwater tasks are important and how AUVs can be applied. These divide AUVs into three main categories.

A. Operations at a safe distance from the sea floor (C or cruising type)
   - observation of the sea floor using sonar,
   - examination of water composition,
   - examination and sampling of floating creatures.
B. Inspections in close proximity to the sea floor and man-made structures (B or bottom reference type)
   - inspection of hydrothermal activity,
   - inspection of creatures on the seafloor,
   - inspection of underwater cables, pipelines and offshore platforms.

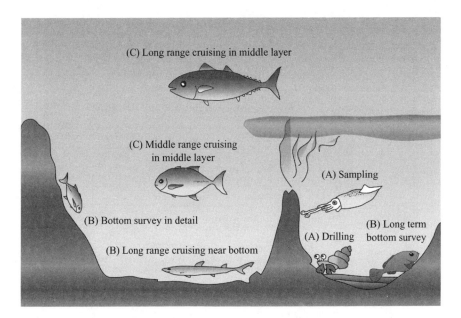

Figure 11.8 AUVs wearing skin-like costumes of fish

C. Interactions with the sea floor and man-made structures (A or advanced autonomy type)
- sampling of substance on the sea floor,
- drilling.

Although C type AUVs are not affected much by the configuration of the sea floor during the mission [7], collision avoidance manoeuvres near the sea floor are among the most important concerns for B type AUVs. These concepts are illustrated in Figure 11.8 where the characteristic shape of the robots are depicted corresponding to their environment. At present, most development has been into C type AUVs because these are the most simple and carry the least risk. However there is much hope and demand for B type vehicles that operate close to the seafloor and so research into this field is also necessary.

One approach to realise a B type robot is to upgrade the sensors of a C type robot and reform it in conformity with the mission requirement. However, this process yields the shift in emphasis from long-range forward propulsion towards high manoeuvrability, control and rapid sensing capabilities of the environmental topology in order to cope with sudden drops and the harsh underwater landscape. The B type robot should be designed based on these requirements. The robot Tri-Dog 1 [8,9] in Figure 11.9 is an example of an AUV which can perform the B type missions.

Acoustic and visual observation of the sea floor by C and B type robots may be the first stage of underwater exploration. The natural and much desired progression is to sample and interact with the environment. However, there has been little research into AUVs capable of sampling [10]. This signifies the difficulties and complexities

252 *Advances in unmanned marine vehicles*

Figure 11.9 AUV 'Tri-Dog 1' surveying top of rock mound for deep caissons constructed at the mouth of Kamaishi Bay

involved in underwater sampling. There is strong competition from ROVs, which are currently relied upon for sampling tasks. However, it is anticipated that within a few years the objects found by C and B type robots will be sampled by A type robots. It is thought that AUVs such as Tri-Dog 1, which can operate in close proximity to underwater structures, can be modified to accommodate such capabilities.

11.3.2 Road map of R&D of AUVs

As AUV technology advances steadily, it becomes necessary to consider what type of underwater tasks AUVs should perform, and to create a road map for future AUV development. The following road map is our idea, which considers only investigation of the sea floor.

2006
- Autonomous locating of hydrothermal chimneys and hydrothermal ore deposits (B type).

2007
- Recording images around the entire circumference of hydrothermal chimneys (B and A types).

2008
- Continuous one month operation by recharging power and exchanging information with an underwater station (C type),

- Autonomous mosaicing of hydrothermally active areas (B type),
- Target object and area specific sampling (A type),
- Sub-bottom observation by pulling streamer cables (C type).

~2010
- CTDO observation to be carried out by AUVs as standard (C type),
- Multiple robots [11,12] observing a region cooperatively (C, B and A types),
- Many deep-tow vehicle operations superseded by AUVs (C type),
- Cooperative missions between AUV and USV (unmanned surface vehicle) (C, B and A types).

~2015
- Researchers to have individual AUVs allocated (C, B and A types).

~2020
- Underwater station maintenance to be carried out by AUVs (A type).

~2025
- Cooperative missions between AUV, USV and UAV (unmanned aerial vehicle) (C, B and A types).

11.4 Acknowledgements

In this chapter, the motion of the robot shown was recorded by KH04-2's navigation at sea, and the road map was developed based on the experience of various groups. This kind of road map can only be laid out as a result of the significant advances in AUV technology and the experiences gained in the field. This has involved the cooperation and input of many people and I would like express my appreciation towards the people involved in our work.

References

1. Ura, T. (2002). Construction of AUV r2D4 based on the success of full-autonomous exploration of Teisi Knoll by R-one Robot. In Proceedings of the AUV Show Case, UK, pp. 23–28.
2. Ura, T. (2002). Development of autonomous underwater vehicles in Japan. *Advanced Robotics*, 6(1), pp. 3–15.
3. Ura, T., T. Obara, S. Takagawa, and T. Gamo (2001). Exploration of Teisi Knoll by autonomous underwater vehicle "R-one Robot". In Proceedings of the OCEANS 2001, Hawaii, USA, vol. 1, pp. 456–461.
4. Ura, T., T. Obara, K. Nagahashi *et al.* (2004). Introduction to an AUV "r2D4" and its Kurosima Knoll survey mission. In Proceedings of the OCEANS'04 (OTO'04), Kobe, Japan, pp. 840–845.
5. Koyama, T., A. Asada, T. Ura, Y. Nose, T. Obara, and K. Nagahashi (2004). 'Bathymetry by new designed interferometry sonar mounted on AUV'. In Proceedings of the OCEANS'04 (OTO'04), Kobe, Japan, pp. 1169–1174.
6. NOAA: http://oceanexplorer.noaa.gov/explorations/04fire/welcome.html

7 Iwakami, H., T. Ura, K. Asakawa *et al.* (2002). Approaching whales by autonomous underwater vehicle. *Marine Technology Society Journal*, 36(1), pp. 80–87.
8 Kondo, H. and T. Ura (2004). Navigation of an AUV for investigation of underwater structures. *Control Engineering Practice*, 12, pp. 1551–1559.
9 Kondo, H., T. Maki, T. Ura, and Y. Nose (2002). Structure tracing with a ranging system using a sheet laser beam. In Proceedings of the UT'04, Taipei, Taiwan, pp. 83–88.
10 Imai, T., T. Ura, and Y. Nose (2002). Semi-autonomous touching of underwater object by unmanned untethered vehicle. In Proceedings of the OCEANS'02, Biloxi, USA, pp. 236–241.
11 Yu, S. and T. Ura (2004). Experiments on a system of multi-AUV interlinked with a smart cable for autonomous inspection of underwater structures. *International Journal of Offshore and Polar Engineering*, 14(4), pp. 274–278.
12 Yu, S. and T. Ura (2004). A system of multi-AUV interlinked with a smart cable for autonomous inspection of underwater structures. *International Journal of Offshore and Polar Engineering*, 14(4), pp. 265–273.

Chapter 12

Guidance and control of a biomimetic-autonomous underwater vehicle

J. Guo

12.1 Introduction

The development of propulsion mechanisms for autonomous underwater vehicles (AUVs) has recently attracted increased attention. However, underwater vehicles that are powered by thrusters and control surfaces which exhibit poor hovering, turning and manoeuvring performance in water currents. Natural selection has ensured that fish have evolved highly efficient swimming mechanisms. Their remarkable swimming abilities can be used as inspiration for innovative designs that improve the man-made systems operating in, and interacting with, aquatic environments.

In the early twentieth century, Breder [1] divided fish swimming movements into two main categories: those that are propelled by their bodies and tails; and, those that oscillate their median and/or paired fins to manoeuvre. The first group is further divided into Anguilliform, Ostraciiform, Thunniform and Carangiform swimmers. The second group has as many as seven styles: Amiiform, Gymnotiform, Balistiform, Rajiform, Tetraodontiform, Labriform and Diodontiform. Typically, the locomotion styles of fish are named by body and/or caudal fin (BCF) locomotion, and median and/or paired fin (MPF) locomotion, respectively. Sfakiotakis *et al.* [2] present a detailed description for BCF and MPF locomotion. The BCF control method is usually suitable for high-speed movement, while the MPF control method is designed for manoeuvring control at slow speeds and hovering motions. The most common BCF locomotion is the Carangiform as less energy is lost in lateral water shedding and vortex formation than with other BCF motions. Numerous studies on the design of biomimetic robots have been based on BCF control and have focused on the motion of a fin as an oscillating foil for propulsion. Carangiform swimmers are propelled by undulating their bodies. The bodily undulations are confined to the last third of the body length. Kelly and Murray [3] used a reduced Lagrangian formulation in which a fish is assumed as a rigid body, with its tail action represented by a point vortex of independently controlled position and strength. Mason and Burdick [4] developed a robotic tester for planar Carangiform locomotion. They proposed an approach based on a highly simplified quasi-static lift and drag modelling of the forces impacting the body and tail of the fish.

Oscillation of the paired pectoral fins to control the motion of biomimetic-autonomous underwater vehicle (BAUV) is based on Labriform locomotion in fish. A review of pectoral fin locomotion in terms of structure, kinematics, and neural control can be found in Reference 5. Fin shapes, motion patterns of pectoral fins are reviewed in Reference 6. It was found that rowing motion provides better manoeuvrability than flapping motion at slow speeds. However, rowing is less efficient than flapping. Second, the high efficiency and performance of the Labriform mode of locomotion in fish indicates that it would be a useful means of propulsion for autonomous underwater vehicles. Finally, fin shape is a key influence on propulsion performance. The pectoral fin motion adopted in this work is a type of rowing motion.

Kato [7] investigated methods for guiding and controlling the movements of a fish robot. He proposed a hydrodynamic model for pectoral fin motion and evaluated the manoeuvrability of a fish robot equipped with a device that moved the pectoral fins. He employed an unsteady vortex–lattice method to model the unsteady forces in the pectoral fin model. The fish robot, with a pair of pectoral fins, swam on a predetermined course and hovered near a specific point in water current under fuzzy control. Later, Kato et al. [8] developed an underwater robot named *PLATYPUS* with four pectoral fins, which performed a rendezvous and docking with an underwater post in still water with satisfactory accuracy. A vehicle named *PILOTFISH* that achieved six-degree-of freedom motion by controlling the amplitude, frequency and centre position of four flexible oscillating fins was constructed [9]. Recently, the design of an underwater vehicle, BFFAUV, which can be used as an experimental platform for propulsion and manoeuvring by utilising four biologically inspired flapping foils was presented in Reference 10.

Control problems for a biomimetic-autonomous underwater vehicle (BAUV) include the following: station keeping, trajectory planning, tracking planned trajectories, manoeuvring, power efficiency and active noise control [11]. This chapter focuses on achieving coordinated control by integrating BCF and MPF swimming modes for waypoint tracking, turning, hovering and braking. There is evidence in nature that some fish use MPF locomotion for swimming slowly and switch to BCF during high-speed swimming [12]. Expectations are that the integrated controller will combine the advantages of both swimming modes to control hovering, accurate positioning and agile turning.

Previous work by the author on the modelling and control of a BAUV is summarised as follows: Chiu et al. [13] simulated the undulatory motion of a flexible slender body; Guo and Joeng [14] developed a method for coordinating body segments and paired fins and, thus, to control the motion of a BAUV; Guo et al. [15] later developed an optimal body-spline to enable a BAUV to swim forward; Guo et al. [16] combined turning motions with forward-swimming motions to design a control system to track way-points; and, Chiu et al. [17] developed a method to estimate hydrodynamic forces induced by pectoral fins and presented simulation results.

In this chapter, a simulation method developed under the assumption of a slender body is presented. The length of this flexible slender body is divided into a number of segments. When the body undulates, a wave passes over it moving from its nose to its tail. Reaction forces due to momentum change, friction and cross-flow drag, are evaluated for each segment. The equations of motion described by the body-fixed coordinate are derived by summing the longitudinal force, lateral force and yaw moment acting on each segment. The modelling of the pectoral fin based on a blade element synthesis scheme for evaluating the hydrodynamic forces of pectoral fin motion is presented. Joint angles are coordinated with reference to the bodily shape of the BAUV in the local coordinate frame. Furthermore, a global control law is proposed for BAUV navigation. The remainder of this work is organised into three sections. Section 12.2 develops a dynamic model of the BAUV by employing the Lagrange equation of motion. The dynamic equation is then combined with the Euler–Lamb equation and fluid viscous forces. Next, Section 12.3 presents the structure of the controller and the local and global controls. Section 12.3 also outlines experimental tests and details the experimental results with a related discussion. Finally, in Section 12.4, conclusions are presented.

12.2 Dynamic modelling

Figure 12.1 illustrates the photograph of the BAUV testbed. The testbed BAUV comprises five components – head, caudal body, tail fin and two pectoral fins. The head, tail fin and pectoral fins are rigid, while a rigid link together with flexible materials support the caudal body. The net mass of the BAUV is 200 kg, and has length 2.4 m, width 0.4 m and height 0.7 m. A Doppler sonar and a heading sensor, as well as a depth metre, are located in the abdominal portion of the BAUV to sense its direction of motion, velocity and depth in the water. An echo sounder is

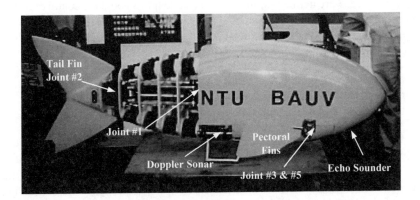

Figure 12.1   The BAUV testbed

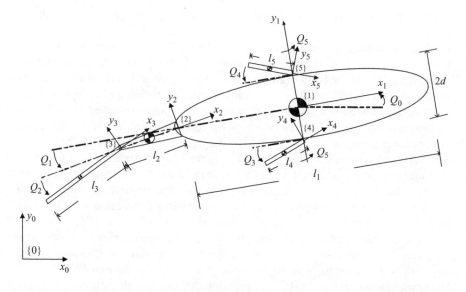

Figure 12.2   Definitions of coordinates of the BAUV

located at the bottom of the head for sensing environmental features. Servo motors control the movements of the caudal peduncle, tail fin and pectoral fins. This section obtains a mathematical model of a BAUV that moves in the horizontal plane. Full dimensional equations of motion of BAUVs can be derived using the procedure presented here.

12.2.1   Rigid body dynamics

The body dynamics of a BAUV are now described. Segment-fixed coordinate frames $x_i y_i z_i, i = 1, 2, 3$ are defined for the body and caudal fins about the centre of mass of each segment (Figure 12.2) and frames $x_i y_i z_i, i = 4, 5$ are defined for the pectoral fins (Figure 12.3). Table 12.1 lists the notations used for the definition of the coordinate systems. The following assumptions are made:

1. all segments are symmetrical in the x–z and x–y planes;
2. the vehicle body is comprised of homogenous materials;
3. the buoyancy is always negated by gravity.

The first two assumptions make the dynamic system symmetric in the x-direction.

x_b, y_b, θ_b as listed in Table 12.1 represent the position and orientation of the first segment (the main hull) described in the space-fixed coordinate frame $x_0 y_0 z_0$. The main hull contains environmental sensing devices and the navigation equipment of the BAUV, and is the primary target of motion control. Homogeneous coordinate transformation matrices at the ith joint in Figures 12.2 and 12.3 are as

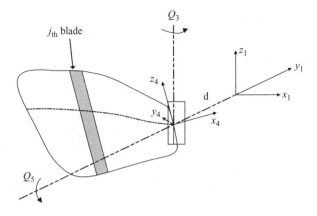

Figure 12.3 Frontal view of the right-side pectoral fin

follows:

$$\mathbf{T}_i^{i-1} = \begin{bmatrix} \mathbf{R}_i^{i-1} & \mathbf{o}_i^{i-1} \\ \mathbf{0} & 1 \end{bmatrix}$$

$$= \begin{bmatrix} \begin{bmatrix} \cos\theta_{i-1} & -\sin\theta_{i-1} & 0 \\ \sin\theta_{i-1} & \cos\theta_{i-1} & 0 \\ 0 & 0 & 1 \end{bmatrix} & \mathbf{o}_i^{i-1} \\ \begin{bmatrix} 0 & 0 & 0 \end{bmatrix} & 1 \end{bmatrix}, \quad i = 1,2,3 \quad (12.1)$$

$$\mathbf{T}_i^1 = \begin{bmatrix} \mathbf{R}_i^1 & \mathbf{o}_i^1 \\ \mathbf{0} & 1 \end{bmatrix}$$

$$= \begin{bmatrix} \begin{bmatrix} \cos\theta_{i-1}\cos\theta_{i+1} & -\sin\theta_{i-1} & \cos\theta_{i-1}\sin\theta_{i+1} \\ \sin\theta_{i-1}\cos\theta_{i+1} & \cos\theta_{i-1} & \sin\theta_{i-1}\sin\theta_{i+1} \\ -\sin\theta_{i+1} & 0 & \cos\theta_{i+1} \end{bmatrix} & \mathbf{o}_i^1 \\ \begin{bmatrix} 0 & 0 & 0 \end{bmatrix} & 1 \end{bmatrix},$$

$$i = 4,5 \quad (12.2)$$

Here $\mathbf{T}_i^{i-1}$ denotes a transformation from the coordinate i to the coordinate $i-1$. $\mathbf{R}_i^{i-1}$, $\mathbf{R}_i^1$ represent the rotational matrices and $\mathbf{o}_i^{i-1}$, $\mathbf{o}_i^1$ are the position vectors of the origins. The subscript is the index that specifies the source coordinate and the superscript specifies the destination, or reference, coordinate. Table 12.2 lists the origins of coordinate systems and the position vector of the centre of mass $\mathbf{p}_i^i$ of the ith joint.

Each segment-fixed coordinate is related to the space-fixed coordinate by,

$$\begin{bmatrix} \mathbf{p}_i^0 \\ 1 \end{bmatrix} = \mathbf{T}_i^0 \begin{bmatrix} \mathbf{p}_i^i \\ 1 \end{bmatrix} \quad (12.3)$$

Table 12.1 Notations used in the BAUV's dynamic model

Notations	Definitions
$\{i\}$	Frame number of the ith joint, $\{0\}$ denotes the space-fixed frame
θ_0	BAUV heading angle with respect to frame $\{0\}$
θ_1	Angle of the caudal peduncle with respect to frame $\{1\}$
θ_2	Angle of the caudal fin with respect to frame $\{2\}$
θ_3	Lead–lag angle of the right pectoral fin with respect to frame $\{1\}$
θ_4	Feathering angle of the right pectoral fin with respect to frame $\{1\}$
θ_5	Lead–lag angle of the left pectoral fin with respect to frame $\{1\}$
θ_6	Feathering angle of the left pectoral fin with respect to frame $\{1\}$
$2d$	Width of the BAUV
l_i	Length of the ith segment
(x_b, y_b, θ_b)	Orientation and heading of the BAUV with respect to $\{0\}$

Table 12.2 Origins of coordinates and positions of centre of mass

Frame	$\mathbf{o}_i^k$, origin of frame i relative to frame k	$\mathbf{p}_i^i$, centre of mass of the ith link relative to the origin of frame i
$\{1\}$	$\mathbf{o}_1^0 = (x_b, y_b, 0)^\mathrm{T}$	$\mathbf{p}_1^1 = (0, 0, 0)^\mathrm{T}$
$\{2\}$	$\mathbf{o}_2^1 = (-0.5l_1, 0, 0)^\mathrm{T}$	$\mathbf{p}_2^2 = (-0.5l_2, 0, 0)^\mathrm{T}$
$\{3\}$	$\mathbf{o}_3^2 = (-l_2, 0, 0)^\mathrm{T}$	$\mathbf{p}_3^3 = (-0.5l_3, 0, 0)^\mathrm{T}$
$\{4\}$	$\mathbf{o}_4^1 = (0, -d, 0)^\mathrm{T}$	$\mathbf{p}_4^4 = (-0.5l_4, 0, 0)^\mathrm{T}$
$\{5\}$	$\mathbf{o}_5^1 = (0, d, 0)^\mathrm{T}$	$\mathbf{p}_5^5 = (-0.5l_5, 0, 0)^\mathrm{T}$

where $\mathbf{T}_i^0 = \mathbf{T}_1^0 \mathbf{T}_2^1 \cdots \mathbf{T}_i^{i-1}$ for $i = 1, 2, 3$, and $\mathbf{T}_i^0 = \mathbf{T}_1^0 \mathbf{T}_i^1$ for $i = 4, 5$. The position of the centre of mass of link i in the space-fixed coordinate, $\mathbf{p}_i^0$, is specified by the first two components in the last column vector of $\mathbf{T}_i^0$. The linear velocity of the centre of mass of link k with respect to the space-fixed coordinate is,

$$\dot{\mathbf{p}}_i^0 = \mathbf{J}_i \dot{\mathbf{q}} \qquad (12.4)$$

where

$$\mathbf{J}_i = \left[\frac{\partial \mathbf{p}_i^0}{\partial q_1} \quad \frac{\partial \mathbf{p}_i^0}{\partial q_2} \quad \cdots \quad \frac{\partial \mathbf{p}_i^0}{\partial q_9} \right],$$

$$\mathbf{q} = \begin{bmatrix} x_b & y_b & \theta_0 & \theta_1 & \theta_2 & \theta_3 & \theta_4 & \theta_5 & \theta_6 \end{bmatrix}^\mathrm{T}$$

represents the vector of independent variables, and $\mathbf{J}_i$ is the Jacobian matrix.

Figure 12.3 illustrates the frontal view of the right-side pectoral fin. The pectoral fin is treated as a number of moving blade elements. Moreover, let $\mathbf{p}_{i,j}^i = [x_{i,j}, 0, z_{i,j}]^T$ denote the centre of mass of the blade element j of the fin i. The centre of mass of the blade element j is located at the middle of the element. In this study, the pectoral fin is divided into ten moving blade elements. The rotational matrices and the positional vectors for each blade element are displayed as follows:

$$\mathbf{R}_i^1 = \begin{bmatrix} \cos\theta_{i-1}\cos\theta_{i+1} & -\sin\theta_{i-1} & \cos\theta_{i-1}\sin\theta_{i+1} \\ \sin\theta_{i-1}\cos\theta_{i+1} & -\cos\theta_{i-1} & \sin\theta_{i-1}\sin\theta_{i+1} \\ -\sin\theta_{i+1} & 0 & \cos\theta_{i+1} \end{bmatrix}, \quad \text{for } i = 4, 5 \tag{12.5}$$

and

$$\mathbf{p}_{i,j}^1 = \begin{bmatrix} \cos\theta_{i-1}(z_{i,j}\sin\theta_{i+1} - x_{i,j}\cos\theta_{i+1}) \\ (-1)^{i+1}d + \sin\theta_{i-1}(z_{i,j}\sin\theta_{i+1} - x_{i,j}\cos\theta_{i+1}) \\ (z_{i,j}\sin\theta_{i+1} + x_{i,j}\cos\theta_{i+1}) \end{bmatrix},$$

for $i = 4, 5$ and $j = 1, \ldots, 10$ \hfill (12.6)

The linear velocity of the centre of the blade element with respect to the fluid is determined using Eqs. (12.5)–(12.7) as follows:

$$\begin{bmatrix} u_{i,j} \\ v_{i,j} \\ w_{i,j} \end{bmatrix} = (\mathbf{R}_i^0)^{-1}(\mathbf{J}_{i,j}\dot{\mathbf{q}}), \quad \text{for } i = 4, 5 \text{ and } j = 1, \ldots, 10 \tag{12.7}$$

where

$$\mathbf{J}_{i,j} = \begin{bmatrix} \dfrac{\partial \mathbf{p}_{i,j}^0}{\partial q_1} & \dfrac{\partial \mathbf{p}_{i,j}^0}{\partial q_2} & \cdots & \dfrac{\partial \mathbf{p}_{i,j}^0}{\partial q_9} \end{bmatrix}$$

and $u_{i,j}$, $v_{i,j}$ and $w_{i,j}$ are the velocity components of the mid-position of blade element parallel to the x_i-, y_i- and z_i-axes, respectively.

Next, the Lagrange equation of motion is obtained based on the concept of generalised coordinates, energy, and generalised forces. Let $\mathbf{T}$ and $\mathbf{U}$ represent the kinetic energy and potential energy of the BAUV, respectively. The Lagrangian is then defined as the difference between the kinetic and potential energy, and $L(\mathbf{q}, \dot{\mathbf{q}}) \equiv T(\mathbf{q}, \dot{\mathbf{q}}) - U(\mathbf{q})$. $U(\mathbf{q}) = 0$ is set because the motion is planar. The general equations of motion of the BAUV can be formulated using the Lagrangian function, as follows:

$$\frac{d}{dt}\frac{\partial}{\partial \dot{q}_i}L(\mathbf{q}, \dot{\mathbf{q}}) - \frac{\partial}{\partial q_i}L(\mathbf{q}, \dot{\mathbf{q}}) = F_i, \quad 1 \leq i \leq 9 \tag{12.8}$$

Here, F_i is the generalised force of the ith generalised coordinate.

The total kinetic energy can be expressed as $T(q,\dot{q}) = \frac{1}{2}\dot{q}^T D(q)\dot{q}$ where $D(q)$ is the inertia matrix.

$$D(q) = \sum_{i=1}^{5}\left[m_i(J_i)^T J_i\right] + \begin{bmatrix} 0 & 0 & 0 & 0 & 0 & 0 & 0 & 0 & 0 \\ 0 & 0 & 0 & 0 & 0 & 0 & 0 & 0 & 0 \\ 0 & 0 & \sum_{i=1}^{5} I_i & \sum_{i=2}^{3} I_i & I_3 & I_4 & I_5 & 0 & 0 \\ 0 & 0 & \sum_{i=2}^{3} I_i & \sum_{i=2}^{3} I_i & I_3 & 0 & 0 & 0 & 0 \\ 0 & 0 & I_3 & I_3 & I_3 & 0 & 0 & 0 & 0 \\ 0 & 0 & I_4 & 0 & 0 & I_4 & 0 & 0 & 0 \\ 0 & 0 & I_5 & 0 & 0 & 0 & I_5 & 0 & 0 \\ 0 & 0 & 0 & 0 & 0 & 0 & 0 & 0 & 0 \\ 0 & 0 & 0 & 0 & 0 & 0 & 0 & 0 & 0 \end{bmatrix}$$

(12.9)

where m_i is the mass of link i and I_i is the inertia of the ith segment.

The equations of motion are

$$\sum_{j=1}^{9} D_{ij}\ddot{q}_j + \sum_{k=1}^{9}\sum_{j=1}^{9} C_{kj}^{i}\dot{q}_k\dot{q}_j = F_i \tag{12.10}$$

where

$$C_{kj}^{i} = \frac{\partial}{\partial q_k} D_{ij} - \frac{1}{2}\frac{\partial}{\partial q_i} D_{kj}, \quad 1 \leq i,j,k \leq 9$$

The BAUV system then can be formulated as a six-input control system using the first three rows of Eq. (12.10).

$$\begin{bmatrix} D_{11} & D_{12} & D_{13} \\ D_{21} & D_{22} & D_{23} \\ D_{31} & D_{32} & D_{33} \end{bmatrix} \begin{bmatrix} \ddot{x}_b \\ \ddot{y}_b \\ \ddot{\theta}_b \end{bmatrix} + \begin{bmatrix} D_{14} & \cdots & D_{19} \\ D_{24} & \cdots & D_{29} \\ D_{34} & \cdots & D_{39} \end{bmatrix} u + \begin{bmatrix} \dot{q}^T C^1(q)\dot{q} \\ \dot{q}^T C^2(q)\dot{q} \\ \dot{q}^T C^3(q)\dot{q} \end{bmatrix} = \begin{bmatrix} F_x \\ F_y \\ \tau \end{bmatrix}$$

(12.11)

where

$$C^i(q) = \begin{bmatrix} C_{11}^i & \cdots & C_{19}^i \\ \vdots & \ddots & \vdots \\ C_{91}^i & \cdots & C_{99}^i \end{bmatrix}$$

The control input $u = \begin{bmatrix} \ddot{\theta}_1 & \cdots & \ddot{\theta}_6 \end{bmatrix}^T$ is the angular acceleration of each joint. The motor controller directly controls the angular velocity of each joint. Moreover, joint torque limit can be transformed into an angular acceleration limit, an angular velocity limit and an angle limit.

12.2.2 Hydrodynamics

The primary driving force for controlling the motion of the BAUV is the interactive force between the hull and fluid. In Reference 14, the authors considered the friction and cross-flow drag in two dimensions and simulated the motion. The forces associated with fluid interaction on the body/caudal fin include friction C_f (in the x_i direction), and cross-flow drag C_d (in the y_i direction). They used the Euler–Lamb equation for the ith segment described in each segment-fixed coordinate frame $x_i y_i z_i$.

$$\begin{bmatrix} F_{x_i} \\ F_{y_i} \\ \tau_{z_i} \end{bmatrix} = \begin{bmatrix} -m_{xi}\dot{u}_i + m_{yi} v_i \dot{\psi}_i - \frac{1}{2}\rho u_i |u_i| C_f S_i \\ -m_{yi} \dot{v}_i - m_{xi} u_i \dot{\psi}_i - \frac{1}{2}\rho v_i |v_i| C_d d_i l_i \\ -I_{ai}\ddot{\psi}_i - (m_{yi} - m_{xi})u_i v_i - \rho \dot{\psi}_i |\dot{\psi}_i| C_d d_i (l_i^4/64) \end{bmatrix}, \quad i = 1,2,3 \tag{12.12}$$

where S_i denotes the wet surface area of the ith segment; d_i represents the height of the cross-section of the ith segment and l_i is the length of the ith segment. Moreover, u_i and v_i are the local velocities parallel to the x_i and y_i axes, respectively, and ψ_i denotes the angle with the x_0 axis; for example, $\psi_3 = \theta_b + \theta_1 + \theta_2$. The terms m_{xi} and m_{yi} represent the added mass of link i with respect to the x_i and y_i axes, and I_{ai} is the added inertia. Furthermore, F_{x_i} and F_{y_i} are the forces that act locally on the ith segment. Additionally, τ_{z_i} is the torque acting on the ith segment. This equation contains only the forces of interaction with the fluid, and does not incorporate rigid body dynamics. Here, the forces acting on the ith segment are described in segment-fixed coordinates. Meanwhile, in Eq. (12.11), the forces are described in space-fixed coordinates, so the local coordinates are transformed into global coordinates by

$$F_i = \begin{bmatrix} \cos\psi_i & -\sin\psi_i \\ \sin\psi_i & -\cos\psi_i \end{bmatrix} \begin{bmatrix} F_{x_i} \\ F_{y_i} \end{bmatrix}, \quad i = 1,2,3 \tag{12.13}$$

$$\tau_i = \tau_{z_i} + \begin{pmatrix} F_{y_i} & -F_{x_i} \end{pmatrix} \begin{pmatrix} \bar{x}_i \\ \bar{y}_i \end{pmatrix}, \quad i = 1,2,3 \tag{12.14}$$

where

$$\begin{pmatrix} \bar{x}_i \\ \bar{y}_i \end{pmatrix} = \mathbf{p}_i^0 - \mathbf{o}_1^0$$

Pectoral fin motions involve the combination of two modes, the lead–lag and the feathering motion. The rotations with respect to the z_1- and y_1-axes are defined as the lead–lag motion and feathering motion, respectively. $i = 4$ denotes the right-side pectoral fin while $i = 5$ represents the left-side pectoral fin in the following derivations.

Figure 12.4 illustrates a sketch of force diagram of the jth blade element of the right-side pectoral fin. The linear velocities $v_{i,j}$ and $w_{i,j}$ are the velocity components defined in Eq. (12.7).

Following the method proposed in Reference 17, a simple model to evaluate hydrodynamic forces acting on a pectoral fin was constructed. The force at the jth blade element of the ith pectoral fin, $\mathbf{L}_{i,j}$, comprises added inertia, the cross-flow

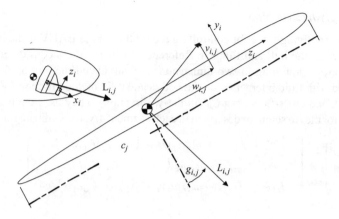

Figure 12.4 Forces at the jth blade element of the pectoral fin

drag acting on the y_i-axis and the lifting force,

$$\mathbf{L}_{i,j} = \begin{bmatrix} 0 \\ -\pi\rho \cdot (c_j/2)^2 \cdot \dot{v}_{i,j} - \frac{1}{2}\rho v_{i,j}|v_{i,j}|c_j \cdot C_d \\ -\frac{1}{2}\rho w_{i,j}^2 c_j \cdot 2\pi(v_{i,j}/w_{i,j})C(\kappa)\cos\gamma_{i,j} \\ \frac{1}{2}\rho w_{i,j}^2 c_j \cdot 2\pi(v_{i,j}/w_{i,j})C(\kappa)\sin\gamma_{i,j} \end{bmatrix},$$

$$i = 4, 5 \text{ and } j = 1, \ldots, 10 \tag{12.15}$$

where $\gamma_{i,j} = \tan^{-1}(v_{i,j}/w_{i,j})$, $\rho, c_j, C(\kappa)$ and C_d denote the water density, the chord length of the jth blade, the Theodorsen function and the cross-flow drag coefficient, respectively. The reduced frequency is defined by $\kappa = c\omega/U$, where c represents the maximum chord length, ω is the oscillating frequency and U denotes the maximum forward speed of the BAUV. This study adopts the values of $C(\kappa) = 0.5$ for relatively high reduced frequency and $C_d = 2.0$ for a plate element.

The force on the jth blade element in Eq. (12.15) is then transformed into the space-fixed coordinate via a rotation matrix.

$$\begin{matrix} \mathbf{F}_{i,j} = \mathbf{R}_i^0 \mathbf{L}_{i,j} \\ \tau_{i,j} = (\mathbf{p}_{i,j}^0 - \mathbf{o}_1^0) \times \mathbf{L}_{i,j} \end{matrix}, \quad \text{for } i = 4, 5 \text{ and } j = 1, \ldots, 10 \tag{12.16}$$

Finally, the total forces that act on $(x_b \ y_b)$ under the body/caudal fin motion and the pectoral fin motion in the horizontal plane are,

$$\begin{bmatrix} F_x \\ F_y \end{bmatrix} = \begin{bmatrix} \sum_{i=1}^{3}(F_i)_x \\ \sum_{i=1}^{3}(F_i)_y \end{bmatrix} + \begin{bmatrix} \sum_{i=4}^{5}\sum_{j=1}^{10}(F_{i,j})_x \\ \sum_{i=4}^{5}\sum_{j=1}^{10}(F_{i,j})_y \end{bmatrix} \tag{12.17}$$

$$\tau = \sum_{i=1}^{3}\tau_i + \sum_{i=4}^{5}\sum_{j=1}^{10}\tau_{i,j} \tag{12.18}$$

where $(\cdot)_x, (\cdot)_y$ denote the x- and y-components of the variable.

Table 12.3 Values of parameters of the BAUV

Parameter	Segment				
	1	2	3	4	5
l_i (m)	0.9	0.4	0.5	0.33	0.33
d (m)	0.1	0	0	0	0
m_i (kg$_f$)	8.786	2.084	0.221	0.1	0.1
I_i (kg$_f$ m^2)	0.5568	0.3939	0.00944	0.001	0.001
m_{xi} (kg$_f$)	0	0	0	0	0
m_{yi} (kg$_f$)	12.8	3	9	2	2
I_{ai} (kg$_f$ m^2)	1	0.0527	0.22	0.1	0.1
S_i (m^2)	0.986	0.328	0.452	0.2	0.2
d_i (m)	0.426	0.31	0.48	0.2	0.2
C_d	1.2	1.2	2	2	2
ρ (kg$_f$/m^2)	100	100	100	100	100

Table 12.4 Values of parameters of the blade elements of the pectoral fin

jth blade of fin i	$x_{i,j}$ (mm)	$z_{i,j}$ (mm)	c_j (mm)
1	33.00	13.02	114.34
2	66.00	26.06	149.08
3	99.00	34.74	166.45
4	132.00	43.42	179.47
5	165.00	49.21	189.61
6	198.00	55.00	196.84
7	231.00	60.79	196.84
8	264.00	69.47	188.16
9	297.00	92.63	144.74
10	330.00	136.06	0.00

Note: c_j is the chord length of the jth blade.

Tables 12.3 and 12.4 specify the parameter values used in the BAUV's model and define the blade elements of the pectoral fins.

12.3 Guidance and control of the BAUV

This section considers a way-point tracking problem for the BAUV. The control action is divided into two levels. The local controller generates the oscillating motion,

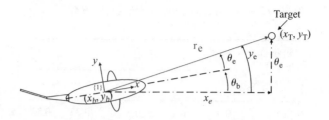

Figure 12.5 The local versus global frames for way-point tracking

and the global controller provides the global tracking precision. Section 12.3.1 presents the guidance problem and joint motion limitations for the controller design. Section 12.3.2 proposes control architectures for the local and global controllers. Finally, Section 12.3.3 presents the results of experiments using the global control functions to coordinate the body/caudal fin and a pair of pectoral fins for the BAUV to swim forward, turn, brake and hover.

12.3.1 Guidance of the BAUV

Figure 12.5 illustrates necessary parameters for formulating a BAUV guidance problem. A posture vector that describes the position and heading direction of the target and the BAUV is defined as (x_T, y_T, θ_T) and (x_b, y_b, θ_b), respectively. A posture error is then defined by (ρ_e, θ_e), where $\rho_e = \sqrt{(x_e)^2 + (y_e)^2}$ and $\theta_e = \theta_T - \theta_b$. The intervals of the posture error are $\rho_e \in [0, \infty)$ and $\theta_e \in [-\pi, \pi)$. Tracking control aims to track the target point by ensuring that (ρ_e, θ_e) converges asymptotically to a null vector.

Because the inputs, angular accelerations, are constrained by the torque output of the motors, the following inequalities define the limitations of angles and the angular velocities.

$$\begin{aligned}
-\frac{\pi}{6} \leq \theta_1 \leq \frac{\pi}{6} \text{ rad}, &\quad -4\pi \leq \dot{\theta}_1 \leq 4\pi \text{ rad/s} \\
-\frac{\pi}{3} \leq \theta_2 \leq \frac{\pi}{3} \text{ rad}, &\quad -4\pi \leq \dot{\theta}_2 \leq 4\pi \text{ rad/s} \\
0 < \theta_3 \leq \frac{\pi}{2} \text{ rad}, &\quad -4\pi \leq \dot{\theta}_3 \leq 4\pi \text{ rad/s} \\
-\frac{\pi}{2} \leq \theta_4 \leq 0 \text{ rad}, &\quad -4\pi \leq \dot{\theta}_4 \leq 4\pi \text{ rad/s} \\
0 \leq \theta_5 \leq \frac{\pi}{2} \text{ rad}, &\quad -4\pi \leq \dot{\theta}_5 \leq 4\pi \text{ rad/s} \\
0 \leq \theta_6 \leq \frac{\pi}{2} \text{ rad}, &\quad -4\pi \leq \dot{\theta}_6 \leq 4\pi \text{ rad/s}
\end{aligned} \quad (12.19)$$

Here, the first two motors, which are the driving motors of body/caudal fin, are assumed to be able to generate sufficient torque to produce any desired angular accelerations sufficiently fast; this assumption is valid for a high reduction gear. Meanwhile, the last four motors are the driving motors for the pectoral fins, and have a limited range of angular acceleration, owing to reduction gear selection, yielding large angular velocities.

12.3.2 Controller design

For BAUVs, no direct control input exists controlling the motion in the x_0-direction. Meanwhile, for the BAUV system, the hydrodynamics are difficult to model accurately, and the joint angle limitation should also be satisfied. This will create a complex control problem.

To solve the difficulties associated with the actuator assignment problem, the proposed control system mimics the locomotion of fish. Basically, fish swim by body/caudal fin oscillation and manoeuvre using their median/pectoral fins. Using this concept, this work divides the control into two levels. The local control method controls the body/caudal fin and pectoral fins to generate the desired motions in the local frames. The global control level coordinates the local controllers and ensures tracking in the global frame.

12.3.2.1 Local controller

This study develops a control method based on BCF locomotion by transferring the oscillations of the body/caudal fin to the oscillation of the body-spline [15]. A body-spline equation was used to represent the centre line of the body by continuously coordinating the joint angles to match the desired shape of the body-spline in the local coordinate frame. Figure 12.6 gives a conceptual view of the shape of the body-spline. In Figure 12.6, $y(x, t)$ denotes the body-spline equation, c_1 and c_2 represent coefficients of the envelope, λ is the body wave length and f_c denotes the oscillation frequency. Moreover, A_m denotes the limits of the envelope at the fish tail. If the fish is in the process of turning, the parameter 'offset' has a non-zero value.

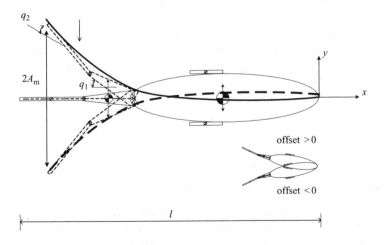

Figure 12.6 Body-spline of a BAUV

A body-spline equation to represent the bodily motion of the BAUV is defined as

$$y(x,t) = G_c(c_1 x + c_2 x^2)(A \sin(kx + 2\pi f_c t) + \text{offset}) \tag{12.20}$$

According to Reference 15, a body-spline for the BAUV is obtained with the coefficients $c_1 = -0.075$, $c_2 = 0.017$ and $\lambda = 3.6$. The operating ranges of the parameter values are determined based on the angular limitations of joint motors. Parameter A represents the oscillation amplitude. The parameter 'offset' determines the offset of the centroid of the fish body during swimming by the BAUV. G_c denotes the parameter of the global controller and is discussed later in Section 12.3.2.2. The maximal swing of the body is set to $A = 1$, $|\text{offset}| = 2$ and $G_c = 1$. θ_1 and θ_2 can be determined by

$$\begin{aligned}\theta_1 &= \tan\left(\frac{\partial}{\partial x} y(x,t)\right)^{-1}\bigg|_{\theta_1} \\ \theta_2 &= \tan\left(\frac{\partial}{\partial x} y(x,t)\right)^{-1}\bigg|_{\theta_2}\end{aligned} \tag{12.21}$$

Velocities and accelerations of the joint angles for body/caudal oscillation can then be obtained based on the first and second derivatives of θ_1, and θ_2, respectively. Inputs and outputs of the body/caudal fin controller are (A, f_c, offset) and $\mathbf{u}_c = (\ddot{\theta}_1, \ddot{\theta}_2, 0, 0, 0, 0)^T$, respectively.

The inputs–outputs of the BAUV swimming behaviour in relation to the bodily motion were examined via simulations using the dynamic model and experimentation using the free running BAUV testbed. Swimming behaviour curves demonstrate the relationship between A, f_c, offset and the surge speed and yawing rate of the BAUV, as illustrated in Figure 12.7. G_c equals 1 for all of the plots in this figure.

Figure 12.7(a) and (b) indicates that parameters A and f_c are both directly proportional to the surge speed. Moreover, the parameter 'offset' is directly proportional to the yawing rate. Thus, surge speed could be modulated by tuning A or f_c and the yawing rate could be modulated by turning 'offset'.

This section designs a controller based on MPF locomotion for turning, braking and hovering by controlling the feathering and lead–lag motions described in Section 12.2.1. This work proposes a simple control method for pectoral fins, as follows:

$$\begin{aligned}\theta_3(t) &= -G_p \cdot \beta[1 - \cos(2\pi f_p t)] \\ \theta_4(t) &= -G_p \cdot \beta[1 - \cos(2\pi f_p t)] \\ \theta_5(t) &= -G_p \cdot \alpha[1 - \cos(2\pi f_p t + \Delta\varphi + \Delta\theta)] \\ \theta_6(t) &= -G_p \cdot \alpha[1 - \cos(2\pi f_p t + \Delta\varphi - \Delta\theta)]\end{aligned} \tag{12.22}$$

where $\alpha, \beta, \Delta\varphi, \Delta\theta, f_p$ and G_p denote the feathering angle, lead–lag angle, braking factor, turning factor, oscillation frequencies of the pectoral fins and enabling function, respectively. G_p approaches 1 when using a full envelope of the pectoral fins. G_p was reduced to zero when no motion was required for the pectoral fins.

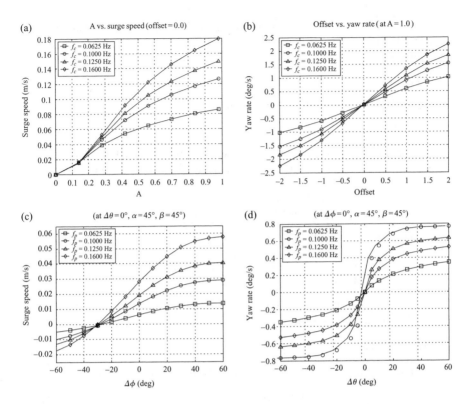

Figure 12.7 Characteristic curves of the body/caudal fin and pectoral fin plotted against surge speed and yawing rate (a) u_1 versus A for different f_cs (b) q_b versus offset for different f_cs (c) u_1 versus $\Delta\varphi$ for different f_ps (d) θ_b versus $\Delta\theta$ for different f_ps

Figure 12.7(c) shows that the BAUV using pectoral fins can move forwards or backwards by tuning a braking factor $\Delta\varphi$. BAUV can hover if $\Delta\varphi$ is set to $-30°$, and can be moved forward if $\Delta\varphi$ exceeds $-30°$. Figure 12.7(d) shows that the BAUV turns by tuning the factor $\Delta\theta$. The pectoral fins can control the BAUV in moving forwards or backwards, turning, braking or hovering by tuning the parameters $\Delta\varphi$, $\Delta\theta$ and f_p from the characteristic curves displayed in Figure 12.7(c) and (d).

12.3.2.2 Global controller

A BAUV moving in the global frame controls its own motion as follows: turning via a caudal fin or a pair of pectoral fins, braking via a pair of pectoral fins. The global controller is thus a switching mechanism that chooses local controllers and enables them to cooperate to achieve the control objectives in the global frame. Enabling functions presented as overlapped saturation functions are proposed in this section.

The input of the global controller is the posture error (ρ_e, θ_e), which was defined in Section 12.3.1. Meanwhile, the outputs are [offset G_c]^T for the body/caudal fin and $[\Delta\varphi \quad \Delta\theta \quad G_p]^T$ for the pectoral fins. Functions used in the global controller include: (1) turning and braking functions and (2) enabling functions. The turning and braking functions are a piecewise continuous function $f(x)$ that satisfies

(1) $\|x\| \cdot f(x) = x \cdot \|f(x)\|$
(2) $\|f(x)\| \leq 1$, if $\|x\| < k$, k is real (12.23)
(3) $\|f(x)\| = 1$, otherwise

For the enabling function that controls the switching between the body/caudal fin and the pectoral fins, a function $S(x; a, b)$ that satisfies

$$S(x;a,b) = \begin{cases} 0 & x < a \\ 2\left(\dfrac{x-a}{b-a}\right)^2 & a \leq x < \dfrac{a+b}{2} \\ 1 - 2\left(\dfrac{x-b}{b-a}\right)^2 & \dfrac{a+b}{2} \leq x < b \\ 1 & x \geq b \end{cases} \quad (12.24)$$

serves as a basic construction function. For example, a smooth trapezoidal function $\Pi(x; a, b, c, d)$ can be constructed by

$$\Pi(x; a, b, c, d) = S(x; a, b) - S(x; c, d) \quad (12.25)$$

and an anti-trapezoidal function $U(x; a, b, c, d)$ has the following form:

$$U(x; a, b, c, d) = 1 - \Pi(x; a, b, c, d) \quad (12.26)$$

Further illustrations of the enabling function can be found in Section 12.3.3. Figure 12.8 shows overall feedback control diagram for the BAUV system.

12.3.3 Experiments

Two experimental examples are illustrated here. The turning experiment involves three way-points around which the BAUV swims. Figure 12.9 presents the results from tracking three way-points arranged such that the BAUV must make a strange turn. These points are (4, 0) m, (4, 2) m and (2, 2) m. When the distance between the BAUV and the waypoint reduces to less than 0.5 m, the waypoint is considered to have been reached. Figure 12.9(b) and (c) plot the heading and offset parameters during the turning experiments. Figure 12.9(d) plots the velocity of the BAUV in the vehicle-fixed coordinate. During the turn, the large offset of the body-spline prevents the full amplitude of the tail oscillation from being achieved. Hence, the magnitude of the velocity declined as the BAUV turned from 0 degree to 180°.

Figure 12.10(a) illustrates the results of a braking experiment in which the BAUV swam to a target at (3.5,0) and hovered at the target position. Moreover, Figure 12.10(b) displays that the surge velocity was varied near the target, while the sway velocity was maintained within a small range. The heading of the BAUV tended to stay at a constant angle near the target as shown in Figure 12.10(c). Figure 12.10(d)

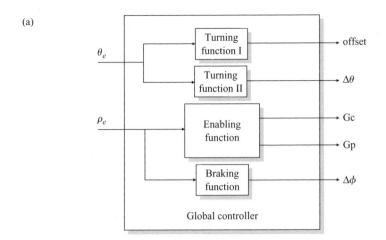

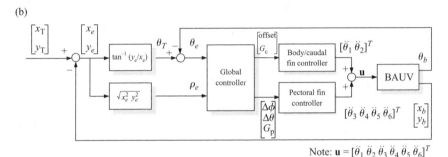

Figure 12.8 *The feedback control diagram for the BAUV system (a) global control structure (b) control system of the BAUV*

illustrates that coordinating the body/caudal fin and pectoral fins could minimise the positioning error. The heading of the BAUV stabilised upon reaching the target. The enabling functions $G_c(\rho_e)$ and $G_p(\rho_e)$ were switched at approximately 25 and 55 s, respectively, as shown in Figure 12.11(a). The positioning errors result from the noises of the position signals of the BAUV derived from the Doppler sensor and the time delay of the motion control system of the joint motors.

Figure 12.11(a) illustrates the enabling functions versus time. Moreover, Figure 12.11(b) shows the brake command $\Delta\varphi$ versus time. The increase of $\Delta\varphi$ with time demonstrated the braking action of the BAUV motion. Figure 11(c) and (d) plot the turning functions 'offset' and $\Delta\theta$ versus time. Positive values of 'offset' and $\Delta\theta$ display counterclockwise turning motion.

In the above experiments, the parameters for the body/caudal fin and pectoral fin control were set to $c_1 = -0.075, c_2 = 0.017$ and $\lambda = 3.6, f_c = f_p = 0.16$ Hz, $A = 1$, offset$_{max} = 2, \alpha = \beta = 45°, \Delta\theta_{max} = 50°$ and $\Delta\varphi_{max} = 15°$. The global controller was designed using the functions described by Eqs. (12.23)–(12.26). The

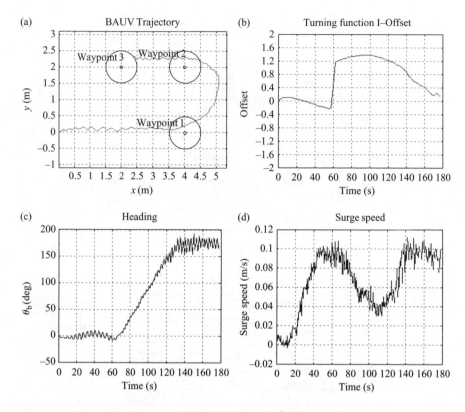

Figure 12.9 Data of a turning experiment for a BAUV tracking three way-points: (a) trajectory (b) offset of the body-spline (c) heading (d) surge speed

components of the global controller are listed as follows:

$$\text{offset} = \text{offset}_{\max} \cdot \tanh(2\theta_e), \quad \Delta\theta = \Delta\theta_{\max} \cdot \tanh(\theta_e)$$

and

$$\Delta\varphi = \begin{cases} \Delta\varphi_{\max} \cdot \sqrt{\rho_e} + \Delta\varphi_{\text{shift}}; & \rho_e \leq 1 \\ \Delta\varphi_{\max} + \Delta\varphi_{\text{shift}}; & \rho_e > 1 \end{cases}$$

The enabling functions were defined as follows:

$$G_c(\rho_e) = \Pi(\rho_e; 0.2\rho_{e,\max}, 0.3\rho_{e,\max}, 0.7\rho_{e,\max}, 0.8\rho_{e,\max})$$
$$G_p(\rho_e) = U(\rho_e; 0.2\rho_{e,\max}, 0.3\rho_{e,\max}, 0.7\rho_{e,\max}, 0.8\rho_{e,\max})$$

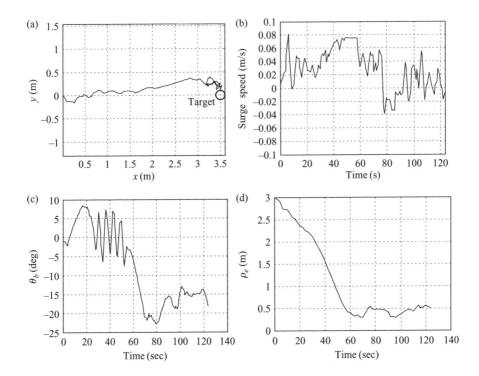

Figure 12.10 Data of an experiment for a BAUV performing braking and hovering at a target position: (a) swimming trajectory (b) surge speed (c) heading (d) position error

12.4 Conclusions

This chapter summarises the development of a method for controlling BAUV motion. The contributions of this study are as follows:

1. A mathematical model of the BAUV equipped with body/caudal fin and two-degree-of-freedom pectoral fins was established. This model was derived using the Lagrange method. This model deals with the horizontal-plane motion because the potential energy is ignored in dynamic model formation. The hydrodynamic modelling is based on drag forces.
2. Parameters related to the swimming performance of the body/caudal fin and pectoral fins were defined and their applications in a guidance and control system were clarified. The forward speed of the BAUV can be controlled by alternating the amplitude and frequency of body-spline equation. Moreover, an offset parameter of the body-spline can be controlled for directly determining the yawing rate. For a pair of pectoral fins, a braking factor that creates phase difference between the lead–lag and feathering angles is used to control the forward speed.

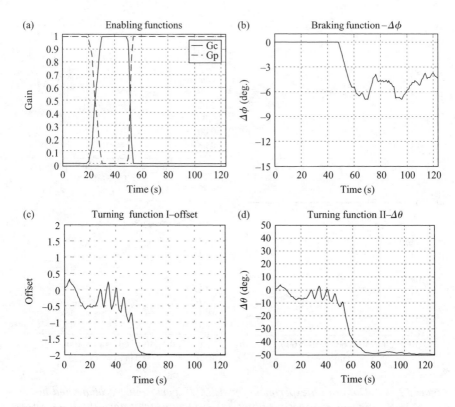

Figure 12.11 Outputs of the global controller for a BAUV performing braking and hovering at a target position: (a) enabling functions versus time in braking (b) $\Delta\varphi$ versus time in braking (c) offset versus time in braking (d) $\Delta\theta$ versus time in braking

Furthermore, a turning factor that defines the relative motion of the two pectoral fins is specified as the key parameter in turning control.
3. A coordinated control method is proposed that combines the body/caudal fin and pectoral fins. The main advantages of the method are that it can support a braking action, a small turning radius and also hovering around a position. Erratic BAUV turns and braking motions were demonstrated using the proposed controller.

Acknowledgements

The author would like to thank the National Science Council of the Republic of China for financially supporting this research under Contract No. NSC91-2611-E002-008. Thanks are also extended to Professors Sheng-Wen Cheng, Forng-Chen Chiu, Jing-Fa Tsai and Jeng-Shiang Kuo for their valuable comments, and Mr Chuan-Hsing Wu for assisting in the experiments.

References

1. Breder, C.M. (1926). The locomotion of fishes. *Zoologica*, 4, pp. 159–256.
2. Sfakiotakis, M., D.M. Lane, and J.B.C. Davies (1999). Review of fish swimming modes for aquatic locomotion. *IEEE Journal of Oceanic Engineering*, 24(2), pp. 237–252.
3. Kelly, S.D. and R.M. Murray (2000). Modeling efficient pisciform swimming for control. *International Journal of Robust Nonlinear Control*, 10, pp. 217–241.
4. Mason, R. and J.W. Burdick (2000). Construction and modeling of a carangiform robotic fish. In *Lectures Notes in Control and Information Sciences*, Springer Verlag, London, UK, pp. 235–242.
5. Westneat, M. W., D.H. Thorsen, J.A. Walker, and M.E. Hale (2004). Structure, function, and neural control of pectoral fins in fishes. *IEEE Journal of Oceanic Engineering*, 29(3), pp. 674–683.
6. Walker, J.A. and M.W. Westneat (2002). Kinematics, dynamics and energetics of rowing and flapping propulsion in fishes. *Integrative Comparative Biology*, 42, pp. 1032–1043.
7. Kato, N. (2000). Control performance in the horizontal plane of a fish robot with mechanical pectoral fins. *IEEE Journal of Oceanic Engineering*, 25(1), pp. 121–129.
8. Kato, N., H. Liu, and H. Morikawa (2004). Biology-inspired precision maneuvering of underwater vehicles. In *Bio-Mechanisms of Swimming and Flying*, N. Kato, J. Ayers, H. Morikawa, Eds., Springer-Verlag, Tokyo, pp. 111–125.
9. Hobson, B., M. Murray, and C. Pell (1999). Pilotfish: maximizing agility in an unmanned-underwater vehicle. Proceedings of the 11th International Symposium on UUST, pp. 41–51.
10. Licht, L., V. Polidoro, M. Flores, F. Hover, and M.S. Triantafyllou (2004). Design and projected performance of a flapping foil AUV. *IEEE Journal of Oceanic Engineering*, 29(3), pp. 786–794.
11. Colgate, J.E. and K.M. Lynch (2004). Mechanics and control of swimming: a review. *IEEE Journal of Oceanic Engineering*, 29(3), pp. 660–673.
12. Archer, S.D. and I.A. Johnston (1989). Kinematics of labriform and carangiform swimming in the Antarctic fish *Notothenia neglecta*. *Journal of Experimental Biology*, 143, pp. 195–210.
13. Chiu, F.C., C.P. Wu, and J. Guo (2000). Simulation on the undulatory locomotion of a fexible slender body. In Proceedings of the 1st International Symposium on Aqua Bio-Mechanisms, Hawaii, pp. 185–190.
14. Guo, J. and Y.-J. Joeng (2004). Guidance and control of a biomimetic autonomous underwater vehicle using body-fin propulsion. *Proceedings of the Institution of Mechanical Engineering Part M: Journal of Engineering for the Maritime Environment*, 218, pp. 93–111.
15. Guo, J., F.C. Chiu, C.C. Chen, and Y.S. Ho (2003). Determining the bodily motion of a biomimetic underwater vehicle under oscillating propulsion. In IEEE International Conference on Robotics and Automation, Taipei, pp. 983–988.

16 Guo, J., F.C. Chiu, S.W. Cheng, and Y.S. Ho (2003). Control systems for waypoint-tracking of a biomimetic autonomous underwater vehicle. In Proceedings of the MTS/IEEE OCEANS 2003, San Diego, CA, pp. 333–339.
17 Chiu, F.C., C.K. Chen, and J. Guo (2004). A practical method for simulating pectoral fin locomotion of a biomimetic autonomous underwater vehicle. In Proceedings of the International Symposium on Underwater Technology, Taipei, pp. 323–329.

Chapter 13
Seabed-relative navigation by hybrid structured lighting

F. Dalgleish, S. Tetlow and R.L. Allwood

13.1 Introduction

In recent years, advances in the automatic creation of large, spatially accurate composite images of underwater scenes have revolutionised marine science, as it is now possible to view large contiguous areas of the seabed as a single image. In order to make these imaging survey tasks possible from an autonomous underwater vehicle (AUV), it is necessary to have a precise seabed-relative navigation and control capability, allowing both fine manoeuvring in the area of interest and the provision of necessary data to robustly create composite images. In previous work, this has been successfully demonstrated using conventionally illuminated optical imaging systems deployed from an ROV, with several systems capable of using the image registration results to navigate in real time, and other systems attempting to reduce the inevitable integration error build up by employing revisiting strategies [1–3]. In this chapter, work is described that has been carried out within the Offshore Technology Centre in which a laser-assisted vision sensor has been deployed on the Centre's AUV '*Hammerhead*' to provide navigational data and enhanced imaging.

The performance of underwater vision systems is affected by the scattering and absorption properties of the water through which they are operating. Other than in shallow water during daytime when ambient lighting may be sufficient, light has to be provided artificially. The conventional solution, which usually consists of a lamp with a wide angle of illumination as a source and a CCTV camera, is limited by:

(i) the absorption of the light as it propagates through varying path lengths of water to the target (and back again), resulting in images of uneven intensity;
(ii) the near-field backscatter of the light, particularly when the lamp and camera are close together.

In the case where the intensity of scattered light is such that it veils the image, the system is said to be 'contrast limited'. An increase in the separation between the camera and the lamp can reduce the near-field backscattered light, but then the system may become limited by absorption where the returning signal is too weak to be detected. In this case, the system is said to be 'power limited'. Increasing the power of the lamp does not significantly improve the situation due to the exponential decay of light intensity with distance. Furthermore, the spreading effect of the light beam and the finite aperture of a camera lens further reduce the useful received light intensity.

The problems referred to above have driven research efforts towards the development of a number of novel underwater viewing techniques, most of which employ green light emitting lasers to coincide with the wavelength of minimum attenuation in seawater. These systems fall into three main categories:

(i) range gated viewing systems
(ii) synchronous scanning systems
(iii) non-synchronous scanning systems.

The range gated viewing system exploits the 'time of flight' of an illuminating light pulse in seawater and requires precise gating of both the light source and the receiver [4,5].

Synchronous scanning or laser line scan (LLS) methods exploit the highly collimated nature of a laser beam by using a narrow field of view (FOV) receiver to track the spot at high speed, thus reducing backscattered and forward scattered light [6,7]. In addition, more sophisticated pulsed laser variations exist [8].

A less expensive and sophisticated alternative is to optically or mechanically form a fan of light as a source and use a standard receiver such as a low light charged coupled device (CCD) camera with a wide angle lens. Image processing techniques can be used to partially remove the forward and backscattered light. These methods are known as non-synchronous scanning systems or laser stripe illumination (LSI) systems, of which at least two systems have been developed for underwater use [9,10].

The basis of the system that was the subject of the research described in this chapter is that of non-synchronous scanning. Although much research has been performed at Cranfield University into image formation, processing and composite imaging algorithms, it has mostly been concerned with two-axis scanning methods. In order to deploy this type of sensor, with single axis scanning, from an AUV platform and to provide realistic images of the seabed, where the cross-axis scan is provided by the scanner and the through axis scan is provided by the forward motion of the vehicle, it is necessary to be able to accurately sense horizontal motion. Therefore, the work in this chapter describes the design and development of a hybrid system that uses extended range laser imaging to obtain simultaneous seabed bathymetric and reflectivity datasets, and a conventional imaging technique to determine horizontal motion by image registration. Results from a series of full-scale dynamically constrained motion trials are presented to determine the utility of such a system onboard an AUV for both navigational purposes and the creation of motion-compensated composite images. The stand-alone navigational results from this sensor, known as

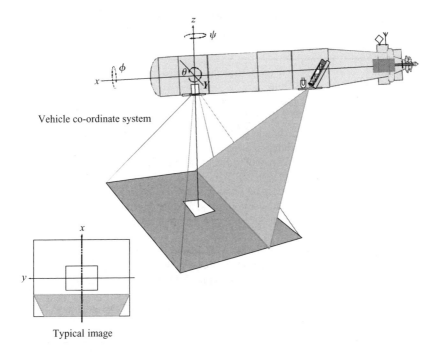

Figure 13.1   Physical arrangement and principles of the laser-assisted vision sensor

laser-assisted vision, are promising for a slow moving platform in low to medium turbid water conditions.

13.2   Description of sensor configuration

The physical arrangement and principles of the AUV-deployed laser-assisted vision sensor deployed on the *Hammerhead* AUV is shown in Figure 13.1.

It can be seen in Figure 13.1 that the bottom third of the FOV is used for the laser stripe bathymetry (and altimeter) and reflectivity profiling, whilst the centre of the FOV is used for the application of the image registration process (region-based tracker) to determine horizontal motion. These subsystems are described in Section 13.3. A summary of the components of the complete system is given in Table 13.1.

13.3   Theory

Apart from the seabed reflectivity information contained in each laser stripe image, bathymetric information can also be recovered at each pixel. This is because there is a fixed geometry between the laser projector and the camera and where the stripe falls in

Table 13.1 Summary of Hammerhead AUV data and image acquisition system with integrated laser-assisted vision

Laser	100 mW (cw) frequency doubled diode-pumped Nd:YAG (532 nm)
	2.2 mrad beam divergence
Light	50 W dichroic halogen (12 V)
Camera	Monochrome $\frac{1}{3}''$ low light CCD, 768 × 576 pixels maximum resolution
	Sensitivity 0.0001 lux (AGC on), automatic control of AGC (on/off)
Scanner	80 Hz resonant scanner, sinusoidal driving signal
Lens	Automatic control of iris, focus and zoom
Compass	TCM2 tilt-compensated electronic compass (20 Hz update rate)
Frame grabber	National Instruments PCI-1409, Image depth is 8 or 10 bit
Data acquisition	National Instruments PCI-6024E
Computer	MicroATX motherboard, AMD 2800+ processor, 1024 MB RAM, 120 GB HD
Software	Windows™ 2000 Pro, LabVIEW 6.1 Full Development System

the image will indicate the distance to the target (seabed). Therefore, for each image, the altitude of the vehicle above the seabed can be determined. Once the distance to the seabed is known, each pixel in the image can be attributed with real world dimensions in the x- and y-directions. A region-based tracker is then used to estimate the translation of a group of pixels between two consecutive images. Knowing the spatial dimensions in both images, as described above, allows the spatial translation between images to be derived. These image shifts are typically several centimetres at 2 knots cruising speed and 4 m altitude.

Table 13.2 summarises the sensor measurements involved in the computation of three-dimensional (3D) seabed-relative motion by the system. It can be seen that one degree of freedom is derived entirely from images (heave), two are provided entirely from additional inclinometers (pitch and roll) and three are derived by fusing visual information with independently measured rotational information. In addition, the use of a pressure sensor is required to measure depth, which is integrated with the altitude estimate from the laser triangulation process to provide bathymetry with reference to the sea surface.

In addition to these dynamic values, there are several static quantities and measurements which also contribute to deriving the solutions to estimate the trajectory of the vehicle. The static quantities are:

 (i) distance between the centre of rotation of the vehicle and the camera;
 (ii) horizontal and vertical angle of view (AOV) of the camera lens in water;
(iii) horizontal and vertical image size.

Table 13.2 Summary of computer vision output parameters and sensor measurements

Sensor	Value	Description	Units
Laser altimeter	Altitude (Z)	Distance from camera to seabed (also bathymetry and reflectivity along the laser stripe for imaging purposes)	Metres
Region-based tracker	dx, dy, {yaw(ψ)}	Vertical and horizontal local image displacements[a]	Pixels
Inclinometers	Pitch (θ) and roll (Φ)	Rotation measured inside the vehicle (distance between sensor centre and optical axis is known)	Radians
Compass	Yaw (ψ)	Tilt-compensated electronic compass	Radians
Pressure	Depth (W)	Ambient water pressure (converted to metres seawater or freshwater)	Metres

[a] Although yaw (ψ) can be solved when two or more tracking regions are applied, this technique will not be considered here.

In order to determine the motion of the camera (or vehicle) over the seabed and produce composite images, it was necessary to create several mathematical models to allow a software implementation. These do not include a dynamic system model for the platform vehicle but are based entirely on measurements from monocular computer vision modules and a tilt-compensated electronic compass (pitch, roll and yaw).

In the following sections, the main software modules for the laser bathymetry and the region-based tracker are described in more detail.

13.3.1 Laser stripe for bathymetric and reflectivity seabed profiling

The laser bathymetry sensor employs the technique of triangulation to measure the 3D coordinates of points on the seabed with respect to a global reference frame. The system uses a laser stripe generator as a source, and a 2D light sensing receiver (CCD camera), the optical axis of which is located at a known disparity angle (α) to the plane of the source. This is shown in Figure 13.2, for the through-axis (longitudinal) cross section, with an illustration of the cross-axis (lateral) profiling principles of the sensor. It should be noted that the laser stripe is generated along the cross-track axis only, with the through-track axis scan being provided by the forward motion of the AUV. The relationship between the range, laser stripe position at a particular point and α is only derived for the case when the laser scan angle is zero. Knowing the laser scan angle corresponding to each point (pixel) along the line,

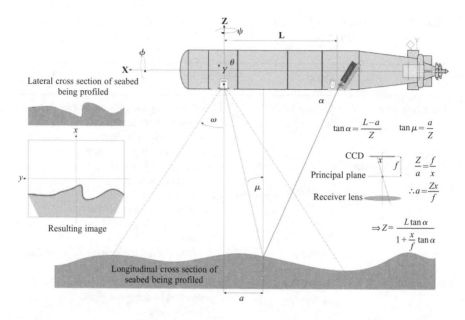

Figure 13.2   Geometry and reference frames

it would be possible to analytically determine the range at each point. However, the physics of projecting a plane of laser light through a plane air–Perspex–water interface produces an undesirable phenomenon which has the effect of varying the sensor baseline (L in Figure 13.2). This has been modelled as part of the calibration process in work relating to similar systems [11,12], but for the purpose of this work a 2D multipoint interpolation method is used to account for this error and other nonlinearities [13].

For each of the range (or bathymetry) points (or pixels) determined from the returning laser signal, there exists a unique reflectivity value, where reflectivity is defined as the ratio of the radiant energy reflected to the total that is incident upon that surface. Albeit a simplification, the reflectivity dataset is generated from the intensity value of the returning signal from the seabed. A light or shiny surface will have a high reflectivity value approaching unity.

However, the general algebraic solution for the laser triangulation process in the case where the scan angle is zero, is shown in Figure 13.2. It is possible to remove the x/f ratio by substituting in the measured pixel position of the laser return (with respect to the centre of the image, x_p) as a proportion of the total amount of pixels in the horizontal axis of the CCD (X_{res}) and the FOV at this range. The resulting expression is

$$Z = \frac{L \tan(\alpha)}{1 + \left[\dfrac{2x_p \tan(\omega) \tan(\alpha)}{X_{res}}\right]} \qquad (13.1)$$

13.3.2 Region-based tracker

In the quest to simplify the solution of determining motion from a region in the centre of monocular image sequences, it is believed that an orthographic projection is acceptable, albeit with the drawback of sacrificing some accuracy.

The application of a single tracking region to determine one correspondence pair allows the solution of the two equations necessary to determine X and Y displacements relative to the seabed. However, more correspondence pairs could be used to derive yaw as well and provide a more robust least-squares estimate of the parameters.

This approach, which can be described as correlation-based window tracking, was implemented using the LabVIEW IMAQ™ image processing and machine vision toolbox, using a pattern-matching technique designed for rapid industrial inspection applications. The algorithm performs an intelligent correlation between an initial stored image region (template) and subsequent image regions where image displacement is measured directly. The displacement is taken from the centre point of the image at each step whereby the difference between the new position and the previous position is taken as the inter-frame displacement. The correlation uses a non-uniform sampling technique where only a few points that represent the overall content of the image are extracted. As well as the position of the match, a confidence value describing the quality of the match is also given [14].

Moreover, an adaptive search strategy based on previous displacements is incorporated to further improve tracking efficiency. A new template is selected every n images to reduce the degradation of the correlation process over time. The complete process, including the integration of the laser altimeter data to transform into real world coordinates, is represented as a flowchart in Figure 13.3, where n is the number of iterations between the extraction of a new template image and m is an index representing the iteration number.

13.4 Constrained motion testing

In order to investigate the effectiveness of the proposed hybrid optical navigation methodology, a set of constrained motion experiments were carried out at the Deep Wave Basin at IFREMER at Brest, France. In these experiments, the *Hammerhead* AUV was mounted beneath a large carriage that was able to move at varying speeds along the 50 m tank of which the depth varied from 10 to 20 m. In addition to the image and vehicle specific data, the measurements necessary to build a ground truth model were also acquired and used in ascertaining the accuracy of the sensor.

13.4.1 Laser altimeter mode

In this mode of operation, the position of the laser within the FOV of the camera was only considered at the scan angle of 0° (i.e., in the centre of the scan, coincident with the optical axis of the camera). Figure 13.4 shows the theoretical curve based on the geometry of the viewing system, with the six points overlaid based on actual measurements from which images were taken in clear filtered seawater. The extraction of the stripe was performed to the nearest pixel.

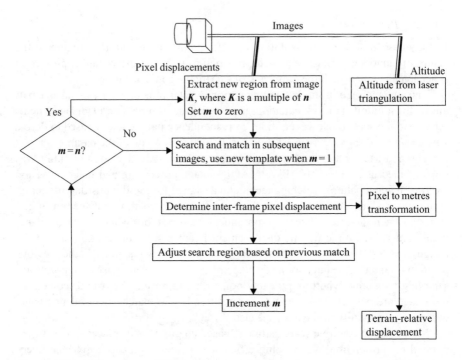

Figure 13.3   Flowchart of algorithm for vision-based navigation

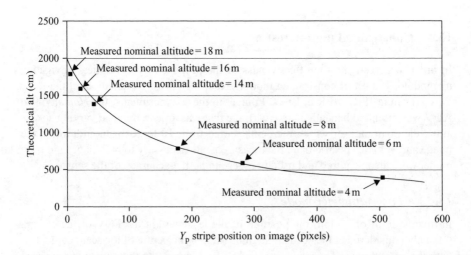

Figure 13.4   Theoretical attitude from stripe position curve with image-derived points overlaid

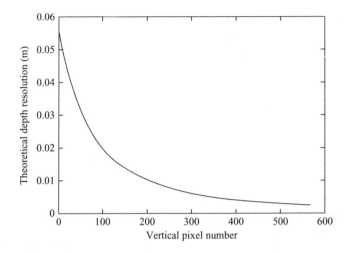

Figure 13.5   Nearest pixel theoretical depth resolution across vertical FOV

It can be seen how the accuracy falls off between 14 and 18 m altitude. This is due to both the nonlinearity of the lens as the stripe approaches the extreme of the FOV and a contribution from measurement errors.

From an analysis of Eq. (13.1), and using nearest pixel stripe extraction, the theoretical depth resolution across the entire vertical FOV can be shown (Figure 13.5). Due to the particular geometry used, the resolution is poor in the far FOV. For example, at 16 m altitude the theoretical depth resolution was over 0.04 m. In addition, due to the laser beam diverging and occupying several pixels at this range, the real resolution is probably worse than it appears. Indeed, from Figure 13.4, it can be seen that the difference between measured altitude and theoretical altitude (at $Y_p = 25$; 16 m nominal altitude) is approximately 0.6 m (4 per cent of total altitude).

Therefore, although errors exist as a result of measurement inaccuracies of the parameters in Eq. (13.1), the dominant error is due to the optical distortion at the edge of the FOV at this distance.

13.4.2 Dynamic performance of the laser altimeter process

The plot in Figure 13.6 shows the derived altitude (using Eq. (13.1) at zero scan angle) whilst traversing the tank bottom at a speed of $0.5 \, \text{m s}^{-1}$ (i.e., approximately 1 knot) and at an altitude of almost 4 m. The total attenuation coefficient of the seawater (α) was estimated at $0.2 \, \text{m}^{-1}$. The range to the tank bottom that was derived from both the motion mechanism measurements and the model for the tank bottom is also shown on the plot.

At this altitude and with the particular system geometry, the theoretical (nearest pixel) ranging resolution (from Eq. (13.1)) was almost 0.02 m. However, from the data in Figure 13.6, the mean RMS error was more than 0.04 m. This error is likely due to measurement inaccuracies in the generation of the ground truth model, and in

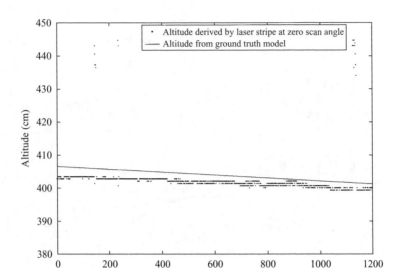

Figure 13.6   Dynamic range sensing laser altimeter mode

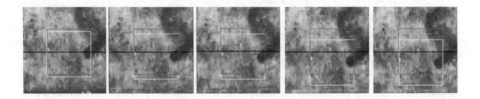

Figure 13.7   Zoomed-in regions of images showing results of correlation process

the three measured parameters in Eq. (3.1) [laser exit angle (α), AOV of lens (2ω) and baseline (L)]. The position of the stripe at the zero scan angle was extracted to the nearest pixel. It can be seen that there are some occasions when there has been a brighter pixel in the image than the stripe, leading to outliers (top left and top right in Figure 13.6). These outliers were removed using a simple post-processing method where a trimmed median filter is used. In this case, the altitude values were sorted into ascending order. Then, the values which come into the lower or upper 5 per cent are discarded. This was sufficient to remove the relatively few outliers in the dataset.

13.4.3   Dynamic performance of region-based tracker

The zoomed-in regions of consecutive histogram-equalised images in Figure 13.7, show the results of the correlation process for a matching region size of 25 × 25 pixels. The images were acquired at 25 Hz and are displayed with the oldest image on the left-hand side and the newest on the right-hand side. The grey square bounds the position of the original template while the white square bounds the position of the most recent match. These image regions were taken from a sequence at a nominal altitude of 8 m and at a speed of 1 m s^{-1}.

A simple underwater calibration of the lens was performed by deploying a scale on the tank bottom and positioning the imaging system at several known altitudes in order to determine the angle of view (AOV). In assuming identical pixel sizes over the CCD array and by neglecting geometric distortion which is more severe at the edges of the image, Eqs. (13.2) and (13.3) can be used to transform from image coordinates to real world coordinates. In this case, the correlation process output measurements (δx(pixels) and δy(pixels)), the resolution (height and width) of the image in pixels (572 by 768 in this case) and the distance between the camera and the seabed (Z) are used to determine the image displacement relative to the seabed.

$$\delta x(\text{m}) = \delta x(\text{pixels}) \left[\tan\left(\frac{\text{AOV}_y}{2}\right) \left(\frac{2Z(\text{m})}{x_{\text{res}}(\text{pixels})}\right) \right] \quad (13.2)$$

$$\delta y(\text{m}) = \delta y(\text{pixels}) \left[\tan\left(\frac{\text{AOV}_x}{2}\right) \left(\frac{2Z(\text{m})}{y_{\text{res}}(\text{pixels})}\right) \right] \quad (13.3)$$

One of the performance objectives of the computer vision applications being investigated in this research was near real-time operation; meaning it was important that they could perform as close to 25 Hz as possible. Therefore, in the evaluation of the region-based tracker, it was very important to monitor the processing time required by the correlation process. It was clear from the nature of the correlation process that the processing time would increase exponentially with the template size. However, it was found that the tracker became sensitive to noise, if the template size was reduced too much. In addition, if the image contrast was low, the tracker performance was reduced. Therefore, because the raw images acquired during these experiments were generally of low contrast, several techniques were used to improve the contrast before applying the tracker. For the results presented in this chapter, histogram equalisation was used. Although not presented, the optimal template size in terms of both robustness to noise and processing efficiency was determined from static and dynamic image sequences. The interested reader can refer to Reference 13 for further details.

The combination of the altitude data in Figure 13.6 and the pixel displacement as determined by the region-based tracker with a 25 × 25 template size, and the use of Eqs. (13.2) and (13.3) produce the trajectory shown in Figure 13.8. The associated Z trajectory data had been filtered using a trimmed median filter to remove outliers greater than 5 per cent. In the absolute accuracy assessment, the vision-derived end position along the major axis of the tank was 27.07 m, whilst according to the ground truth model it should have been 27.89 m. This represents an error of 3 per cent. Along the minor axis of the tank, where the system was supposedly rigidly constrained, there is an apparent shift of 0.31 m from the centre position, over the duration of the run. Whilst during the experiments every effort was made to align the camera

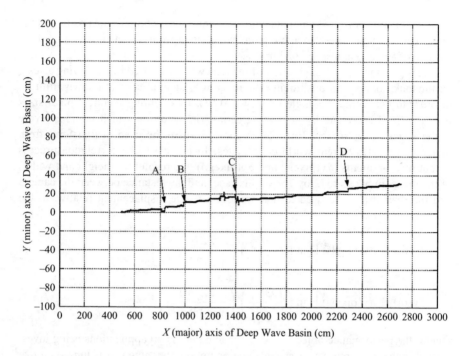

Figure 13.8 *Dead-reckoned position of camera port for a dynamic trial over 22 m (nominal altitude = 4 m, towing carriage speed = $0.5\,m\,s^{-1}$)*

with the longitudinal axis of the *Hammerhead* AUV and also to align the vehicle with the longitudinal axis of the tank, it is likely that a small bias existed. Indeed, the mean pixel displacement in this (lateral) direction was 0.14 pixels, indicating that there was a slight offset between the heading of the mounted AUV hull and the direction in which the towing carriage moves. Also, in order to avoid violating the assumption of orthography, it was necessary to align the camera axis to be perpendicular to the bottom of the tank. It is likely that a portion of the longitudinal (major axis) error was attributable to a bias of this nature.

In addition to biases due to alignment, which are a major limitation of this type of sensor, there are also random errors experienced at each iteration due to random image noise, and more seriously the effect of outliers which appear as sudden jumps in the trajectory. These can be seen at points A, B, C and D, which are overlaid in Figure 13.8.

13.4.4 Dynamic imaging performance

The acquired images are processed to produce continuous 2D reflectivity or bathymetry, images or 3D range images of the tank bottom. Such images derived from laser stripes are attractive in that they are of high contrast, are optically flat (even intensity) and can overcome many of the problems associated with producing

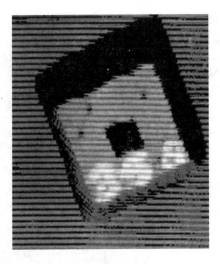

Figure 13.9   The 2D composite waterfall image

large seabed images by mosaicking. However, in the absence of image overlap, the spacing between the consecutively extracted laser stripes can be determined by the region-based tracker. It is therefore possible to reference the composite images with relative positional information.

A 2D waterfall composite image of a concrete block is shown in Figure 13.9. In this case, the laser stripe region (which is several pixels wide due to laser beam divergence and forward scattering effects) from each image is extracted using an irregular region of interest. The composite image is created by stitching together consecutive stripe regions to form a continuously evolving image. Note that the final composite image is viewed from above; an orthographic view. Although this method incorporates more image information into the composite images, the disadvantages are that the image files are quite large and accurate measurements cannot be easily made as the stripe region width is based on an intensity threshold rather than a fixed number of pixels.

The 2D intensity waterfall images shown in Figure 13.10 consist of 1200 individual images taken at a nominal altitude of 6 m and a speed of $0.5\,\mathrm{m\,s}^{-1}$ where forced sway and roll were applied to the mounting onto which the *Hammerhead* AUV was fixed. The length of tank covered was approximately 24 m. In the composition of the right-hand side waterfall, the displacement information between each frame as calculated by the region-based tracker has been used to translate the position of the newly extracted laser stripe region onto the continuously evolving composite image. The left-hand side waterfall does not integrate this displacement information, and is therefore not spatially accurate. It can be seen from the motion-corrected waterfall that there have also been some spurious matches or outliers. These instances are more apparent when looking at the black PVC pipeline section (which should be straight). As the region passes over the pipeline, a phenomenon known as the aperture problem

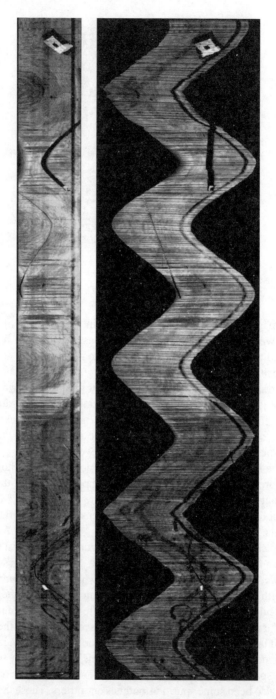

Figure 13.10 The 2D intensity waterfall images: (left) uncompensated for translation; (right) compensated for translation

occurs.[1] The tracker extracts a template which is at the edge of the pipeline. In the next iteration, a good match (greater than 60 per cent confidence level) can be made in several positions, as the correlation process searches up the edge at almost identical regions. Indeed as the pipeline width is also greater than the width of the tracking region in the FOV of the camera, the aperture problem is occurring in two dimensions in some instances, and this has resulted in the distortion of the image of what is a straight pipeline. This result is somewhat paradoxical, in that the expected drop-out of the quality of these measurements was expected to occur when traversing over a featureless seabed which lacks texture, not an environment rich in visual information such as in this case. Manual examination of the images at the other points where outliers occur reveals areas of little or no intensity variation.

Regardless of the errors made in the automatic creation of the waterfall images shown in Figure 13.10, the dead-reckoned end position is different from the measured end position by 0.19 m in the major (longitudinal) axis and 0.12 m in the minor (lateral) axis. However, range variations due to both the uneven tank bottom and the secondary heave motion associated with the pendulum motion of the towing carriage have not been scaled out.

13.5 Summary

It has been shown by experiment that the vision sensor as detailed in this chapter, whilst simultaneously acquiring bathymetry and reflectivity profiles, can derive robust altitude data at extended ranges (4–8 m) over extended image sequences using laser triangulation alone to an accuracy of between 1 and 5 per cent. In addition, it is possible to utilize the unused portion of the CCD image to determine horizontal motion by means of a region-based tracker. The combined measurements from the region-based tracker and the laser triangulation sensor can produce a stand-alone accuracy of around 97 per cent of the distance travelled at an altitude of 4 m in filtered seawater. This result is subject to degradation as the turbidity of the water increases, the velocity and altitude increase and the planar nature of the tank bottom is reduced. Finally, as well as deriving navigational information, the motion detected can be used to modify contrast enhanced composite waterfall images such that they are spatially representative of the actual seabed scene.

13.6 Acknowledgements

This work was funded by the EPSRC. We would like to acknowledge the help we received from IFREMER for work in the Deep Wave Basin which was partially funded by the EU under the Access to Research Infrastructure action of the Improving Human Potential Programme.

---

[1] The aperture problem in computer vision occurs when the motion component parallel to an edge is lost. Only the component normal to the edge can be detected.

References

1 Gracias, N. and J. Santos-Victor (2000). Underwater video mosaics as visual navigation maps. *Computer Vision and Image Understanding*, 79(1), pp. 66–91.
2 Negahdaripour, S. and X. Xu (2002). Mosaic-based positioning and improved motion-estimation methods for automatic navigation of submersible vehicles. *IEEE Journal of Oceanic Engineering*, 27, pp. 79–99.
3 Fleischer, S.D. (2000). Bounded-error vision-based navigation of autonomous underwater vehicles. Ph.D. thesis, Stanford University, Stanford, CA.
4 Fournier, G.R., D. Bonnier, J. Luc Forand, and P.W. Pace (1993). Range-gated underwater laser imaging system. *Optical Engineering*, 32(9), pp. 2185–2190.
5 Swartz, B., R. Morton, and S. Moran (1993). Diver and ROV deployable laser range gated underwater imaging system. In Proceedings of the MTS Underwater Intervention '93, New Orleans, LA, USA, 18–21 January.
6 Strand, M.P. (1995). Underwater electro-optical systems for mine identification. In Proceedings of Autonomous Vehicles in Mine Countermeasures, Monterey, CA, 4–7 April.
7 Coles, B. (1997). Laser line scan systems as environmental survey tools. *Ocean News Technology*, 3, pp. 22–24.
8 Moore, K.D., J. Jaffe, and B. Ochoa (2000). Development of a new underwater bathymetric laser imaging system: L-Bath. *Journal of Atmospheric and Oceanic Technology*, 17(8), pp. 1106–1117.
9 Tetlow, S. and R.L. Allwood (1995). Development and applications of a novel underwater laser illumination system. *Underwater Technology*, 21(2), pp. 13–20.
10 Langebrake, L.C., S.A. Samson, E.A. Kaltenbacher, E.T. Steimle, J.T. Patten, C.E. Lembke, R.H. Byrne, K. Carder, and T. Hopkins (2000). Sensor development: progress towards systems for AUVs/UUVs. In Proceedings of the MTS/IEEE OCEANS 2000, Providence, RI, USA, 12–14 September.
11 Chantler, M.J., J. Clark, and M. Umasuthan (1997). Calibration and operation of an underwater laser triangulation sensor: the varying baseline problem. *Optical Engineering*, 36(9), pp. 2604–2611.
12 Moore, K.D. (2001). Intercalibration method for underwater three-dimensional mapping laser line scan systems. *Journal of Applied Optics*, 40(33), pp. 5991–6004.
13 Dalgleish, F.R. (2004). The development and evaluation of a laser-assisted vision sensor for AUV navigation. Ph.D. thesis, Cranfield University, UK.
14 National Instruments (2000). IMAQ Vision Concepts.

Chapter 14

Advances in real-time spatio-temporal 3D data visualisation for underwater robotic exploration

S.C. Martin, L.L. Whitcomb, R. Arsenault, M. Plumlee and C. Ware

14.1 Introduction

We do not know how to display effectively and assimilate in real-time the quantitative data generated by the huge variety of sensors on remotely operated underwater robotic vehicles. Underwater robots now perform sampling, manipulation and high-resolution acoustic and optical survey missions in the deep ocean that previously were considered impractical or infeasible [1–4]. Over 1000 robotic uninhabited undersea vehicles (UUVs) are presently in operation worldwide [5]. For example, Reference 6 reports a 1997 survey in which the *Jason* and *Argo II* underwater vehicles were deployed to survey a $2\,km^2$ shipwreck site at 4100 m depth in the Pacific. Over a year of post-processing of the raw survey data resulted in data products including a detailed bathymetric and debris map of the site; site mosaics; 135,774 electronic still images (each covering about $7\,m \times 10\,m$); and hundreds of hours of conventional and high-definition video.

Underwater robotic vehicles now also enable scientists to deploy sensors at full ocean depths with unprecedented precision. For example, Reference 7 reports the highest resolution bathymetric survey then performed in the deep ocean, in which the *Jason* underwater vehicle was deployed to precisely survey a $20 \times 50\,m$ second century B.C.E. shipwreck site at about 800 m depth in the Mediterranean. This survey data comprised millions of sonar pings, thousands of electronic still images (each covering about $3\,m \times 4\,m$) and hundreds of hours of broadcast quality video. This survey was performed by flying the *Jason* vehicle in a precise grid pattern under closed-loop control, at an altitude of 3–4 m above the wreck site, while scanning the site with a pencil beam 675 kHz sonar and an electronic still camera.

Visibility on the ocean floor is poor, often only a few metres. The human operators must mentally construct a composite three-dimensional (3D) map and image of the overall survey site based on smaller visual images and sonar scans.

14.1.1 The need for real-time spatio-temporal display of quantitative oceanographic sensor data

Despite the fact that remotely operated robotic vehicles have displaced inhabited submarines as the leading method of US deep-ocean oceanographic research, their human–computer interfaces (HCI) have remained largely unchanged since the invention of underwater robots in the mid-1970s. At present, much of the quantitative data obtained with these systems is automatically logged for post-processing, but is not easily available in real-time, thus vitiating both data utility to human operators and mission utility to science.

Given that deep-sea oceanographic research is frequently conducted at sites for which little *a priori* information is available, we argue that rapid real-time presentation of sensor data could significantly improve our ability to explore the unstructured environment of the benthic floor.

This chapter reports the development and validation of a system, which provides an interactive 3D graphical user interface, that displays the quantitative spatial and temporal sensor data that is presently available to pilots and users only as numerical displays. This chapter builds and extends upon the work previously presented in the 2005 IEEE International Conference on Robotics and Automation, Barcelona [8]. Our goal is to address the following three classes of sensor data:

1. *Navigation sensor data* comprising vehicle state data (position, orientation and velocities), way-points, targets, vehicle trajectories and survey tracklines.
2. *Scientific sensor data* comprising sensors such as electronic cameras; sidescan, pencil-beam and multibeam sonar; water column velocity profilers; physical oceanography sensors such as conductivity, temperature and depth (CTD), transmissometer; fluorimeter; and the like.
3. *Vehicle sensor and system status data* such as power levels, payload, endurance, operational mode and error conditions.

The goal of this system is to (i) graphically display real-time numerical data in a spatially natural and concise format; (ii) provide temporal context, when required, of previous sensor values, for example, vehicle position, trajectory, survey footprint or a CTD anomaly in a buoyant plume; (iii) indicate data precision and reliability and (iv) indicate data 'staleness'.

The remainder of this chapter is organised as follows: Section 14.2 describes the implementation of the system. Section 14.3 describes the system's capability to replay previously acquired data. Section 14.4 reports comparisons of this real-time system to a laser scan. Section 14.5 reports a preliminary field trial and Section 14.6 concludes and outlines future work.

14.2 System design and implementation

The system has three parts: a navigation subsystem, a real-time data display subsystem and a subsystem to geo-reference sensor data. Figure 14.3 shows the process by which a single sonar ping is geo-referenced for display in GeoZui.

14.2.1 Navigation

The navigation subsystem employed in this project is DVLNAV, an interactive program for precision 3D navigation of underwater vehicles (Figure 14.1). DVLNAV employs a bottom-lock Doppler sonar and a North-seeking gyro to compute accurate vehicle *XYZ* displacements in a true North coordinate grid. It is also equipped to initialise the bottom track to external sources such as long-baseline acoustic navigation systems, global positioning systems (GPS) or manual input. The software provides an interactive user interface for real-time display of all sensor and navigation information. The system is currently deployed on six operational underwater robotic vehicles and on the *Alvin* inhabited oceanographic submersible. Recent sea trials with this navigation system are reported in Reference 9.

14.2.2 Real-time spatio-temporal data display with GeoZui3D

The visualisation component of the system is provided by a customised version of GeoZui3D (Geographic Zooming User Interface 3D). GeoZui3D is a platform for

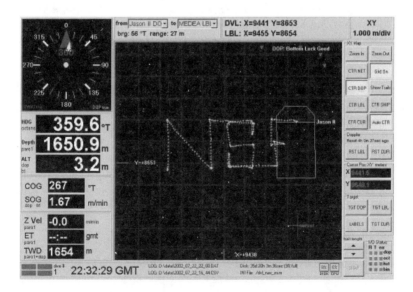

Figure 14.1 Navigation: screen shot showing the DVLNAV Doppler-based navigation and closed-loop tracking performance of the new Jason 2 *during sea trials on 22 July, 2002 at 1650 m depth on the Juan de Fuca Ridge [9]*

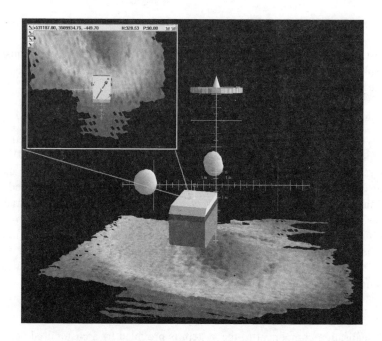

Figure 14.2 GeoZui3D: example of two interlinked views of a vehicle and the bathymetry that has just been acquired. In this example, the smaller plan view is always centred on the vehicle. This figure shows data playback of data from the survey of a 750 B.C. shipwreck in the Eastern Mediterranean conducted by the Jason 1 *ROV, Jason Dive 251, 19 June, 1999, 00:16 GMT to 01:17 GMT [17]*

research into visualisation and interaction with 3D geospatial and temporal data [10, 11]. The interface is illustrated in Figure 14.2. The following describes the features relevant to vehicle visualisation and control.

14.2.2.1 Real-time input rendered to a 3D display

GeoZui3D receives real-time geospatial data from UDP data packets over the local network. Different kinds of data may be received on one or more ports. The following real-time objects are presently supported:

1. dynamically generated bathymetric surfaces;
2. vehicles and their tracks;
3. targets displayed as spheres, cubes, cones or cylinders;
4. scalar 3D sensor values displayed as tubes with colour and/or size mapping.

To allow for real-time surface visualistion, resolved 3D soundings from the vehicle's Imagenex881A scanning pencil-beam sonar are integrated into a dynamically growing surface using a weighted spatial averaging method. The surface is composed of a tree of regularly spaced grids. Each grid contains 100 by 100 square cells with a cell size predetermined by the user through a configuration file. As the surface grows, new grids are added to accept soundings that fall outside existing grids. The grids are organized by a tree data structure: each leaf node can contain up to 100 cells (10 × 10) and each non-leaf node can contain up to 100 (10 × 10) tree nodes. The spherical objects in Figure 14.2 represent targets that can be thought of as 3D Post-It® notes, saving geo-referenced textual comments. Clicking on a target reveals the comment. More real-time objects are planned as future work.

14.2.2.2 View control and linked views

Conceptually, in GeoZui3D the centre of the 3D workspace is a point located just behind the user's monitor screen, and this workspace centre provides a focus for interactions. Rotation is accomplished using the 3D widgets shown in Figure 14.2 and scaling is done about the workspace centre at a carefully calibrated rate of 8 times magnification per second. Vehicle position and orientation information is shown using a 3D model geo-referenced in the 3D scene. Clicking on the model causes the view to be linked to the vehicle's movement as though from a camera mounted on the vehicle [12]. Clicking on some other part of the scene 'unmounts' the camera and changes the view's frame of reference back to the static scene.

GeoZui3D has specific capabilities for integrating multiple simultaneous views of the same 3D scene. Multiple windows can be created and attached to static or moving objects to support different tasks. For example, an 'over the shoulder viewpoint' above and behind a vehicle may be best for steering it through an environment that has already been mapped. A simultaneous top–down overview can show a wider context. Visual 'tethers' show the relationship of the detailed forward-looking view window to the overview window. Such views can be set up either through user interactions, or through scripting.

14.2.2.3 Time control

In addition to displaying real-time data, the data are recorded to allow for instant replay of the vehicle's position. For example, GeoZui3D's time control bar may be moved back, say, 3 min to review the vehicle's position at that time, then real-time viewing of the vehicle's current position may be instantly resumed.

14.2.3 Real-time fusion of navigation data and scientific sensor data

Commonly used oceanographic sensors such as sonar, CTD and ADCP provide raw sensor data via a diverse variety of hardware interfaces (RS232, RS422, RS485, Ethernet and analog are presently the most common), data formats and sampling intervals. In present day remotely operated underwater vehicles, sensor data (e.g., sonar range and angle, conductivity, temperature, depth, altitude, speed) are typically displayed in real-time as a number on an instrument panel, or as a sensor-specific display in sensor instrument coordinates. These raw data are normally logged and time-stamped in real-time. In subsequent off-line post-processing, the data are fused with logged vehicle navigation data and sensor calibration data to geo-reference the sensor data values.

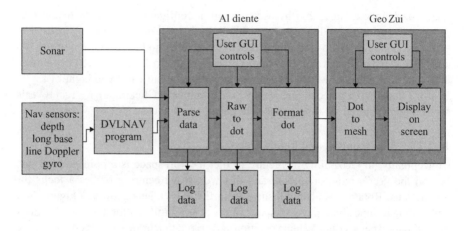

Figure 14.3 Flow diagram displaying how a sonar ping is analysed, geo-referenced then displayed in GeoZui

Plotting and interactive display programs (such as GeoZui3D) generally cannot accept raw sensor data; they require processed geo-referenced data.

We have developed a computer program, called 'Al-dente', to combine sensor, navigation and calibration data in real-time, under user control, and to transmit the resulting geo-referenced sensor data via Ethernet to GeoZui3D for interactive display (Figure 14.3). Al-dente has a graphical user interface that enables the user to control its operations. The program utilises the following data:

1. *Vehicle sensor calibration data input:* Vehicle instrument calibration and sensor information is loaded from a user-selectable initialisation file. Vehicle information includes a definition of the vehicle's coordinate system, and the 6-DOF position and orientation of each sensor as mounted on the vehicle. Sensor information includes calibration, setting and communication parameters. Sonar data can be filtered in realtime based upon a maximum and minimum depth value that is entered into the user-selectable initialisation file.
2. *Raw navigation data input:* Al-dente is presently configured to receive vehicle navigation data via UDP packets in a standardised format from the DVLNAV navigation program [9].
3. *Raw oceanographic sensor data input:* Al-dente is designed to interface directly to certain sensors such as the Imagenex 881A scanning pencil-beam sonar (Figure 14.4). A graphical interface enables the user to examine and control the sonar's frequency, pulse sequence and all other internal sonar parameters. Raw real-time sensor values are then processed and logged in real time.
4. *Incrementally processed geo-referenced sensor data output:* Al-dente combines the sensor, navigation and calibration information to compute the exact geo-referenced coordinates and UTC time of each sensor reading. The incrementally processed data is transmitted to GeoZui3D and is logged to disk.

Figure 14.4   Photograph of Imagenex 881A pencil-beam sonar shown as mounted on the JHU ROV

The program has two modes of operation:

(a) *Real-time mode:* Provides real-time centralisation of sensor, navigation and calibration data to compute geo-referenced data. Each sensor datum is consolidated upon receipt, logged and forwarded to GeoZui3D via UDP for display.
(b) *Replay mode:* Previously logged sensor, navigation and calibration data can be loaded, geo-referenced and transmitted to GeoZui3D via UDP. This process can be performed either as a batch process or incrementally.

For the representative case of processing sonar bathymetric data from an Imagenex 881A scanning sonar, the system works as follows:

(a) At program start, Al-dente loads a file containing all vehicle calibration data.
(b) In real-time, Al-dente receives navigation data packets via UDP broadcast from the vehicle navigation system.
(c) Al-dente configures the sonar internal parameters and initiates sonar pings as specified by the user.
(d) Upon receipt of each sonar ping data packet, Al-dente uses the most recent vehicle 6-DOF navigation position, the previously loaded calibration data, and the scanning-angle and range of the sonar ping to compute the 3-DOF position and approach vector of the sonar reflection. Al-dente also provides a conventional polar-scan and a grid plot, as shown in Figure 14.5, of the scanning sonar depicting colour-mapped return intensities and bottom profile.

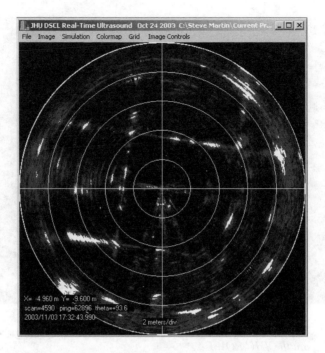

Figure 14.5   Al-dente provides a conventional polar plot depicting colour-mapped return intensities, bottom profile and range rings

(e) The spatially fused sonar data is transmitted to GeoZui3D via UDP, and logged to disk.

For a detailed discussion of high-resolution sonar bathymetry, the reader is referred to Reference 13. Although it is hard to depict in static images, the real-time spatially accurate display of survey bathymetric and vehicle trajectory data provides the user with an instant comprehension of the progress of the survey, the completeness of the bathymetric sonar coverage and the quality of the data.

Our preliminary impression is that this system provides the user with spatio-temporal awareness superior to that of conventional ROV instrumentation displays.

14.3   Replay of survey data from Mediterranean expedition

Al-dente has the capability to load log files from previous vehicle deployments, and to replay the logged data as if it were real-time data. In replay mode, the logged data is loaded and fused by Al-dente, and then incrementally transmitted to GeoZui3D. This enables the user to interactively review previously logged underwater surveys. Figure 14.2 shows a screen shot of GeoZui3D as it received geo-referenced data from a survey of a 750 B.C. deep-water shipwreck conducted by the *Jason 1* ROV in the

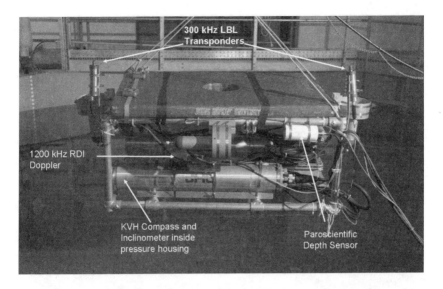

Figure 14.6   The JHU ROV – preliminary experimental test-bed for the new real-time spatio-temporal 3D interface

Eastern Mediterranean in 1999 [7]. The figure clearly depicts the central cargo pile, and the shapes of individual Amphora can be seen in the bathymetry.

14.4   Comparison of real-time system implemented on the JHU ROV to a laser scan

To validate the accuracy of the data displayed in GeoZui3D several bathymetric surveys of the Johns Hopkins University Hydrodynamics Test Tank [14] were conducted using the real-time system. The resulting sonar bathymetry was compared to data obtained from an entirely separate laser scan of the test tank floor performed by Cullinan Engineering Inc, a commercial survey firm. Figure 14.7 depicts the JHU Hydrodynamics Test Tank and Figure 14.10 identifies major bottom features in the tank.

14.4.1   Real-time survey experimental set-up

The real-time system has been implemented on the Johns Hopkins University Remotely Operated Vehicle (JHU ROV), a research testbed vehicle developed at Johns Hopkins University. The JHU ROV is equipped for full 6-DOF position measurement. Vehicle heading, roll and pitch (and their time derivatives) are instrumented with a three-axis KVH ADGC gyro-stabilised magnetic compass and a Phins North-seeking three-axis fibre-optic gyro. Depth is instrumented by a Paroscientific depth sensor. A 1200 kHz Doppler sonar provides *XYZ* velocity measurements, and in combination with the gyro and depth sensor, enabling Doppler navigation

Figure 14.7 Photograph of 43,000 gallon JHU Hydrodynamics Test Tank. The tank measures 7.6 m in diameter and 4.27 m in depth

via the DVLNAV software program [9]. Vehicle *XYZ* position is instrumented with a 300 kHz time-of-flight acoustic navigation system. An Imagenex 881A scanning sonar (280 kHz–1.1 MHz) provides bottom sonar ranges. Figure 14.6 shows the JHU ROV and identifies the major elements of its sensor suite [14].

Figure 14.8 depicts GeoZui3D real-time interface showing the vehicle and test-tank bathymetry generated in real time. Sonar data from an Imagenex 881A sonar and navigation data from DVLNAV were fused by the Al-dente program in real time and transmitted for display by GeoZui3D.

14.4.2 Laser scan experimental set-up

The test tank was drained following the completion of the real-time sonar surveys. Cullinan Engineering, a survey company was contracted to perform a laser scan of the test tank using a Leica Geosystems HDS2500 laser scanner.

The complete laser scan of the tank interior was comprised of multiple individual scans. These individual scans were spatially referenced using common reference points distributed throughout the tank in the form of targets and fiducial spheres. The individual scans were joined into point clouds using a least-squares adjustment routine. Once this was complete, the individual points clouds were combined into a complete topographic map of the test tank.

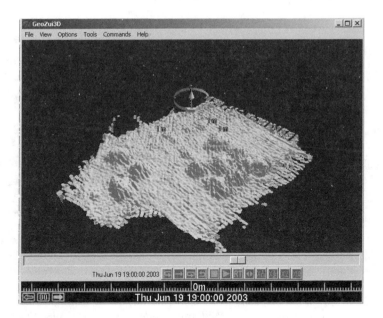

Figure 14.8   Real-time spatio-temporal interface showing the test-tank bathymetry. Sonar data from an Imagenex 881A sonar and navigation data from DVLNAV was fused by the Al-dente program and transmitted for display by GeoZui3D

14.4.3   Real-time system experimental results

Using the system installed on the JHU ROV several bathymetric surveys were conducted of the JHU test tank (Figure 14.7). The survey path was designed to have maximum coverage of the tank bottom and approximately uniform sonar returns per section of the tank. At the beginning of each survey DVLNAV was reset to the most recent 300 kHz time-of-flight acoustic navigation fix. The relative position of the vehicle was then calculated based upon velocity measurements from the Doppler sonar and in combination with the three-axis gyro, these enable Doppler navigation via the DVLNAV software program [9].

The sonar data shown in Figure 14.8 were obtained with an Imagenex 881A profiling sonar configured to run at a frequency of 800 kHz, maximum range of 5 m, pulse length of 60 μs, absorbtion of 1.65 dB/m and gain of 9 dB. The sonar's scanning parameters were set to a sector width of 78°, train angle of −9°, and step size increment of 1.2°. GeoZui was set up to have spatial averaging parameters: grid size of 0.01 m and beam shape of radius 0.05 m. The survey comprised 71,795 sonar pings and had a total track length of 32.16 m.

14.4.4   Laser scan experimental results

Five laser scans of the test tank were performed with a Leica Geosystems HDS2500 laser range scanner, which provides laser ranges with an accuracy of ±4 mm. The laser

Figure 14.9 Mesh of composite laser scan data formed by Cullinan Engineering. Clearly depicted are the large boulders shown in Figure 14.10

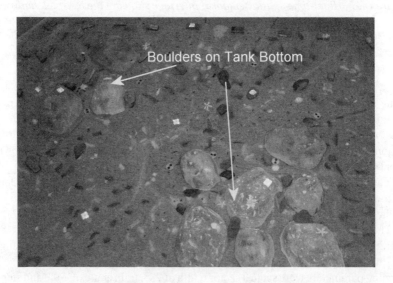

Figure 14.10 Photograph of bottom of JHU Hydrodynamics Test Tank with the identification of major bottom features

scanner was placed on the catwalk, which is situated around the JHU Test Tank, to perform these scans. The catwalk is 10 m from the centre of the tank bottom. The composite laser scan, shown in Figure 14.9, is comprised of over 3.5 million laser range points. This figure clearly depicts the large boulders shown in Figure 14.10.

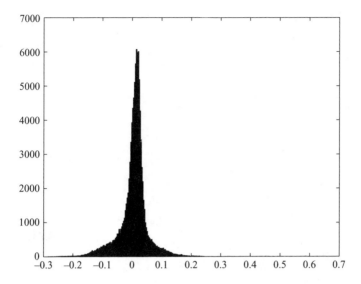

Figure 14.11   Histogram of depth difference between real-time sonar bathymetry and high-resolution laser scan of the test-tank bottom features

14.4.5   Comparison of laser scan to real-time system

A comparison of the laser and sonar survey data is complicated by the fact that the two systems acquire 3D spatial data in independent Cartesian frames of reference. To compare the quality of the sonar bathymetry to the laser scan bathymetry, we (a) clipped bathymetric data to include only the tank floor features, that is, excluding the tank wall data; (b) computed a best fit 6-DOF rigid body transformation between the two sets of bathymetry and (c) computed the histogram of depth differences between the sonar and laser bathymetry maps (Figure 14.11).

Figure 14.12 displays the 3D plot of the error between ground truth laser scan to a set of bathymetry from the real-time system. The standard deviation of the bins with sonar data is 0.0469 m, and the mean absolute error is 0.0041 m. Given that the intrinsic resolution of the sonar used at 5 m is 0.01 m, the observed accuracy of the real-time sonar bathymetric map is reasonable. While it is noted that careful post-processing of navigation and sonar data may yield improved bathymetric accuracy [13], the purpose of this system is to present reasonably accurate 'first cut' bathymetric data in real time to the vehicle pilots and observers, which we have achieved.

14.5   Preliminary field trial on the *Jason 2* ROV

To evaluate the feasibility of employing this real-time spatio-temporal data system on an actual full-ocean depth underwater vehicle, we installed and tested the system on the *Jason 2* ROV [15,16]. *Jason 2* is a 6500 m ROV developed by the Woods Hole Oceanographic Institution (WHOI) for oceanographic research. *Jason 2* is operated

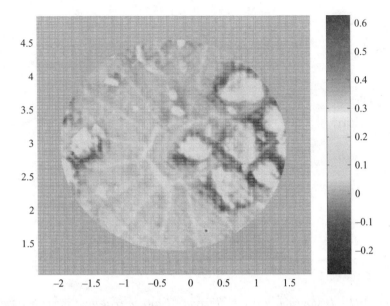

Figure 14.12   Two-dimensional plot of the error between the mesh of the ground truth laser scan and the mesh bathymetry of real-time system

for the oceanographic research community by WHOI's National Deep Submergence Facility[1] (NDSF) as part of the U.S. University-National Oceanographic Laboratory System[2] (UNOLS). Our objectives were to install our real-time spatio-temporal data system on *Jason 2*, and to demonstrate the feasibility of this system during shallow water engineering trials.

The *Jason 2* ROV is equipped for full 6-DOF position measurement. Vehicle heading, roll and pitch (and their time derivatives) are instrumented with a Octans three-axis fibre-optic gyro. Depth is instrumented by a Paroscientific depth sensor. A 1200 kHz Doppler sonar provides *XYZ* velocity measurements, and in combination with the gyro and depth sensor, enabling Doppler navigation via the DVLNAV software program. An Imagenex 881A scanning sonar (280 kHz–1.1 MHz) provides bottom bathymetry (Figure 14.13). Figure 14.14 shows the *Jason 2* ROV and identifies the major elements of its sensor suite.

The *Jason 2* was deployed in Woods Hole harbour. Under closed-loop control it performed a 180° spin (for sonar calibration) and four tracklines spaced 2–4 m apart and approximately 10 m in length to obtain bathymetry of the sea floor. These tracklines encompass a total distance travelled of 57 m, 83,848 sonar pings, covering an area of 200 m^2.

Figure 14.13 depicts the GeoZui3D real-time interface displaying the *Jason 2* vehicle, bathymetry generated in real time during these dock trials, and the

---

[1] http://www.whoi.edu/marops/vehicles
[2] http://www.unols.org/

3D data visualisation for underwater robotic exploration 307

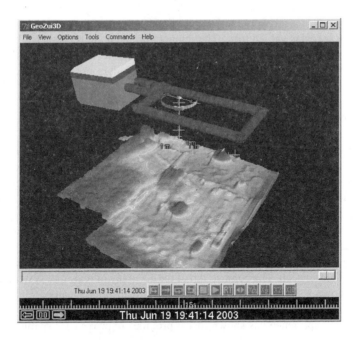

Figure 14.13   Screen shot of GeoZui3D showing bathymetry of Jason 2 *from 2 to 10 m long tracklines during dock trials on 15 October, 2004 at 3 m depth off the dock at the Woods Hole Oceanographic Institute*

Figure 14.14   Photograph of Jason 2 – *Woods Hole Oceanographic Insitute deep ocean scientific vehicle with the identification of the major elements of its sensor suite*

temperature (as measured by the Doppler Sonar's temperature sensor) displayed in real time as a scalar valued tube. The temperature data are represented by the tube attached to the vehicle and double as a trackline of the vehicle during the survey. Clearly depicted in Figure 14.13 are a number of boulders and rocks on the sea floor.

14.6 Conclusions and future work

This chapter has reported the design and development of a real-time human–computer interface to enable a human operator to more effectively utilise the large volume of quantitative data (navigation, scientific and vehicle status data) generated in real time by the sensor suites of underwater robotic vehicles. The system provides an interactive 3D graphical interface that displays, under user control, quantitative spatial and temporal sensor data presently available to pilots and users only as alpha-numerical and 2D displays.

The system has been experimentally evaluated based upon a comparison to a ground truth laser scan. The comparison of the accuracy of real-time system to a laser scan has shown a standard deviation of 0.0469 and absolute mean of 0.0041 m. This demonstrates that the system accurately displays data within the capabilities of the sensors. Our experience has shown that it provides improved 3D spatial awareness to the user.

Although it is difficult to depict in static images, the real-time spatially accurate display of survey bathymetric and vehicle trajectory data provide the user with an instant comprehension of the progress of the survey, the completeness of the bathymetric sonar coverage, and the quality of the data. The system has been tested on the Woods Hole Oceanographic *Jason 2* ROV. These data have shown the feasibility of using the system with at sea oceanographic survey operations.

Our preliminary impression is that this system provides the user with spatio-temporal awareness superior to that of conventional ROV instrumentation displays. In the future, we plan to add several additional object types to Al-Dente and GeoZui3D to support additional scientific sensors, support for multiple vehicles, and additional spatial and temporal controls. We hope to test this system at sea in actual underwater vehicle oceanographic survey operations.

Acknowledgements

We are grateful to the National Science Foundation for support under OCE-0112737 (Whitcomb) and 0081292 (Ware), and to NOAA for support under NA17OG2285 (Ware). Stephen Martin is supported under a National Defense Science and Engineering Graduate Fellowship.

The laser scan of the JHU test-tank was obtained in collaboration with Dr Hanumant Singh (and his students) of the Woods Hole Oceanographic Institution.

The survey data depicted in Figure 14.2 depict survey data obtained by Whitcomb and collaborators with the *Jason 1* ROV on an expedition to the Eastern Mediterranean in June 1999, on which the chief scientists were Robert Ballard, Lawrence Stager (archaeology) and Dana Yoerger (engineering) [17].

The authors are grateful to the staff of the National Deep Submergence Facility of the Woods Hole Oceanographic Institution for their collaboration in installing and testing the reported system on the *Jason 2* ROV during engineering dock trials in October 2004.

References

1 Robigou, V., K. Stewart, and R. Ballard (1993). Hydrothermal vent sites in the guayamas basin revisited by an ROV: high precision bathymetric, geological and biological mapping. *EOS Transactions of the American Geophysical Union*, 74, p. 573.
2 Sulanowska, M.M., S.E. Humphris, and J.C. Howland (1996). Detailed analysis of the surface morphology of the active TAG hydrothermal mound by mosaicking of digital images. *Transactions of the American Geophysical Union*, 77, p. 768.
3 Tivey, M.A., H.P. Johnson, A.M. Bradley, and D.R. Yoerger (1998). Thickness of a submarine lava flow determined from near-bottom magnetic field mapping by autonomous underwater vehicle. *Geophysical Research Letters*, 25(6), pp. 805–808.
4 Whitcomb, L.L., D.R. Yoerger, H. Singh, and J. Howland (1999). Advances in underwater robot vehicles for deep ocean exploration: navigation, control, and survey operations. In J. Hollerbach, and D. Koditschek, Eds., *Robotics Research – The Ninth International Symposium*, Springer-Verlag, London, Chapter 13, pp. 439–448.
5 Ocean News and Technology Press Staff (1998). *Remotely Operated Vehicles of the World*. Ocean News and Technology Press, Palm City, FL.
6 Williams, R.A. and R. Torchio (1998). *M.V. Derbyshire Surveys – UK/EC Assessors Report: A Summary*. Department of the Environment, Transport and the Regions, Eland House, Bressenden Place, London SW1E 5DU, London, 1998. ISBN 1-85112-075-0, Abbreviated version available online from the U.K. Government at http://www.dft.gov.uk/stellent/groups/dft_shipping/documents/pdf/dft_shippin_pdf_505280.pdf.
7 Ballard, R.D., A.M. McCann, D.R. Yoerger, L.L. Whitcomb, D.A. Mindell, J. Oleson, H. Singh, B. Foley, J. Adams, and D. Picheota (2000). The discovery of ancient history in the deep sea using advanced deep submergence technology. *Deep Sea Research Part 1*, 47(9), pp. 1591–1620.
8 Martin, S., L.L. Whitcomb, R. Arsenault, C. Ware, and M. Plumlee (2005). A system for real-time spatio-temporal 3-D data visualization for underwater robotic exploration. In Proceedings of the IEEE International Conference on Robotics and Automation, Barcelona, Spain, pp. 1628–1635.
9 Kinsey, J.C. and L.L. Whitcomb (2004). Preliminary field experience with the DVLNAV integrated navigation system for oceanographic submersibles. *Control Engineering Practice*, 12(12), pp. 1541–1549.

10 Arsenault, R., C. Ware, M. Plumlee, S. Martin, L.L. Whitcomb, D. Wiley, A. Shorter, T. Gross, and A. Bilgili (2004). A system for visualizing time varying oceanographic 3D data. In Proceedings of IEEE/MTS Oceans'2004, Kobe, pp. 743–747.

11 Ware, C., M. Plumlee, R. Arsenault, L. Mayer, S. Smith, and D. House (2001). GeoZui3D: data fusion for interpreting oceanographic data. In Proceedings of IEEE Oceans'2000, pp. 1960–1964.

12 Plumlee, M. and C. Ware (2003). Integrating multiple 3D views through frame-of-reference interaction. In Proceedings of the International Conference on Coordinated and Multiple Views, London, pp. 34–43.

13 Singh, H., L.L. Whitcomb, D. Yoergar, and O. Pizarro (2000). Microbathymetric mapping from underwater vehicles in the deep ocean. *Computer Vision and Image Understanding*, 79(1), pp. 143–161.

14 Kinsey, J.C., D.A. Smallwood, and L.L. Whitcomb (2003). A new hydrodynamics test facility for UUV dynamics and control research. In Proceedings of IEEE/MTS Oceans'2003, San Diego, CA.

15 Johnson, H.P. (2003). Probing for life in the ocean crust with the lexen program. *EOS, Transactions of the Americal Geophysical Union*, 84(12), pp. 109–116.

16 Whitcomb, L.L. J.C. Howland, D.A. Smallwood, D.R. Yoerger, and T.E. Thiel (2003). A new control system for the next generation of US and UK deep submergence oceanographic ROVs. In Proceedings of the 1st IFAC Workshop on Guidance and Control of Underwater Vehicles, GCUV'03, 9–11 April, pp. 137–142.

17 Ballard, R.D., L.E. Stager, D. Master, D. Yoerger, D. Mindell, L.L. Whitcomb, H. Singh, and D. Piechota (2002). Iron age shipwrecks in deep water off Ashkelon, Israel. *American Journal of Archaeology*, 106(2), pp. 151–168.

Chapter 15

Unmanned surface vehicles – game changing technology for naval operations

S.J. Corfield and J.M. Young

15.1 Introduction

In 1718 Captain Woodes Rogers, the new Governor of New Providence in the Bahamas and a man with a distinguished former 'career' as a privateer, managed to secure the two exits from Nassau harbour with his small force of two Royal Navy warships and two sloops. If the buccaneers, pirates and privateers of Nassau could have united, the small Royal Navy task force could have been overcome. However, the King had offered a pardon to the residents of Nassau and any potential alliance of like-minded rogues had crumbled as a result. Henry Jennings, the unofficial mayor of Nassau and others had already been to Bermuda to accept the pardon and, with the risk to his entrepreneurial activities now clear, Edward Teach had left for the Carolinas before Woodes Rodgers' arrival.

To disrupt this modest blockade and escape with the stolen goods in his possession, Captain Charles Vane fell back on a historical precedent. The magazine of his most recent prize, an ex-French brigantine was filled and its guns loaded. It was aimed towards Woodes Rogers' warship *Rose* and her companion sloop *Shark*, set on fire and let loose. As the brigantine fire ship came towards them, its cannons exploded and the Royal Navy ships were forced to cut anchors and take evasive action. Captain Vane's sloop slipped out of the harbour under a sky lit up by the final explosion of the fire ship's magazine.

Of course, this tale is not as well known as Howard's and Drake's use of fire ships against the Spanish Armada in 1588, but it is a tale that illustrates perfectly how the clever, low-cost conversion and use of existing vessels as unmanned systems can provide a competitive advantage.

15.2 Unmanned surface vehicle research and development

The last 15 years have seen a great deal of well-publicised research and development of unmanned underwater vehicles (UUVs) throughout the world. The naval requirements for stealthy reconnaissance in the littoral and the possibilities of driving down the costs of civil seabed survey operations have combined with the intrinsically interesting technical challenges of producing cost effective UUVs to generate a vibrant research community.

At the same time, work on unmanned surface vehicles (USVs) has continued largely unnoticed in the background. USVs are not a modern invention and various projects have been conducted since the Second World War, leading to a number of radio controlled USVs being built and operated. These early USV systems were primarily used as gunnery and missile target systems. Some remote mine sweeping applications were developed in the 1950s and 1960s and other experimental monitoring systems and attack systems have been reported from time to time.

A step change in development occurred at the end of the 1970s when a number of European navies started to develop a new generation of mine countermeasure (MCM) systems [1,2]. These were designed to use multiple radio-controlled drones running in front of a manned MCM ship which was controlling the drones. The key benefits of this type of system were envisaged to be the increased stand-off afforded to the manned platform, and hence the reduced risk to the crew, and the ability for a single manned platform to operate several drones simultaneously. This provided the basis for faster rates of area search or sweep in accordance with the modern in-stride concepts of mine clearance.

The first of the German Troika MCM systems went into service in 1980. The Danish SAV Class drones went into service in 1991, with further units following in 1996 and MSF Mk1 drones added in 1998. With these modern precedents for the use of remotely operated surface vehicles in operational roles, it is not really clear why USV research and development for a wider range of operational roles has not been of higher importance. Certainly there have been and still are significant philosophical differences in how different organisations view the potential of unmanned systems in different operational roles. While the benefits of being able to insert unmanned systems into high-threat environments without endangering human operators are well known and increasingly appreciated, serious consideration of the use of USVs within a wider range of roles has ultimately awaited recent developments in advanced distributed communications, command and control systems. Experience with other unmanned systems has contributed to a better understanding of how technology developments since the end of the 1980s can be best integrated to produce unmanned systems that will be both reliable and effective within the maritime battlespace.

A recurring important issue in any modern navy is how to reduce equipment procurement and running costs while maintaining existing capability levels and, in some cases, increasing them. This is particularly the case for modern multi-role systems where flexibility is a key factor and the systems have to possess good performance in a wide range of operational tasks in order to be considered for procurement in the first place. In modern navies, which employ sophisticated platforms and combat systems, the costs of training and manpower are generally very significant. Any genuine reduction in the need for trained personnel to meet operational requirements can represent a significant through life cost saving and thus, where the use of unmanned systems can potentially provide the required operational capability, there is now serious consideration about how such systems can be developed and fielded.

Since the end of the 1990s, a number of significant research and development programmes on advanced USVs have been conducted [3,4]. The SAIC/Navtec Owl and Owl II programmes in the USA have investigated the role of small (3 m) USVs in surveillance and harbour protection operations and these programmes have been successful in shedding light on many of the technical issues and potential solutions in key USV technical areas, including long range command and control and in-theatre launch and recovery. The US Spartan USV programme has focused on the adaptation of standard 7 m and 11 m rigid inflatable boats (RIBs) to produce viable semi-autonomous multi-role USVs. An important aspect of the Spartan programme is the development of modular mission payloads that can be easily and rapidly integrated onto the main Spartan platforms to meet the requirements for roles which include anti-terrorism protection, force protection, shallow water anti-submarine warfare, shallow water mine countermeasures, discrete surveillance and surface attack.

The attack role is a relatively recent development. The advantages and disadvantages of allowing remotely operated systems to deploy weapons have been debated at length within the robotics research community over the years. It is fairly obvious that where unmanned systems can reliably act as low-cost force multipliers with the ability to both detect and prosecute targets, this capability would potentially represent a significant military capability enhancement. When US Predator unmanned aircraft systems (UAVs) deployed Hellfire anti-tank missiles under remote control conditions during operations in Afghanistan in 2001, it was clear that a number of military and philosophical barriers were being broken down and that the wider use of weapons from unmanned systems is now both inevitable and near term. In the USV world the evidence for this is clearly instantiated in the current Israeli Protector USV which, like Spartan, uses a RIB as the hull platform. Protector is fitted with electro-optic sensors, radar, a GPS system and an inertial navigation system like other USV prototypes; but it also includes a stabilised 12.7 mm machine gun and is seen as a possible replacement for conventional manned fast patrol craft.

15.3 Summary of major USV subsystems

15.3.1 The major system partitions

A USV system comprises two major partitions: the USV drone system and its associated onboard subsystems; the mission command and control subsystem. It should be noted that the drone system can comprise multiple drone units and that the mission command and control system can be geographically and functionally extended and distributed to include local control stations, in-theatre tactical area command stations and remote strategic command stations operated from many thousands of miles away.

During the design of a USV system, the functional split between the mission command and control system, located at some host facility (e.g., a land base, ship base, etc.), and the USV drone system is not an *a priori* defined split. If it is decided that the majority of the command and control functionality, including sensor data interpretation, should be conducted by an operator at a host facility workstation, the design of the USV drone system will tend towards the use of relatively simple remotely controlled drone units. However, if it is decided that the majority of the command and control functionality should be built into the drone on-board processing subsystems and that the host facility workstation will only provide high-level mission plans to the drones, then the USV drone units will tend towards advanced technology autonomous systems.

A common characteristic of many robotics research programmes has been to assume that the greater the amount of mission command and control functionality that can be built into the remote drone systems, the better. In reality, careful consideration should be given to this design decision since it greatly affects the complexity, reliability and cost of the overall USV system solution. The availability of advanced sensors, high-capacity LAN/WAN networks, satellite-based data communications and high-accuracy positioning systems with global coverage means that, at present, very advanced semi-autonomous USV capabilities can be built which use the best aspects of computer-based processing in combination with the unrivalled interpretative capabilities of the human brain. It is believed that this combination of human–computer interaction to produce optimal system performance over a wide range of USV mission conditions is likely to remain a key feature of USV systems over the next 20 years.

15.3.2 Major USV subsystems

The major USV subsystems are the:

(a) hull (or multi-hull)
(b) auxiliary structural elements
(c) engines and associated propulsion subsystems
(d) energy/fuel subsystems
(e) on-board communications, command and control subsystem which includes onboard processing subsystems for the USV's:
 - own sensor systems
 - engine management systems
 - payload sensor systems
 - navigation, guidance and control (NGC) subsystems
(f) host facility based mission command and control workstation
(g) host-to-drone communications infrastructure.

15.3.3 Hulls

The development costs of specialised unmanned vehicle bodies, airframes or hulls tend to be very high. Hence, during initial proof of principle phases of many unmanned vehicle programmes, existing platforms from conventional systems are used or modified to provide the basis for the unmanned system platforms. To date, USVs have been developed around a number of different existing hull types including jetski hulls, RIBs, custom tri-hulls, patrol craft mono-hulls and swath catamaran hulls.

As with any marine vehicle design, the choice of USV hull depends on the nature of the desired USV operation and its associated seakeeping requirements, the performance flexibility required, the numbers, types and masses of the different payload sensors/equipment that the USV is to host and, of course, cost constraints.

At this stage of USV development, the emphasis appears to be firmly on the development of small USVs which can be transported into theatre by ships, aircraft or possibly submarines. This emphasis has resulted in the use of commercial jetski, modified jetski and RIB hulls within a number of important USV development programmes. These hulls are readily available, robust and relatively low cost. At present, more developments appear to be using RIB hulls than jetski hulls since these are already operated and handled by navies and can accommodate much larger fuel tanks to provide high endurance while still retaining a high payload capacity (Figure 15.1).

Since USV drones are not subject to the same type of limitations due to human operator discomfort under adverse sea conditions as conventional boats, there is clearly scope to examine the possible benefits offered by a number of alternative types of hull configuration [5]. Any relaxation of environmental constraints may provide opportunities for increased speed and manoeuvrability performance that can be

Figure 15.1 A RIB-based USV conversion (Reproduced by permission of QinetiQ Ltd)

better exploited by alternative hull types rather than more conventional configurations. While a number of studies are addressing advanced SWATHs, planing hulls, hydrofoils, wave piercing hulls, very slender vessel hulls, surface effect hulls and wing-in-ground effect hulls, no definitive views have yet emerged on how such hull configurations impact on overall USV drone system effectiveness when considered in conjunction with all the rest of the USV system elements.

15.3.4 Auxiliary structures

Any USV will need to have auxiliary topside structures to protect on-board systems, to house or mount on-board sensors and to provide the USV with protection from environmental conditions and enemy threat systems. For any ship, but especially for small USVs, a key feature of any such auxiliary structure is its strength to weight ratio. The forces acting on a USV superstructure can be very high, particularly since one of the advantages of a USV is that it can operate over a wider range of sea conditions than a comparable manned system. The use of advanced lightweight composite materials is an important aspect of overall USV platform design since the mass of superstructure elements will affect the overall hydrostatic and hydrodynamic stability of the USV. It will also impact on the payload capacity of the USV. Glass fibre, carbon fibre and Kevlar reinforced composite auxiliary structures would normally be considered standard on modern USVs.

In military applications, many USV mission concepts require the above water signature of the USV to be low in order to minimise the probability of detection, classification and targeting by enemy forces. Various advanced stealth composite technologies can be applied to the superstructure of a USV and it is likely that the use of such materials will be increasingly evident in future USVs.

15.3.5 Engines, propulsion subsystems and fuel systems

Current developmental USVs tend to use standard commercial off-the-shelf (COTS) engines and propulsion systems. While clearly there may be scope for the use of hybrid engines where a low-noise electric motor based propulsion mode is used to give a greater degree of stealth, this aspect of USV propulsion system design is currently not considered to be the most critical.

Where jetski-based systems have been developed, the COTS marine engines, which tend to use unleaded petrol as the fuel, and the pumped waterjet propulsion systems are usually retained. Typically the engines used in these types of system will be 1000 cm^3 or below. Increasing the size of the fuel tanks allows endurance to be increased to typically in the order of 48 h, giving a range in the order of 200 nm. In RIB-based designs, standard COTS outboard engines are normally retained, currently giving an endurance of 8 h and a range in the order of 150 nm.

In shallow water applications, steering functionality is often provided by turning the propellers or diverting the water jets. The conversion of these types of engine, propulsion system and steering systems to semi or fully autonomous control can be accomplished relatively easily through the use of computer controlled actuators operating on conventional cable and rod arrangements.

The use of standard COTS engines and propulsion systems reduces procurement cost and ensures low maintenance costs and good availability of spares. The use of standard marine fuels ensures that standard fuel handling procedures and facilities can be used.

Two of the major limitations of current USVs are:

1. The need for relatively frequent engine maintenance in comparison with desired mission endurances.
2. Endurance limitations due to relatively small fuel tank capacities and current lack of on-line USV refuelling infrastructure.

To address the first limitation, it is considered that further research is required into low maintenance small marine diesel engines for USVs. To address the second issue, research and development work is required on automatic ship based refuelling systems such that USVs can dock temporarily with refuelling facilities deployed over the side of a refuelling ship to receive rapid injections of fuel. In advanced heterogeneous USV swarm concepts, it is envisaged that unmanned fuelling drones could be deployed to remotely refuel on-task USVs in a way that maximises their mission availability.

15.3.6 USV autonomy, mission planning and navigation, guidance and control

Current USV system architectures are generally based on hierarchical functional schemes which can equally accommodate autonomous USV functionality and operator-based USV remote control functionality. In these schemes, steering, engine control functions and payload operation are implemented within the lowest levels of the hierarchy, with guidance functions, navigation functions, reactive planning functions and tactical planning functions occurring at higher levels.

At present, most prototype USV systems use pre-programmed way-points as the basis for tactical mission planning, track definition and drone navigation. Methods for making on-line operator changes to way-points have been proven. Various intelligent systems techniques are potentially available for local replanning of way-points in response to fault conditions detected autonomously by onboard sensors and processing systems but at present there have not been many published examples of their use within a USV system.

USV navigation is generally based on the use of GPS or DGPS as the baseline positioning system. This is integrated with a ring laser gyro or fibre optic gyro system and self-calibrating magnetic compass system to provide high accuracy interpolated dead-reckoning data and heading data. If required, USV velocity through the water can be provided by electromagnetic log systems, acoustic log systems or simple mechanical logs.

USV guidance systems are usually based on heading guidance algorithms and two-dimensional track following algorithms which use position data, heading data, velocity data and angular rate data as inputs. These algorithms may be based on conventional PID control techniques or may include more advanced elements such as neurofuzzy techniques and state flow techniques to cope with complex guidance tasks and to mitigate system nonlinearities.

A key issue associated with USV operation is that of collision avoidance. Whereas the nature of UUV operations means that the risk of collision with ships is normally relatively low, with USVs transiting in open waters the risk of collisions with other surface vessels could be high and the consequences serious. Currently, USV collision avoidance tends to rely on operator intervention in response to possible collision threats identified by the operator using USV video camera pictures relayed back to the operator control console. Autonomous collision avoidance systems for USVs are currently very immature. The achievement of reliable, autonomous, all-weather marine obstacle detection and characterisation, that is robust to clutter in a USV's environment, is critical if the capability for autonomous USV collision avoidance is to be realised.

At the current stage of USV development, there are few dedicated USV visual, infrared, lidar or radar processing systems designed for use in autonomous USV collision avoidance applications. All weather detection and characterisation of obstacles will require multi-modal sensing capabilities combined with advanced filtering, data fusion and scene interpretation processing. While many of the required sensors and processing techniques have been developed for other manned and unmanned systems, their application for USVs and for collision avoidance in the marine environment has received relatively little research and development. In particular, information on the impact of operating sensors near the air–sea interface, the impact of different weather conditions, the impact of night operation and the characterisation of clutter as well as obstacles is very sparse.

15.4 USV payload systems

A number of abovewater surveillance, underwater search, mine countermeasures and attack payloads have been integrated to developmental USVs. The surveillance payload systems deployed and demonstrated to date include: low light cameras, video cameras, infrared imagers, laser range finders and radars. The underwater search and mine countermeasures sensors deployed and demonstrated to date include: underwater cameras, forward looking sector scan sonars, a Submetrix 2000 sonar, an AQS-14/20X sidescan sonar, an AQS-24 sonar, a Thomson-Marconi SAS TSM 2054 sidescan sonar.

Mini-dyad mine sweep systems and pipe noise makers have been deployed and demonstrated using USVs as tow platforms. These demonstrations have illustrated how synergistic benefits can be obtained when payload systems are integrated with USVs. In the case of shallow water mine sweeping, the shallow draft of USVs allows them to access areas where conventional minesweepers cannot go. The mine sweeping systems can also be used in shallower areas since the tow lengths are able to be shortened without incurring any greater risk to personnel.

To date, attack payloads have been limited to a Mini-Typhoon 12.7 mm gun deployed on the Rafael Protector USV. However, a future payload module under consideration for the US Spartan USV is said to include a 7.62 mm Gatling gun and possibly Hellfire or Javelin missiles.

15.5 USV launch and recovery systems

No definitive solutions for successful, damage-free USV launch and recovery from a range of host surface ships under a wide range of environmental conditions have yet been defined or proven. These systems are very important elements of the USV integration process which impact on the availability of USVs for operational tasks, the number of USVs that can be hosted and the number that can be operated at any given time.

While current launch and recovery systems include the use of A-frames, davits or cranes in conjunction with support boats or the use of stern ramps, various other near-term concepts have been identified by various researchers and industrial organisations. Most of these have yet to be demonstrated with representative USV drones.

Possible future deployment mechanisms include, for example:

(a) over-the-side chute
(b) advanced ship's davit system
(c) stabilised single attachment point crane
(d) stern docking gate with slipway ramp
(e) towed docking frame
(f) air drop from helicopter or aircraft.

Possible future recovery mechanisms include, for example:

(a) capture and lift using a net deployed from a crane (or helicopter)
(b) automatic docking to a towed frame followed by reeling-in to a stern slipway ramp on a host ship
(c) use of docking cones/probes followed by crane or lifting frame lift
(d) remote docking to a towed refuelling cone without recovery onto the host ship
(e) use of dedicated USV support boats equipped with lifting stern docking areas.

Future developments designed to facilitate rapid, safe USV launch and recovery will require autonomous or semi-autonomous USV docking capabilities and new types of stabilised handling system. Both of these developments will contribute to the realisation of a simple high-level operator command process for USV launch and recovery which initiates appropriate automatic sequences of actions by the USV drones, host ship and the launch and recovery systems.

15.6 USV development examples: MIMIR, SWIMS and FENRIR

15.6.1 The MIMIR USV system

In the 1980s, a number of small USV systems were developed at the Admiralty Underwater Warfare Establishment at Portland (UK) to support mine countermeasures

320 *Advances in unmanned marine vehicles*

Figure 15.2   MIMIR (Reproduced by permission of QinetiQ Ltd)

operations. One of these systems used a remotely controlled tri-hull drone to tackle floating mines by approaching the mines and attaching disposal charges to them. Several years after project completion, one of the tri-hulls was reused as the platform for an experimental USV system known as MIMIR (Figure 15.2) [6].

The MIMIR EV1 system was designed to investigate the feasibility of using USVs in underwater search and survey roles, the use of USVs within a networked information system and the potential cost/benefits of such systems. The aim was to maximise shallow water access and data acquisition opportunity. The overall MIMIR concept was to use multiple, low-cost, networked USVs operated from a mobile land-based command centre as nodes within a wireless data network. The data from the USV payloads could be transmitted to the mobile command centre or directly to third party systems using standard COTS networking technologies.

The current MIMIR EV1 is 3 m long and 1.5 m beam. It has a 9 hp single cylinder diesel engine which allows it to operate for in excess of 8 h at speeds of 3–4 knots. The navigation system uses a DGPS as its primary sensor together with a fluxgate compass. It hosts a Submetrix sonar system and a video camera. The sonar data are pre-processed onboard the USV using dedicated hardware and the sonar image data are passed to the sonar display console at the mobile base station in real time using a COTS wireless LAN communications link. USV health data and tracking information are also passed to the operator console using this link. Video data from the camera are passed to the mobile base station using a dedicated COTS video link.

MIMIR EV1 was initially operated in either a manual operator-in-the-loop mode or an autonomous mode. Following various in-water trials, it was decided that a semi-autonomous mode based on guidance functionality implemented at the mobile command centre workstation rather than within the onboard computer systems provided the basis for appropriately robust command and control.

The MIMIR EV1 mobile base station allows a MIMIR EV1 operator to carry out on-site mission planning, monitoring and control. It is based on a commercial four wheel drive vehicle, but the PC based command workstation and its associated communications subsystems can easily be relocated into a boat or building. MIMIR EV1 tracks and position are displayed using a standard low-cost COTS ECDIS system.

15.6.2 The SWIMS USV system

Following the experience of working with MIMIR EV1, the lessons learnt were applied almost immediately in response to a Royal Navy Urgent Operational Requirement for a shallow water influence minesweeping system (SWIMS) to support MCM operations in 2003 during the early phases of Operation TELIC in Iraq [7]. The expertise developed by the MIMIR team allowed an extremely rapid development, build and test programme to be achieved.

In addition to the short timescales, key requirements of the SWIMS programme were that the resulting USVs should be easily transportable and deployable, simple to operate and supportable in theatre. Following project start, a survey of COTS minesweeping equipment was conducted in parallel with a mine threat assessment. As a result of this activity, a number of sweep systems were identified as potential candidates for the shallow water sweep application. These were examined using QinetiQ's Total Mine Simulation System and environmental data sets for the operational area. As a result of the assessment process, mine sweeping equipment supplied by ADI Limited, comprising a mini-dyad system coupled with an acoustic pipe noise making subsystem, was selected as the SWIMS payload. Based on QinetiQ's previous surveys of possible small boats for use in shallow water minesweep towing operations, existing 8.8 m Combat Support Boat (CSB) Mk IIs built by RTK Marine and operated by the British Army were identified as the preferred options for towing the sweep system. These boats are powered by twin waterjets driven by two Yanmar diesel engines. It was further determined that an increased speed of advance could be achieved if two CSBs were used in parallel to conduct the sweep operation and that an even greater speed of advance could be achieved if the CSBs were converted into USVs since this would allow certain human related operational limitations to be overcome.

In December 2002, the UK Defence Procurement Agency tasked QinetiQ to adapt its proprietary USV technology to provide a USV control system that could be fitted on a non-interference basis to twelve CSBs. An outline system requirement for the control system was developed in conjunction with the Royal Navy team assigned to the SWIMS project. This outline requirement comprised eight main elements:

(a) To remove the need for a human CSB driver.
(b) To provide full control over:
 - port and starboard throttles,
 - port and starboard water jet buckets,
 - rudder.
(c) To provide a heading control mode.
(d) To achieve a minimum operating range of 5 nm ahead of a lead mine hunter.

(e) To provide the USV system operator with visual feedback of:
- port and starboard engine speed,
- port and starboard water jet bucket positions,
- rudder angle,
- port and starboard engine alarms.
(f) To provide visual feedback of USV position, speed and heading.
(g) To provide an HCI with track overlay and route planning facilities.
(h) To provide built in safety systems including:
- a set of predefined USV actions when the command signal is out of range,
- system functions and interlocks to ensure safe transfer between manual and automatic USV control modes.

The CSB partition of the control system was implemented using National Instruments hardware. Compact Fieldpoint industrial controllers with an embedded real-time operating system were used to host the control system and provide the input and output interfaces. Five analogue outputs were used to drive four linear actuators (port and starboard throttles and port and starboard water jets) and the CSB hydraulic steering pump. Port and starboard engine tachometer data were input to the controller via frequency-to-voltage converters and two analogue inputs. Port and starboard engine alarms were input to the controller via two digital inputs. RS232 channels were used for input of P-Code GPS data, compass data and the UHF data link. Within the on-board control system, the critical track following guidance mode was implemented using heading control in conjunction with a track bounding algorithm. This mode proved to be very successful and accurate track following to within ± 5 m of the planned baseline track was achieved operationally. The compact, modular controllers provided a high degree of physical robustness. Rapid application development and system prototyping was supported by a good set of integrated software tools (Figure 15.3).

The operator partition of the control system was developed using National Instruments Labview and a custom vehicle control console (Figure 15.4). Labview supported rapid prototyping, redesign and reconfiguration of the HCI as the system requirements for the user control functions, vehicle control console and charting system matured in response to developing SWIMS operational requirements and feedback from the Royal Navy SWIMS team. The charting system was implemented using the COTS Tsunamis Navigator Pro system. It was used to show the operator the position, heading, speed and track of the CSB together with the planned mission tracks.

The initial SWIMS demonstrator system was available within three weeks following contract placement. System demonstration, refinement, testing and operator training was conducted over the next three weeks. A team of eight Royal Navy personnel had been assembled for SWIMS operation and their rapid training programme comprised CSB familiarisation, trials conduct and remote control system operation, maintenance and fault finding (Figure 15.5).

In parallel with the test and training programme, 14 further CSB conversion kits were completed, ready for delivery to Bahrain and integration with the CSBs which were being transported separately. Base station fits to all host mine countermeasures

Unmanned surface vehicles 323

Figure 15.3 SWIMS USV drone carry on/carry off conversion kit (Reproduced by permission of QinetiQ Ltd)

Figure 15.4 SWIMS USV operator console and displays (Reproduced by permission of QinetiQ Ltd)

vessels and conversion kit fits to the 11 CSBs assigned to the SWIMS operation were completed within 3 days. Setting to work and testing of all systems was completed within a further 3 days. Formal sea tests, system acceptance and handover to the Royal Navy SWIMS team were completed within 2 weeks of equipment arrival in Bahrain.

Figure 15.5 SWIMS USV drone undergoing testing in Portland Harbour (Reproduced by permission of QinetiQ Ltd)

Following handover, operational rehearsals were conducted in conjunction with other MCM assets. SWIMS systems and personnel were transported to the rehearsal area on a Royal Fleet Auxiliary vessel and the team were then transferred to a Royal Navy Single Role Minehunter where the operator control system was set up. During the rehearsals, two SWIMS systems and sweeps were operated ahead of a lead mine hunter within track segments, this configuration providing protection for the lead mine hunter and allowing simultaneous hunting and sweeping operations to be conducted from the mine hunter. An important activity during the rehearsals was the validation of new safe operating procedures including those for:

- transferring from manual to remote control,
- transferring sweep systems between CSBs,
- underway refuelling,
- recovery of damaged systems.

Full operational deployment took place at the end of March 2003. The SWIMS system was deployed from HMS *Brocklesby* on a continuous basis to conduct precursor sweeps ahead of the main mine hunting fleet. Operational ranges in excess of 9 km ahead of the lead mine hunter were achieved. On task time was limited only by the fuel capacity of the CSBs – fuel capacity being a characteristic limitation on USV operations. To mitigate this problem, part of the temporary modification process included the installation of enlarged fuel tanks. In practice, cyclical replacement of empty CSBs with fully refuelled CSBs was used during the operation. During this first ever operational deployment of a USV system by the Royal Navy, it was confirmed that a 120 m wide swept corridor over a path length of 27.7 nm was achieved in less than 24 h.

The SWIMS development project is considered to be significant for several reasons beyond purely technical and operational reasons. First, the speed of system development achieved following concept acceptance and contract placement was significantly faster than the development timescales for other maritime unmanned systems. The conversion kit approach allowed flexible use of existing CSB assets and the carry on/carry off nature of the modifications allowed easy reversion back to the normal manned CSB configuration. Second, the conversion and re-role of existing assets demonstrated that the temporary mission focused implementation of multiple USV systems could be achieved at relatively low cost. This temporary status also avoided incurring any significant on-going support costs. Third, it demonstrated that the development to in-service acceptance timeframe for initial capability USVs can potentially be very short indeed. Hence, it may be that a cost-effective model for some future USV procurements will involve low-cost conversion of existing assets to achieve initial in-service semi-autonomous capability followed by modular incremental capability development in response to specific mission requirements.

15.6.3 The FENRIR USV system and changing operational scenarios

In future naval doctrine papers, an increasing emphasis is placed on littoral water and very shallow operations, coordinated operations in association with other maritime and air platforms and reduced cost of operations. The potential use and reuse of families of modular, flexible USVs in a broad range of roles and scenarios offers a new means to satisfy some of the emerging complex and contradictory requirements.

The FENRIR USV family is a generic set of small USV systems which is intended to act as the focus for advanced USV concept development and a variety of different network enabled capability experiments. Typically, FENRIR systems are being investigated to assess their ability to contribute to network enabled operations. For example, FENRIR derivatives are being assessed in the context of using USVs to deliver small UUVs and rapidly deployable sensors into theatre over relatively large stand-off distances. A major limitation of small shallow water reconnaissance UUV systems is the time and energy taken to transit to and from mission areas (Figure 15.6). USVs offer the potential for discrete long-range deployment of UUVs, allowing the UUVs to maximise their underwater on-mission duration and also, under some circumstances, providing a facility for relaying UUV data to mission control centres via satellite communication systems.

Other important areas of investigation are associated with FENRIR-based attack concepts. While some of these concepts envisage the use of single USVs in attack roles, other concepts involve the use of heterogeneous USV swarms communicating coordination commands and tactical information between each other using wireless data links. It is envisaged that these swarms could possess a range of behaviours which would allow them to implement a number of different warfighting roles, including barrier protection, blockade, harbour disruption, attack and counterattack of fast patrol boats, attack of major ship assets and shallow water attack of submarines.

326 *Advances in unmanned marine vehicles*

Figure 15.6 FENRIR rapid deployment into a shallow water area (Reproduced by permission of QinetiQ Ltd)

Typically, a number of key research issues arise when considering FENRIR systems. These include:

(a) Can cooperative multi-USV attacks be effective in meeting engagement goals?
(b) What range of USV payload systems is required for mission success within a heterogeneous USV swarm?
(c) What levels of USV attrition can be expected during operations and can these be mitigated at reasonable cost?
(d) Do radical USV designs have any significant advantages in warfighting scenarios?
(e) Do USVs offer a game changing way ahead for the conduct of some naval operations?

The answers to these are not yet known. However, it is clear that a number of nations are now developing multi-role USVs for near-term use alongside their existing naval assets. The relatively low implementation costs of USV systems together with the ability to integrate subsystems developed for other unmanned land, sea and air applications into USV systems mean that it is likely that some answers will emerge over the next 5–10 years.

15.7 The game changing potential of USVs

USVs have experienced a resurgence of interest over the past 5 years. With the advent of distributed computing, high-bandwidth wireless data networks, smart sensors and the increased capability of intelligent control systems, USVs now offer navies the possibility of game changing operational capabilities. The naval systems research and development communities have started to investigate how these capabilities might be realised and it appears clear that USV research and development will continue to grow over the next 10 years.

While the use of unmanned ships is not likely to be feasible or actually required in the near or medium term, the deployment and use of multiple small USV drones from mother ships is now a near-term reality. Many countries see the potential of USVs as force multipliers and as alternative ways of meeting complex operational requirements in a cost-effective manner. Several known major ship programmes are including allowances for the accommodation, deployment, operation and recovery of USVs in their plans.

The use of USVs to access and operate in high-risk operational areas instead of manned platforms is likely to be attractive not only to the personnel who would otherwise have to carry out the required operations, but also to planning staff who potentially will have access to a wider range of available low-cost assets and politicians who will appreciate not having to explain why personnel are being placed at risk.

In the near and medium terms it is likely that the greatest performance and benefits will arise from the incremental development and use of USVs in advanced semi-autonomous configurations which are based around existing platforms. At present, there does not seem to be any specific need to move towards fully autonomous USV systems. This level of capability has already been shown to be both expensive and technically difficult to produce within the computer hardware and software of other unmanned systems and there is no reason to expect that it will be any different for USVs. An advanced semi-autonomous mode allows USV operators to interact at a high level with remote USV drones while making a significant functional contribution in terms of scene and situation assessment. This will ensure that near term USV systems will possess the high levels of system effectiveness, reliability and robustness that will be required to convince navies that USVs provide a long-term game changing potential that can meet the needs of operations within the complex, ever changing maritime environment of the twenty-first century.

References

1 Saunders, S., Ed. (2004). Mine warfare forces, *Janes Fighting Ships*, pp. 177–178, 268–269, 706–707.
2 German Navy studies future mine warfare (1993). *Janes International Defense Review*, 26(8), p. 597.
3 Brown, N. (2004). More than just a remote possibility: USVs enter the fray, *Janes Navy International*. 109(1), pp. 14–19.

4 Palmer, W. and R. Brizzola (2004). West Bethesda helps implement unmanned vehicle technology. *Wavelengths*, pp.8–11 Naval Surface Warfare Centre, Carderock, February.
5 Newborn, D. (2004). Planing HYWAS integrated node unmanned surface vehicle (PHIN-USV), *Wavelengths*, p. 11 Naval Surface Warfare Centre, Carderock, August.
6 Corfield, S. (2002). Unmanned surface vehicles and other things. Unmanned underwater vehicle showcase 2002, Conference Proceedings, pp. 83–91 25–26 sprearhead Exhibtions, Southampton, September.
7 Ray, S. and M. Leaney (2004). The use of unmanned surface vehicles (USVs) to support extreme littoral water operations during operation TELIC. Underwater Defence Technology Europe 2004, Conference Proceedings CD-ROM, Session 10A-3, Nexus Media Ltd, Nice, June.

Chapter 16

Modelling, simulation and control of an autonomous surface marine vehicle for surveying applications Measuring Dolphin MESSIN*

J. Majohr and T. Buch

16.1 Introduction and objectives

In recent years, autonomous surface craft have been developed in particular for marine research and surveying exploration [1–3] as well as for the rescue of human life at sea. To provide a rescue vehicle, an autonomous *Rescue Dolphin* was developed to rescue people in distress [4]. The rescue system automatically triggers the alarm to the ship's management in case of 'man overboard', independently of ship manoeuvres. It moves fast towards the distressed person, and safeguards the distressed person until recovery by ship.

Market research has confirmed the demand for such a vehicle for shallow water surveying and oceanography. It should not have a detrimental impact on the surrounding environment and should operate with high-precision positioning and tracking in shallow waters. Conventional research and surveying vessels only operate up to a minimum depth of approximately 5 m. They cannot be applied in shallow waters, rivers or in areas which are difficult to access. At present, small vehicles such as rubber dinghies and ship's tenders are used for the purpose. As they do not provide ample space and stability, the valuable measuring equipment is always in danger of being damaged or even lost. Each measuring task required the construction of a special carrier which mostly was only a provisional arrangement. Such vehicles have further disadvantages: a precise track control is impossible, and the sensors do not work very reliably due to lack of stability and the influence of the surrounding environment. Until now it has been common to carry out research tasks on manned craft. This has

---

* Protected by Law No. 29905735.6.

proven to be very costly and time-consuming as well as fairly dangerous in special areas of operation (catastrophes, polar regions).

The development of the unmanned autonomous surface vehicle Measuring Dolphin was carried out within the framework of the German cooperation project MESSIN in the period from 1998 to 2000 sponsored by the BMBF (German Federal Ministry of Education, Research and Technology), project No. 03F0212C [5].

The main task of the project MESSIN consisted of the development and the testing of a prototype of the autonomous surface vehicle Measuring Dolphin which could be applied with high accuracy of positioning and track guidance and under shallow water conditions as a carrier of measuring devices. Fields of application include depth surveying, current and current profile measuring in port entrances and rivers, sediment research, extraction of samples for biological investigations and measuring in drinking water areas.

The important objectives are [6,7]:

- autonomously operating (i.e., unmanned) watercraft;
- automatic control towards a given target or automatic control following a pre-determined surveying guidance programme with metre range precision to travel over a specified water area;
- highly precise positioning and navigation as well as automatic track guidance and control with a low track deviation;
- minimal environmental influences, regardless of what they are (craft's flow distribution, magnetic fields, exhaust gasses, etc.);
- providing the carrier's motion data toward all axes;
- telemetry for transmission of measured data and important functional data to an operation(s) base (on shore or mother ship), for the setting of the parameters and steering of measuring equipment (e.g., sonar) and computers from shore;
- radio control for docking and undocking and retrieval of the Measuring Dolphin in critical situations;
- sufficiently lengthy deployment of the vehicle;
- readiness for rapid deployment by effective slip-operations and simple handling of the equipment carrier;
- development of the Measuring Dolphin as a modular system to be able to solve the various measuring tasks effectively at the lowest costs.

The systems for positioning and modelling, simulation and control of the MESSIN are described in this chapter. Moreover, some aspects of the development of the hull and the propulsion as well as the manoeuvring equipment are also taken into consideration.

16.2 Hydromechanical conception of the MESSIN

As a small shallow-draught vehicle, with a maximum of loading capacity for the storage of measuring equipment and a minimum of movement in rough seas (roll, pitch, yaw) was needed, we decided on a catamaran-type craft (see Figure 16.1).

Modelling, simulation and control of an autonomous surface marine vehicle 331

Figure 16.1 MESSIN in operation; the rudder propeller is represented in the left upper corner [8]

The catamaran design allows the construction of an independent craft fulfilling important safety requirements regarding capsizing and sinking [3]. Using the SWATH principle, both hulls are almost completely under water and therefore cut beneath the sea waves. The MESSIN moves relatively little at sea, and thus damage of the measuring equipment is avoided.

Using a model on the scale 1 : 2 the model errors are negligibly small. Seagoing-, resistance- and propulsion tests for the hydromechanical optimisation of the hull were undertaken at the Model-Testing Tank Establishment (SVA) Potsdam in the middle of 1998 [9]. The results encouraged the project team to carry on, and a full-scale test model of the MESSIN with few construction changes over the 1 : 2 scale model was built. The test model was equipped with propulsion and manoeuvring equipment and the necessary electrical/electronic control so that it could be manoeuvred.

In December 1998, manoeuvring tests were carried out with the self-driven prototype of the MESSIN at the SVA Potsdam [10]. These tests established that the craft has sufficient dynamic yaw stability on the surveying tracks with mostly constant courses. The deployment of a rudder propeller with counter-rotating propellers on each hull for propulsion and steering allows the necessary performance for good manoeuvrability. This confirms the results of an earlier investigation at the University of Rostock in which the theory of the slender body was applied on catamarans and tests with a catamaran model by the rotating-arm technique were carried out [11]. The hulls of the MESSIN are manufactured from highly durable glass-fibre material. The bows of the hulls are interchangeable so that a better handling of

Table 16.1   Main parameters of the MESSIN

Model	Catamaran hull
Movement parameters	
Minimum speed	$0.25\,\text{m s}^{-1}$
Maximum speed	ca. $2\,\text{m s}^{-1}$
Minimum turn radius	$3\,\text{m}$ ($2\,\text{m s}^{-1}$, max. payload)
Mission span	
Hybrid power supply	10 h
Power supply by battery	3 h
Masses	
Mass of the MESSIN	ca. 250 kg for full equipment
Maximum payload	ca. 100 kg
Dimensions	
Length × width × height	3300 mm × 1800 mm × 1500 mm
Draught	
Minimum (minimum equipment)	200 mm
Maximum (maximum payload)	400 mm
Electrical parameters	
Voltages	24/12 V
Power of drives	2 × 400 W
Power of the generator aggregate	1.2 kW
Power supply for measurement	300 W

the depth sensors and sonar system can be guaranteed. Good access is provided to the equipment holds in the hulls. Main parameters of the MESSIN are shown in Table 16.1.

16.3   Electrical developments of the MESSIN

A main part of the project was the development and construction of the complete electronic and electrical equipment of the MESSIN [12,13] so that it can fulfil the requirements for an autonomous surveying craft. The following components were developed:

- system for high-precision positioning and navigation;
- automatic track guidance and control system, modelling and identification of the motion behaviour of the MESSIN;
- hierarchical steering system including steering equipment for rudder machine and propellers;
- hybrid energy supply;
- electrical supply management and distribution;
- data telemetry control;
- motion reference unit;

- hydroacoustic depth sensor and ahead-looking sensors;
- integration of navigational and electrical components.

The development was carried out on the basis of standard components of automatic control and maritime devices. Special-purpose tools such as the hydroacoustics sensors or the generator aggregate were developed as prototypes. The selection of the electrical and electronic devices was carried out with regard to achieving low energy consumption. Thus, special emphasis was laid on the development of the power generation and the energy management. The energy supply was also formed as a modular construction system. The basic equipment consists of low-cost lead-acid accumulators which were positioned in the two hulls. The battery set can be replaced optionally by a generator aggregate in one of the hulls. An internal combustion engine developed for maritime applications generates an electrical power of 1.2 kW by means of a generator. For special applications, there are accumulators on the basis of silver–zinc cells, which have five times the capacity of lead-acid accumulators at the same mass. An energy management system checks the power consumption and when required switches out unimportant systems or reduces the driving speed in order to guarantee a secure return of the MESSIN.

16.4 Hierarchical steering system and overall steering structure

Departing from the task for the development of a universal system of an autonomous measuring craft, the equipment and the degree of automation has to be fitted to the measurement task. Hierarchical systems offer the opportunity on unit basis to increase the reliability and decrease the probability for a total loss. At the lowest level of the hierarchical system the reliability of the units is at the highest level, whereas the complexity and the variability increases to higher levels in the hierarchy. At higher levels, temporary system failures (software crashes) must be tolerated.

The following hierarchy levels were implemented on the prototype of the MESSIN:

1. Analogue controller units for rudder angle and number of revolutions of propeller with overload protection for every catamaran hull.
2. Radio data communication system with programmable controllers (PLCs) function including a hand-held terminal on the shore for service and check of the function of the MESSIN as well as navigation sensors.
3. Autopilot system for the automatic course control.
4. On-board computers for the control and check of the autonomous operation in combination with a control PC (shore component) as well as for data recording during the test.

The overall steering structure for the unmanned MESSIN is presented in Figure 16.2. First, the blocks with white background represent the lowest control level with the manual control elements and the control circuits for drive and rudder servo. A redundancy is established by a double equipment of propeller drive and rudder gear as well as the relevant automatic controller units. Second, the components that are shaded light-grey within the radio remote control, inclusive of PLCs, are implemented in the second level. The hardware supervision of the errors, status signals and the watchdog as well as the voltage supervision also occurs here. All important status information, error reports and measured values for the control of the movement state are transmitted to the hand-held terminal. The latter is used for the simple external control of rudder angle and propeller revolutions in the case of emergency recovery. In addition, the blocks of the navigation sensors belong to the second level. The level for automatic operation mode is represented by the block marked in dark grey. The autopilot carries out automatic course tracking, the initiation of a standard manoeuvre as reversal manoeuvre (U-turn) and the way-point navigation. Thus, the operating capability for surveying tasks is increased. The shaded blocks depict the highest level of the control hierarchy with the on-board computer and the on-shore control computer. The on-board computer has the highest complexity and flexibility, and is arranged at the fourth level. It is used to control and check the autonomous operation mode, to carry out complex manoeuvres for the target searching and target tracking, for collision avoidance as well as complicated docking manoeuvres. In addition, the system is able to react with different emergency programs to breakdowns of single components. The on-board computer can be employed for complex mission planning, track planning and track control in connection with an on-shore computer. Using redundant radio links, the communication to the on-shore unit can be guaranteed. The different radio systems for the service, the parameterisation and correction of position data are also represented in Figure 16.2.

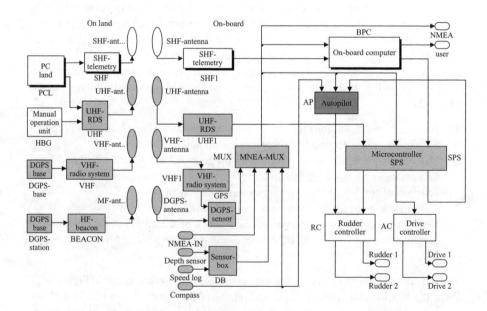

Figure 16.2 Overall guidance structure – MESSIN

The hierarchical principle is continued in the functionality and the operation modes. Using the remote control on shore, one of the following three steering modes can be selected:

- The first steering mode is the manual on-shore remote control, which is operated by use of a joystick for propeller revolution and rudder angle. Propulsion and rudder are controlled via a UHF radio data transmission system (RDS).
- The second mode of operation is the automatic operation mode which uses the functionality of the autopilot. The autopiloting via remote control is carried out by the programming of a predetermined course on the handset and transmission of the latter via the UHF RDS to the on-board autopilot which steers the rudder.
- The third operation mode is the steering program. Automatic control to a preset destination or automatic track guidance and control of a predetermined surveying track profile by the on-board computer can be realised. The setting of several steering programs and parameters of the on-board PC and other systems is carried out by the UHF RDS in connection with the on-shore computer. By means of an electronic sea chart system installed on the on-shore computer, the movement of the MESSIN can be monitored in the operation area. In addition, mission and/or guidance planning occurs here. Fully automatic control should be the normal mode of operation. The other two types of control are only intended to move the MESSIN to a surveying area and to recover it in case of accident.

Important navigational data (course, speed, depth) needed for the guidance of the MESSIN are transmitted via UHF RDS to the hand-held control and represented there. The complete structure (see Figure 16.2) also contains the SHF data telemetry for the transmission of measuring data from on board to shore and for the setting of the parameters for the measuring equipment.

The navigational sensors are represented in Figure 16.2.

- First, a DGPS satellite navigation receiver with the necessary VHF radio system for the transmission of the correctional data from its own reference station.
- Second, the receiving of the correctional data from a DGPS reference station via a HF beacon receiver.
- Additionally, an echo-sounder, a speed log and a compass. To check the speed condition of the vehicle, an impeller-log and a newly developed electromagnetic two-component log are used. As a compass, a magnetic compass supported by a yaw rate gyro is used which shows a very smooth course behaviour.
- Newly developed hydroacoustic depth sensors are used to exactly determine the depth of water (underwater acoustic transducer, vertical) and to recognise obstacles ahead (stones, poles, embankments) regarding their distance and bearing (acoustic transducer oblique port, starboard on each bow). These data are deployed for the anti-collision program of the on-board computer to start anti-collision manoeuvres.

16.5 Positioning and navigation

A DGPS positioning system for high precision (metre range), high availability (100 per cent, not 99.99 per cent!) and high reliability was to be developed and tested. These requirements result from the use of positional information for the safety relevant track control of the unmanned, autonomous MESSIN and the continuous marking of the measurement data with the position at that time.

Many surveying zones in the planned areas of activity of the MESSIN in shallow water are characterised by the close proximity to the shore, to trees, shrubs and buildings and the vehicle with its antennae lies relatively low on the water. There are no optimal conditions for the reception of the satellite signals as there would be in the open sea, instead shadow- and multi-path effects have to be expected. From the outset of the project, there was a requirement to support the DGPS system with a second back-up positioning method and to link both sets of position information via an algorithmic filter forming an integrated navigation system. High-cost support systems such as inertial navigation systems had to be left out from the start because they far exceeded the cost framework of the entire craft. In other words, intelligence-intensive low- and medium-cost solutions had to be found.

As the user is offered a modular concept for positioning which takes into consideration different costs and performance criteria depending on the type of measuring task, various navigational components from low-cost to high-end were manufactured and tested (see Table 16.2).

Wide-ranging tests of the low-cost configuration (Garmin 12-channel DGPS receiver with integrated antenna) in conjunction with the support sensor NHG100 [14] were conducted. The NGH-100 motion sensor, which hitherto has been used on land vehicles, measures yaw rates in the vertical and longitudinal axes by gyroscopes and acceleration component in the longitudinal axis of the vehicle by an acceleration sensor. On the basis of these measurement values, the tracking of the vehicle is carried out if the GPS receiver fails.

Before experiments at sea, several accuracy tests were carried out on a car. The reference positions were provided by the DGPS reference system based on

Table 16.2   Modular concept for positioning systems

Costs and positional accuracy	GPS technology	Motion sensors	Software solution
Low cost; 3–10 m (DGPS)	Garmin GPS 35	NGH-100 [14]	
Middle price segment; submetric (DGPS)	Trimble AgGPS 122 (yellow box)		OFSP on-line filter for estimation (model matching)
High-end (Reference system), <2 cm (RTK)	Ashtech Z-Surveyor	MRU-H [15]	

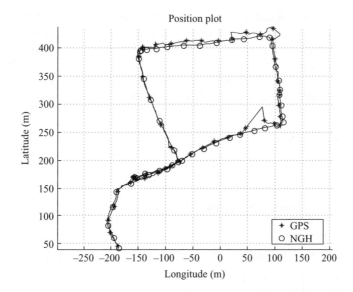

Figure 16.3   Position plot (m) under bad reception conditions

the Ashtech Z-Surveyor consisting of base and rover stations (3∗12-channel GPS 2 frequency receiver) with which the post-processing PDGPS-position by means of the RTK-mode (real-time carrier phase differential surveying) can be established to centimetre accuracy. The evaluation of test trips showed that the NGH-100 in integrated use (NGH-100 + Garmin) in optimal reception conditions resulted, as expected, in a less accurate position compared to the pure use of DGPS. The standard deviation was around 3 m. By contrast, the use of the integrated position solution provides advantages when bad reception conditions prevail in a chosen surveying zone (breaks in DGPS signals, multi-path and increased shadow effects) (see Figure 16.3).

During most of the surveying tests, the MESSIN was equipped with the DGPS-AgGPS Trimble 122 combining GPS and Beacon receiver in a water-protected case. The application of the integrated antenna for the GPS and the Beacon reception is advantageous because the space on the MESSIN is very limited and the minimum distance cannot often be maintained.

16.6   Modelling and identification

The identification of a mathematical model of the steering behaviour of the MESSIN formed the basis for the development and dimension setting of the course and track control systems. The identification was realised by means of manoeuvring tests of the full-scale model at the SVA Potsdam conducting zig-zag and turning manoeuvres [10]. Linearized course and track models of different order for the parameter identification were employed, where functional dependences of the coefficients on the craft's speed in the result of identification are considered. Only a second-order course model of Nomoto [16] and a fourth-order track model [17] shall be demonstrated in the following derivation. The use of these models for a twin-hull ship is permissible [11]. The developed state models are required for direct implementation into the corresponding identification tool of the MATLAB development system [18]. The use of linearised models allows the adaptation to the MESSIN characterized by other design parameters (length, width) with simple methods, such as look-up tables.

16.6.1 Second-order course model [16]

$$F_S(s) = \frac{\text{course } \psi(s)}{\text{rudder angle } \delta(s)} = \frac{K_S}{s(1 + sT_S)} \quad (16.1)$$

where K_S is the turning ability coefficient, T_S is the sum of time constants given by $T_1 + T_2 - T_{D1}$ (time constants corresponding to course model of third order, Eq. (16.8)).

The corresponding state equations give:

$$\dot{\bar{q}} = \begin{bmatrix} 0 & 1 \\ 0 & k_1 \end{bmatrix} \cdot \bar{q} + \begin{bmatrix} 0 \\ k_2 \end{bmatrix} \cdot (\delta + k_3)$$

$$y = q_1$$

$$k_1 = -\frac{1}{T_s}; \quad k_2 = \frac{K_S}{T_S}$$

$k_3 = \delta_{R0}$ bias-value

$q_1 = \psi$ course angle

$q_2 = r$ course angular velocity

initial values of the model are

$$q_1(0) = \psi(0)$$

$$q_2(0) = r(0)$$

16.6.2 Fourth-order track model [17]

The transfer function relates the cross-deviation of the straight track to the rudder angle and is calculated by the course and yaw angle transfer functions. The track angle results from the difference between course angle $\psi(t)$ and yaw angle $\beta(t)$ [17]

$$\phi(t) = \psi(t) - \beta(t) \quad (16.2)$$

The cross-deviation from a straight track (track deviation) then results from the relationship

$$\frac{dY_B}{dt} = v \cdot \sin \phi(t) \quad (16.3)$$

where v is the velocity of the ship.

On the condition that the changes of the track angle are small, the sine function in Eq. (16.3) can be replaced by its argument, which yields

$$\frac{dY_B}{dt} = v \cdot \phi(t) = v \cdot [\psi(t) - \beta(t)] \tag{16.4}$$

According to the Laplace transform of Eq. (16.4), one obtains:

$$s \cdot Y_B(s) = v \cdot [\psi(s) - \beta(s)] \tag{16.5}$$

All terms are divided by the rudder angle, so that the following transfer functions results

$$\frac{Y_B(s)}{\delta(s)} = \frac{v}{s} \cdot \left[\frac{\psi(s)}{\delta(s)} - \frac{\beta(s)}{\delta(s)} \right] \tag{16.6}$$

The following parameters are then introduced:

$$F_\psi(s) = \frac{\psi(s)}{\delta(s)} = \text{transfer function relating course } \psi(s) \text{ and rudder angle } \delta(s)$$

$$F_\beta(s) = \frac{\beta(s)}{\delta(s)} = \text{transfer function relating yaw angle } \beta(s) \text{ and rudder angle } \delta(s)$$

$$F_{Y_B}(s) = \frac{Y_B(s)}{\delta(s)} = \text{transfer function relating track deviation } Y_B(s) \text{ and rudder angle } \delta(s)$$

Thus, one obtains

$$F_{Y_B}(s) = \frac{v}{s} \cdot [F_\psi(s) - F_\beta(s)] \tag{16.7}$$

This representation of the track deviation transfer function has the advantage that both the course- and the yaw angle transfer functions are explicitly contained and transfer functions of other arrangements can be inserted.

The third-order Nomoto transfer function is inserted for the course transfer function [16]

$$F_\psi(s) = K_S \frac{(1 + sT_{D1})}{s(1 + sT_1)(1 + sT_2)} \tag{16.8}$$

The following transfer function is used for the yaw angle transfer function:

$$F_\beta = K_\beta \frac{(1 + sT_{D2})}{(1 + sT_1)(1 + sT_2)} \tag{16.9}$$

Inserting Eqs. (16.8) and (16.9) into Eq. (16.7), the following fourth-order transfer function for the track deviation is obtained:

$$F_{Y_B}(s) = v \cdot \frac{[s^2(-K_\beta T_{D2}) + s(K_S T_{D1} - K_\beta) + K_S]}{s^2(1 + sT_1)(1 + sT_2)} \quad (16.10)$$

The track deviation transfer function is characterised by a double integral behaviour, a second-order delay with the time constants T_1 and T_2, and a second-order differential behaviour with the time constants T_{D1} and T_{D2}. One of the two zeros of the track deviation transfer function is usually situated in the complex right half plane so that a non-minimum phase characteristic (all-pass network) is available [17].

The following model of state space can be derived from this transfer function in observer normal form:

$$\dot{\bar{q}} = \begin{bmatrix} 0 & 0 & 0 & 0 \\ 1 & 0 & 0 & 0 \\ 0 & 1 & 0 & k_1 \\ 0 & 0 & 1 & k_2 \end{bmatrix} \cdot \bar{q} + \begin{bmatrix} k_3 \\ k_4 \\ k_5 \\ 0 \end{bmatrix} \cdot (\delta + k_6) \quad (16.11)$$

$$y = q_4$$

$$k_1 = -\frac{1}{T_1 T_2} \qquad k_2 = -\frac{T_1 + T_2}{T_1 T_2}$$

$$k_3 = \frac{v K_S}{T_1 T_2} \qquad k_4 = v \frac{K_S T_{D1} - K_\beta}{T_1 T_2}$$

$$k_5 = -\frac{v K_\beta T_{D2}}{T_1 T_2}$$

$k_6 = \delta_{R0}$ (bias-value)
$q_4 = y_B$ (track deviation)
q_1, q_2, q_3 (mathematical quantities)

Initial values for this model are

$q_3(0) = \psi(0)$ (initial value for the course angle)
$q_1(0), q_2(0), q_3(0) = 0$ (initial value for the mathematical quantities)
$\bar{q}$ state vector which consists of physical ones and purely mathematical quantities.

During identification, the use of models of different order is very advantageous. The precision of the parameter identification of the more complex models can be increased substantially if the estimated parameters of the models of low order are used as initial values. The parameters of the models mentioned above are determined by the prediction error method (PEM), which is a part of the Identification Toolbox IDENT-Tool of the MATLAB© development program [18].

A selection of the estimated parameters of the Nomoto-course model, second order (Eq. 16.1) represented as functions of the vehicle's speed is shown in Figure 16.4 and that of the track deviation model (Eq. 16.10) is shown in Figure 16.5. Using

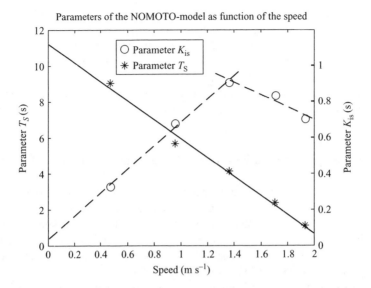

Figure 16.4 Parameters of the Nomoto-course model

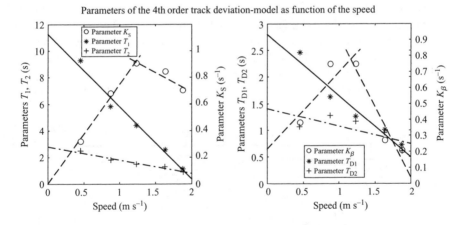

Figure 16.5 Parameters of the fourth order track deviation model

these results, the controller equipment can be adapted to the changing, speed related manoeuvring characteristics of the craft.

Refering to the Nomoto-course model second order (see Figure 16.4), the evaluation shows an increasing growth in the turning ability coefficient K_S (that is approximately linear in relation to speed) to around $1.4 \, \text{m s}^{-1}$ (3 knots) and followed by a slight reduction. The latter effect is considered to be caused by the influence of the edges of the water tank. This was not found in the case of free sea tests. The inertia time constant T_S reduces, depending on the speed, in a linear way. High turning ability coefficients K_S and small inertia time constants T_S, corresponding to a high coefficient $K_S * T_S^{-1}$, reveal a good control property of the MESSIN.

The results of the parameter estimation of the track deviation are represented in Figure 16.5. Both delay and differential constants are reduced with the increase of the speed. The higher parameter is more dependent on the speed than the lower. Both amplification factors K_S and K_β increase up to one specific speed value again (approximately $1.2\,\mathrm{m\,s^{-1}}$) and decrease thereafter.

16.7 Route planning, mission control and automatic control

The block diagram of the track guidance and control system of the MESSIN is represented in Figure 16.6. The planning of routes and measuring scenarios are realised with the aid of the electronic sea chart (ESK). On the one hand, the simple way-point navigation describing the route as a polygonal sequence is used. On the other hand, circular arcs with high accuracy are required for surveying tasks. Further on, the handicap and checking of the speed of the MESSIN is necessary.

To unify these different planning and supervisory functions, homogeneous elements for track planning were employed. This planning element, the so-called track section, consists of a straight line with defaulted length and direction as well as a circular arc with defaulted radius, which follows tangentially the original straight line and closes with the end point of the track section. This end point results from the set of the course of the straight line and the desired course difference to port or starboard. Further sets for the track section are the beginning and/or the final speed and the permissible track deviation. The track sections are combined with each other on the condition that the initial position, the initial direction and the initial velocity

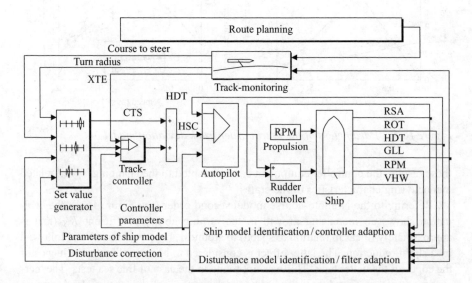

Figure 16.6 Block diagram of the track guidance and control system of the MESSIN

match the final position, the final direction and the final speed of the previous section. Thus, the length of the straight line segment can be made zero so that a pure circular arc results. The minimum radius of the circular arcs, however, depends on the manoeuvering qualities of the MESSIN and cannot be made zero (that is only possible if a dynamic positioning system is used). Therefore, a circular arc is necessary for every course-change. The zero track section is employed to start a surveying field. It is dynamically planned on the basis of the current position, the current course, the current speed as well as the default settings for the turning radius and the permissible cross-track deviation. The unit 'route planning' supplies the unit 'track check' with the sets of the track sections as a list. The unit 'track check' checks the current position and the current speed with the data of the current track section, sets the desired values of course, turning radius and speed and initiates the transition on the next track section if the final position of the current track section is achieved. The 'set value generator' provides the time-controlled set values, as well as the model of the ship's dynamics in order to calculate the 'wheel over points' and the disturbance signal such as sea current. The track deviation, the set points and the actual values are introduced into the cascade control, which is composed of the track distance, course, rate of turn and the speed controllers. The actual control of the set values and the disturbance values occurs here. The unit 'identifier' determines the current ship parameters, the optimal controller parameters and the disturbance parameters such as sea current and sea waves. Here, the identification can be carried out as a combination of off-line identification, speed adaptation and on-line identification of the disturbance signals.

The track guidance system introduced on the basis of track sections and the subsequent track control with a cascade automatic controller is very flexible, it permits a highly precise track guidance along typical forms of surveying tracks. Additionally, tasks of dynamic positioning can be solved by this track control system.

In addition to the cascade control, which is designed on the basis of the classical root locus procedure [5,19], a direct design method of digital controllers in continuous time developed recently was applied to the the course control system of MESSIN [20–23]. Unlike the classical design procedures for sampling controllers, the latter allows the representation of the exact process behaviour between sampling times, which is important for continuous ship processes with stochastic disturbances for a better design of the controller. By means of the present MATLAB Toolbox SISO direct [24,25] a H2 automatic controller with two degrees of freedom (2-DOF) is designed for the course control of the MESSIN. Design using two independent controllers has the advantage that one is able to determine the properties of the inner loop, that is, stability, robustness and disturbance rejection, while the the other is designed with the only aim to improve the follow-up performance. This system is better than the conventional single loop control scheme because here it is difficult to find a trade-off between good compensation of stochastic disturbances and a good follow-up behaviour.

Figure 16.7 shows the block diagram of the 2-DOF sampled-data system. The craft to be controlled consists of two separate blocks with the following transfer functions, based on the Nomoto course model of second order, Eq. (16.1):

$$F_1(s) = \frac{r(s)}{\delta(s)} = \frac{K_S}{1 + sT_S}$$

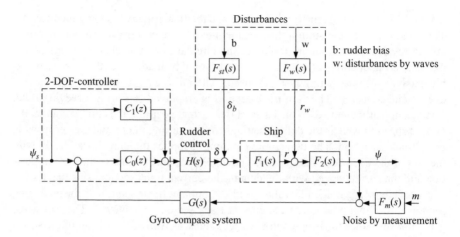

Figure 16.7   Control loop of the sample data system of the MESSIN

where r is the rate of turn and

$$F_2(s) = \frac{\psi(s)}{r(s)} = \frac{1}{s}$$

The craft is acted upon by a wave disturbance. The latter is simulated as a stochastic signal at the output of a forming filter with transfer function [26]

$$F_w = \frac{r_w(s)}{w(s)} = \frac{K_w s}{s^2 + 2\lambda\omega s + \omega}, \quad r_w(s) = \text{rate of turn}$$

excited by zero mean Gaussian white noise $w(s)$.

The control loop also includes the actuator (rudder with own control system) $H(s)$ and a gyro-compass as dynamic negative feedback. The compass measuring system and the complete rudder machine of the ship each can be modelled as a first-order dynamic system. The digital controller consists of the feedback controller $C_0(\varsigma)$ inside the loop and the reference controller $C_1(\varsigma)$. The feedback controller is optimised to ensure minimal course deviation and rudder activity under stochastic wave disturbances, while the reference controller shall ensure a quick follow-up response when the desired course is changed.

In Figure 16.8 the designed 2-DOF-controller is compared with the existing PID-controller by means of the rudder time behaviour and the course follow-up behaviour at a course ramp. Thus, cases with and without stochastic disturbances are considered. The advantages of the former are a better follow-up-course behaviour and a decrease of rudder deflections.

16.8   Implementation and simulation

The prototype of the MESSIN consists of the essential units and function blocks for the realization of the control modes manual control, automatic and program control.

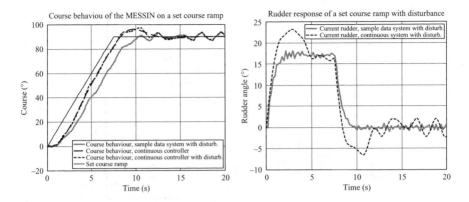

Figure 16.8   Comparison of a time-continuous controller with a sample data controller

The analog measuring and set values as well as the status signals were introduced via external radio remote control with integrated programmable controllers (PLCs). In addition, the autopilot can be operated via external remote control. A PC104 board installed in a waterproof case is used as the on-board computer. It is combined with a synchronous RS232 interface in the PLC. The program mode provides the opportunity that the on-board computer can default the set points and the operating functions for the propulsion, rudder and the course controller. In addition, the on-board computer has one or several NMEA-183 interfaces, which connect the navigation sensors and the autopilot. Selected NMEA data are transmitted via the synchronous RS232 interface to the shore station. In general, the software of the on-board computer was inserted into the programming language Matlab/Simulink [35]. The real-time programming was carried out using the toolboxes 'Realtime-Workshop' and 'RealLink32', the later Matlab Toolbox 'xPCTarget'. In this way, it was possible for the first time to implement the entire design process, the off-line simulation and the generation of a practicable real-time program for the on-board computer to Matlab/Simulink. The entire controlling algorithm of the MESSIN was completely represented with Simulink blocks. The NMEA interfaces were merely linked by Simulink into specific function blocks as a C program. In addition to the generation of the real-time program, xPCTarget also permits the communication and the on-line setting of the parameters of the on-board computer from a host PC. Thus, an extensive check and installation of the software was possible.

In the lab tests, the tool Matlab/Simulink was used alongside with xPCTarget. The MESSIN was programmed as a nonlinear model with Simulink and implemented by means of xPCTarget as a real-time program in an industrial PC (IPC). All analog, digital and serial process interfaces for the navigation sensors and the actuator organs of the MESSIN were provided by the hardware of the industry PC. Thus, the entire track control process could be simulated using 'Hardware in the loop'. Both the on-board computer and the ship simulator MESSIN could be operated and parameterised by the host PC. The compass course was introduced over a turntable

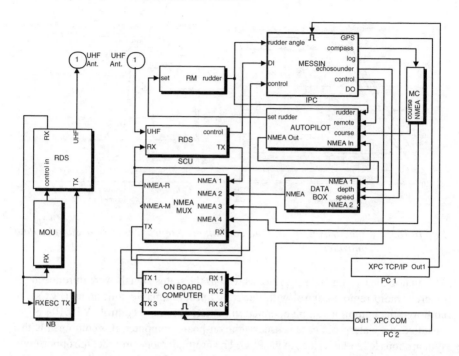

Figure 16.9   Simulation system SIMMESS

and therefore the magnetic compass turned correspondingly. So the original data processing of the compass with the rate of turn sensor could be included into the simulation. Figure 16.9 shows the complete simulation system, called SIMMESS, for the MESSIN. The shaded blocks are the original units of the shore components and of the on-board components. The notebook (NB) containing the electronic sea chart (ESC), the manual operation unit (MOU) as well as the radio data communication system (RDS) with UHF antenna communicates with the shore control unit (SCU). The following on-board components were included as original units in the simulation: the radio data communication system with integrated PLC, the rudder machine (RM) with the analogue actual rudder angle, the magnetic compass on a turntable (MK), the NMEA interface multiplexer (NMEA-MUX), the date box for log and ultrasonic depth finder, the autopilot and the on-board computer. The service and checking of the simulation was carried out by host PC's (PC1 and PC2). PC1 communicates to ship simulator for MESSIN via xPC-TCP/IP-Link. PC2 communicates to on board computer via xPC-RS232-Link. The simulation system SIMMESS proved to be very effective for the design and initiation of the MESSIN.

16.9 Test results and application

After successfull laboratory tests, the MESSIN was subjected to different sea tests. To check the sea and manoeuvre qualities all manoeuvring data were stored in the board

PC and evaluated on shore. The motion data were measured by the highly precise Surveying GPS System Ashtech Z-Surveyor and the motion reference sensor MRU-H [15]. With these data, the models of movement of the MESSIN were verified under free voyage conditions. The Trimble DGPS receiver AgGPS 122 was employed for the following tests of the surveying manoeuvres and of the automatic track control, achieving an absolute accuracy in the 5 m area. The relative accuracy was to the greatest extent of less than 1 m.

The following figures show examples of the manoeuvring experiments. A very good manoeuvrability is achieved with the rudder propeller. A radius of the turning circle of approximately 6 m is already achieved in the case of a rudder angle of 30° and maximum speed (Figure 16.10). A rudder angle sector of ±90° can be employed for special manoeuvres, such as turning on the point. The manoeuvring qualities of the MESSIN when under full payload (total mass 350 kg), are summarised in Table 16.3.

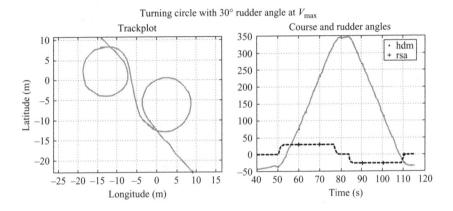

Figure 16.10 Turn cycle manoeuvres

Table 16.3 Manoeuvring quantities of the MESSIN when under full payload (total mass 350 kg)

Manoeuvre	Value	Comment
Radius of turning circle, rudder angle: 30°	6.25 m	Voyage stage 5, propeller rpm: 800 min^{-1}
Radius of turning circle, rudder angle: 60°	1.50 m	Voyage stage 5, propeller rpm: 800 min^{-1}
Radius of turning circle, rudder angle: 75°	0.25 m	Voyage stage 5, propeller rpm: 800 min^{-1}
Run out manoeuvre	20 m	Voyage stage 0, propeller rpm: 0 min^{-1}
Stop manoeuvre, machine backwards	4 m	Voyage stage R5, propeller rpm: -800 min^{-1}

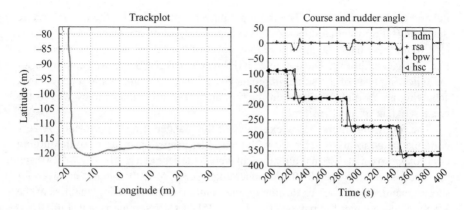

Figure 16.11 Ninety degree course-change manoeuvres with autopilot AP 11

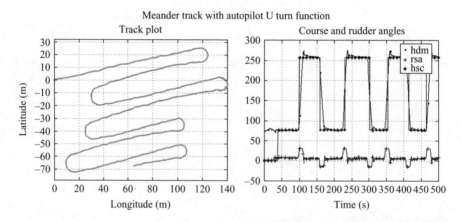

Figure 16.12 U-turn manoeuvre

Later, the on-board autopilot AP 11 (SIMRAD) was tested. Figure 16.11 shows a 90° manoeuvre.

The U-turn manoeuvre can be used for simple surveying tasks as a function of the autopilot. Figure 16.12 shows an example. The unequal turning radii in Figure 16.12 result from the eastern wind prevailing during the test.

The tracks controlled by the on-board computer achieved an accuracy of about 0.5 m under a sea current condition of $0.5\,\mathrm{m\,s}^{-1}$.

After its successful tests, the MESSIN was applied for different measurement tasks. Figure 16.1 shows the MESSIN during a depth measurement in the mud flats of the river Jade. A surveying ultrasonic depth finder was installed as a payload. When employing the MESSIN for current field measurements near port entrances and mole systems [27–33], an ADCP (acoustic Doppler current profiler) current sensor [34] was installed, which is suitable for current profile measurement in individual layers. Here the advantage of the MESSIN, that is, its small influence on the current field showed to its full extent. Additionally, the current measurement was also possible in shallow water because of its small draught.

The tests and measuring voyages of the MESSIN certified the very good manoeuvering and sea qualities. The construction similar to SWATH leads to a good sea attenuation. In the case of calm-rippled sea (significant wave height: 15 cm, wave period: 5 s, maximum speed) and a drifting vehicle the pitch attenuation amounted to approximately 15 dB. The rolling and pitch angle amplitudes lay in the range of 1–2°. Heave movements showed the high values typical for SWATH vehicles with an amplitude of approximately 10 cm during a period of approximately 5 s.

References

1 Robot for exploration of the sea: The catamaran DELFIM and the submarine boat INFANTE (in German), newspaper of the Portuguese pavilion on the EXPO 2000, the Sociedade Portugal, 2001, p. 7.7.
2 Encarnacao, P., A. Pascoal, and M. Arcak (2000). Path following for autonomous marine craft. In Proceedings of the 5th IFAC Conference on Manoeuvring and Control of Marine Craft (MCMC 2000), Aalborg, Denmark, 23–25 August, pp. 117–122.
3 Vaneck, T., C. Rodriguez-Ortiz, M. Schmidt, and J. Manley (1996/1997). Automated bathymetry using an autonomous surface craft. *Journal of the Institute of Navigation*, 43, pp. 407–417.
4 Hahne, J., T. Buch *et al.* (1995). Automatic man overboard rescue equipment (Rescue Dolphin) (in German), Final report, Institute for Safety Technology/ Traffic Safety, Registered Association, Rostock-Warnemünde, Project No. 513/95, Federal Ministry of Economy and Labour (Bundesministerium für Wirtschaft und Arbeit, Germany), 1995.
5 Lampe, B., J. Majohr *et al.*: Joint projekt Messdelphin (MESSIN), Subprojekt 3: Development and system integration of the components navigation, automatic control, communication and energy supply (in German), Final Report, 207 p., Rostock-Warnemünde, 26.02.2001, University of Rostock, Faculty of Electrical Engineering and Information Technologies, Project No. 03F0212C, Federal Ministry of Education, Research and Technology (Bundesministerium für Forschung und Bildung, BMBF, Germany).
6 Majohr, J., T. Buch, and C. Korte (1999). Conception of the navigation and the automatic control of the catamarane-measuring devices carrier Meßdelphin (MESSIN) (in German) *Ortung und Navigation*, 2, pp. 73–82.
7 Majohr, J., T. Buch, and C. Korte (2000). Navigation and automatic control of the Measuring Dolphin (MESSIN). In Proceedings of the 5th IFAC Conference on Manoeuvring and Control of Marine Craft, MCMC 2000, Aalborg, Denmark, 23–25 August, 405–410.
8 Hahne, J. *et al.* (2001). Joint project Messdelphin (MESSIN), Subprojekt 2: System definition as well as design of the boat body including propulsion, measure plat form, launching and lowering technology (in German), Final

Report, Institute for Safety Technology/Traffic Safety, Registered Association, Rostock-Warnemünde, 26 February.
9 SVA Potsdam (1998): Resistance and sea tests with the model of a small catamaran (in German), Report of the SVA Potsdam, No. 2402, June.
10 SVA Potsdam (1998). Manoeuvring tests with the original of a small catamarane measuring dolphin (MESSIN) (in German). Report of the SVA Potsdam.
11 Friedrichs, K. (1970). Transverse forces and moments as a result of motion of catamarans and investigation of their dynamic yaw stability and steering behaviour (in German). *Schiffbauforschung*, 9(1/2), pp. 25–40.
12 Buch, T. and J. Majohr (2000). Presentation of the catamarane MESSIN (in German), In Proceedings of the 6th-*Shipping College*, Institute of Shipping, Warnemünde, 5. p.
13 Majohr, J., T. Buch, and C. Korte (2001). Conception of navigation and automatic steering of the Measuring Dolphin (in German). In Proceedings of the 9th IFAC Symposium Control in Transportation Systems, 13–15 June, Braunschweig, Germany, 6 p.
14 Operation manual, Navigates NGH-100, Autonomous navigation sensor module (in German), Perform Tech, version 2.00 d, 13 November 1997, Hildesheim, 1997, 4 p.
15 Data sheet MRU (motion reference unit)-H, Seatex AS, Trondheim, Norway, 1997.
16 Nomoto, K., T. Taguchi, K. Honda, and S. Hiramo (1957). On the steering qualities of ships. *International Shipbuilding Progress*, 4, pp. 354–370.
17 Majohr, J. (1985). Mathematical model concept for course- and track control objects of ships (in German). *Schiffbauforschung*, 24(2), pp. 75–89.
18 Matlab, Ident-Tool, The Mathworks, Inc., 1999.
19 Unbehauen, R. (2002). *System Theory* (in German), Part 1, Oldenbourg-Verlag, Munich, Vienna.
20 Ladisch, J. (2001). Digital control in continuous time-process oriented design of controller for maritime applications (in German). Master's thesis, University of Rostock, Germany.
21 Rosenwasser, E.N. and B.P. Lampe (2000). *Computer Controlled Systems: Analysis and Design With Process Oriented Models*. Springer, London.
22 Rosenwasser, E.N., K.Y. Polyakow, and B.P. Lampe (1996). Design of a optimal ship course controller by parametric transfer functions (in German). *Automatisierungstechnik*, 44(10), pp. 487–495.
23 Ladisch, J., K.Y. Polyakow, B.P. Lampe, and E.N. Rosenwasser (2004). Optimal design of 2-DOF digital controller for ship course control system. In Proceedings of the IFAC Conference on Control Applications in Marine Systems, CAMS 2004, 7–9 July, Ancona, Italy, pp. 257–262.
24 Polyakow, K.Y., E.N. Rosenwasser, and B.P. Lampe (1999). Direct SD-a toolbox for direct design of sampled data systems. In Proceedings of the IEEE International Symposium CACSD'99, Kohala Coast, Island of Hawaii, USA, pp. 357–362.

25 Polyakow, K.Y., E.N. Rosenwasser, and B.P. Lampe (2001). Optimal design of 2-DOF sampled data systems. In Proceedings of the 13th International Conference on Process Control, Strbske Pleso, SK.
26 Fossen, T. (2002). *Marine Control Systems*. Marine Cybernetics AS, Trondheim, Norway.
27 Korte, H., H.-D. Kachant, J. Majohr, T. Buch et al. (2001). Conception of a modern manoeuvring prediction system for ships (in German). In Proceedings of the 2nd Braunschweiger Symposium Automatisierungs- und Assistenzsysteme für Transportmittel, 20–21 February, Braunschweig, Germany.
28 Korte, H., T. Buch, and H.-D. Kachant (2001). Manoeuvring prediction of ships with including external current information (in German). In Proceedings of the 10th Symposium on Maritime Electronics, 6–7 June, University of Rostock, Germany, pp. 89–92.
29 Majohr, J., T. Buch, C. Korte, and M. Wulff (2001). Sea current ADCP-measures by the surveying catamarane MESSIN in Puttgarden (Isle of Fehmarn/Germany) (in German), In Proceedings of the Hydrografentag, German Hydrographic Society, Part III. 18–20 June, Potsdam, Germany, pp. 1–8.
30 Majohr, J. and H. Korte (2002). Conception of a modern manoeuvre prediction system for ships-inclusion of the current information by ADCP current profile measures with the surveying catamarane MESSIN before the entrance of the ferry port Puttgarden (in German). *Ortung und Navigation*, 1, pp. 35–55.
31 Korte, H., H.D. Kachant, J. Majohr, T. Buch, T. et al. (2001). Concept of a modern manoeuvre prediction system for ships. In Proceedings of the IFAC Conference on Control Applications in Marine Systems, CAMS, No. WA 4.3, 18–20 July, Glasgow, UK.
32 Korte, H., J. Majohr, C. Korte, and M. Wulff (2003). ASFOSS-current information via AIS (in German). In Proceedings of the 9th Schiffahrtskolleg, 5–6 November, Institute of Shipping, Warnemünde.
33 Korte, H., J. Ladisch, M. Wulff, C. Korte, and J. Majohr (2004). ASFOSS-current information system using AIS. In Proceedings of the IFAC Conference on Control Applications in Marine Systems, CAMS, 7–9 July, Ancona, Italy, pp. 179–184.
34 RD Instruments (1989). *Acoustic Doppler Current Profilers – Principles of Operation: A Practical Primer*.
35 Matlab™. The MathWorks, Inc., Natick, MA, USA (http://www.mathworks.com/).

Chapter 17

Vehicle and mission control of single and multiple autonomous marine robots

A. Pascoal, C. Silvestre and P. Oliveira

17.1 Introduction

The last decade has witnessed a tremendous progress in the development of marine technologies that provide scientists with advanced equipment and methods for ocean exploration and exploitation. Recent advances in marine robotics, sensors, computers, communications and information systems are being applied to develop sophisticated technologies that will lead to safer, faster and far more efficient ways of exploring the ocean frontier, especially in hazardous conditions. As part of this trend, there has been a surge of interest worldwide in the development of autonomous marine robots capable of roaming the oceans freely, collecting data at the surface of the ocean and underwater on an unprecedented scale. Representative examples are autonomous surface craft (ASC) and autonomous underwater vehicles (AUVs). The mission scenarios envisioned a call for the control of single or multiple AUVs acting in cooperation to execute challenging tasks without close supervision of human operators. Furthermore, it should be possible for users who are not necessarily familiar with the technical details of marine robot development to do mission programming and mission execution tasks. Thus the need to push the development of methods for reliable vehicle and mission control of single and multiple autonomous marine robots.

The chapter addresses the topics of marine vehicle and mission control from both a theoretical and a practical point of view. The presentation is rooted in practical developments and experiments carried out with the *Delfim* and *Caravela* ASCs, and the *Infante* and *Sirene* AUVs. Examples of mission scenarios with the above vehicles working alone or in cooperation set the stage for the main contents of the chapter.

17.2 Marine vehicles

This section provides a brief description of representative marine vehicles that will be used to motivate mission scenarios and control design techniques in the chapter. The selection includes the *Sirene* and *Infante* AUVs and the *Delfim* and *Caravela* ASCs. Except for *Sirene*, all the vehicles were designed and built by consortia of Portuguese companies and research institutes.

17.2.1 The Infante AUV

Figure 17.1 shows the *Infante* AUV, designed and built by the Instituto Superior Técnico through its Institute for Systems and Robotics. The AUV is the result of a major redesign of the *Marius* AUV [1,2], aimed at obtaining open loop vertical plane stability, increased manoeuvreability, and adequate performance even at low speeds. See Asimov Team [3] and Silvestre [4] and the references therein for descriptions of the vehicle and illustrative mission scenarios.

The vehicle is equipped with two stern thrusters for propulsion and six fully moving control surfaces (two stern rudders, two bow planes and two stern planes) for vehicle steering and diving in the horizontal and vertical planes, respectively. The maximum rated speed of the vehicle with respect to the water is 2.5 m/s. At a cruising speed of 1.3 m/s, the estimated mission duration and range are 18 h and 83 km, respectively. The maximum depth of operation is 500 m. Its main particulars are as follows: length overall: 4.5 m; beam of hull: 1.1 m; beam overall, including bow and stern planes: 2.0 m; draft of hull: 0.6 m; frontal area: 0.7 m^2. Currently, its scientific sensor suite includes a Doppler log, a sidescan sonar, a mechanically scanning pencil beam sonar, a conductivity–temperature–depth (CTD) recorder, a fluorometer, a Plankton sampler and a video camera. In a representative mission, the vehicle performs lawn mowing manoeuvres at different depths to collect scientific data in the water column. Given its good stability properties, the vehicle is also a

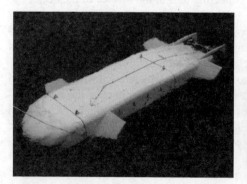

Figure 17.1 The Infante AUV. (Left) Vehicle being deployed. (Right) Vehicle at sea, in the Azores

good platform for manoeuvring at a fixed depth and collecting acoustic data off the seabed for bathymetry mapping and sea-bottom classification purposes.

17.2.2 *The* Delfim *ASC*

The *Delfim* is an ASC that was designed and built at the Instituto Superior Técnico. The research and development efforts that led to the development of *Delfim* were initiated in the scope of a European project that set the goal of achieving coordinated operation of an AUV and an ASC in order to establish a fast direct communication link between the two vehicles and thus indirectly between the AUV and a support vessel. This concept has proven instrumental in enabling the transmission of sonar and optical images through an acoustic communications channel optimised to transmit in the vertical. See Reference 3 and the references therein for a brief description of the project and the major milestones achieved. Over the past few years, the *Delfim* ASC has also been used extensively as a stand-alone unit, capable of manoeuvring autonomously and performing precise path following, while carrying out automatic marine data acquisition and transmission to an operating centre installed on board a support vessel or on shore.

The *Delfim* craft is a small catamaran 3.5 m long and 2.0 m wide, with a mass of 320 kg, see Figure 17.2. Propulsion is ensured by two bladed propellers driven by electrical motors. The maximum speed of the vehicle with respect to the water is 2.5 m/s. The vehicle is equipped with on-board resident systems for navigation, guidance and control, and mission control. Navigation is done by integrating motion sensor data obtained from an attitude reference unit, a Doppler unit and a DGPS (differential global positioning system). Transmissions between the vehicle and its support vessel, or between the vehicle and a control centre installed on-shore are achieved via a radio link with a range of 80 km. The vehicle has a wing-shaped central structure that is lowered during operations at sea. Installed at the bottom of this structure is a low drag body that can carry acoustic transducers, including those

Figure 17.2   *The* Delfim *autonomous surface craft*

356 Advances in unmanned marine vehicles

used to communicate with submerged craft. For bathymetric operations, the wing is equipped with a mechanically scanning pencil beam sonar.

17.2.3 The Sirene *underwater shuttle*

The *Sirene* AUV is an underwater shuttle designed to automatically position a large range of benthic stations on the seabed down to depths of 4000 m. The vehicle and respective systems were developed by a team of European partners coordinated by IFREMER, in the scope of the MAST-II European project Desibel (New Methods for Deep Sea Intervention on Future Benthic Laboratories) that aimed to compare different methods for deploying and servicing benthic stations. The reader will find in Reference 5, a general description of the Desibel project. See also Reference 6 for a theoretical study of the guidance and control systems of *Sirene* and Reference 7 for a description of its mission control system.

The *Sirene* vehicle, shown in Figure 17.3, was designed as an open-frame structure 4.0 m long, 1.6 m wide, and 1.96 m high. Its dry weight is 4000 kg and its maximum operating depth is 4000 m. The vehicle is equipped with two back thrusters for surge and yaw motion control in the horizontal plane, and one vertical thruster for heave control. Roll and pitch motion are left uncontrolled, since the metacentric height is sufficiently large (36 cm) to provide adequate static stability. An acoustic link enables communications between the *Sirene* vehicle and a support ship for tele-operation purposes. At the core of the vehicle navigation system is a long baseline (LBL) positioning system developed by IFREMER [5]. *Sirene* was designed as a prototype vehicle to transport and to position accurately benthic laboratories at predetermined targets on the seabed. See Figure 17.4 (right), which depicts the vehicle carrying a representative benthic lab that is cube shaped and has a volume of 2.3 m^3. In a typical mission (Figure 17.4, left) the Sirene vehicle and the laboratory are first coupled together and launched from a support ship. Then, the ensemble descends in a free-falling trajectory (under the action of a ballast weight) at a speed in the range of 0.5–1 m/s.

Figure 17.3 The Sirene *underwater shuttle*

At approximately 100 m above the seabed, *Sirene* releases its ballast and the weight of the entire ensemble becomes neutral. At this point, the operator onboard the support ship instructs the vehicle to progress at a fixed speed (along a path defined by a number of selected way-points) until it reaches a vicinity of the desired target point. At this point *Sirene* manoeuvres to acquire the final desired heading and lands smoothly on target, after which it uncouples itself from the benthic laboratory and returns to the surface. Tests with the prototype vehicle were carried out off the coast of Toulon, France, in 1997.

17.2.4 The Caravela 2000 autonomous research vessel

The *Caravela* (Figure 17.5) is a long-range autonomous research vessel developed by a consortium of industrial partners (Rinave and Conafi) and research institutes

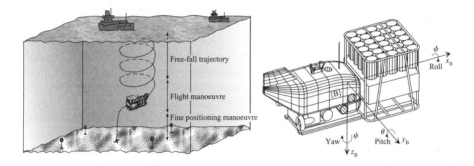

Figure 17.4   *The* Sirene *underwater shuttle. (Left) Laboratory deployment mission. (Right) The shuttle carrying a benthic laboratory*

Figure 17.5   *The* Caravela *autonomous research vessel. (Left) A scale model. (Right) The vehicle hull in the shipyard*

(IMAR and IST/ISR) in Portugal, under the scientific leadership of the IMAR/Department of Oceanography and Fisheries of the University of the Azores [8]. Conceptually, the *Caravela* bears great likeness to the *Delfim* ASC in that it can operate in a fully autonomous mode without constant supervision of a human operator. However, its rugged construction, endurance and high payload capacity, make it perfect for carrying out missions in the open seas for extended periods of time, carrying in its torpedo shaped keel a full array of scientific equipment and acoustic sensors. The *Caravela* was designed to be fully autonomous but capable of responding to commands issued from land or any sea platform via a remote RF/satellite communication link. This link provides a data channel for receiving mission sensor data from the vehicle and for sending operator-generated commands to the vehicle to redirect its mission if required. At the heart of the *Caravela* vessel is an integrated navigation, guidance and control system that allows it to follow predetermined paths with great accuracy. The vessel is both a testbed to try out advanced concepts in vehicle/mission control and radar-based obstacle avoidance and a demonstrator of technologies for the marine science community. The estimated range of operation of *Caravela* is 700 nautical miles. The propulsion system consists of two electrically driven propellers at the stern of the vehicle. The hull houses two diesel generators that charge a pack of batteries. The main particulars of *Caravela* are the following. Length overall: 10 m; beam of hull: 2 m; draft of hull (without mast or keel): 2.3 m; mast height: 3 m; keel height: 2.5 m; 'torpedo' underwater: 4.5 m length/1.2 m diameter.

Development of the *Caravela* was motivated by the need to reduce the cost of operations and improve the efficiency of oceanographic vessels at sea. Conventional oceanographic vessels require a large support crew, are costly to operate, and their availability is often restricted to short periods during the year. However, a large number of oceanographic missions consist of routine operations that could in principle be performed by robotic vessels capable of automatically acquiring and transmitting data to one or more support units installed on shore. In the future, the use of multiple autonomous oceanographic vessels will allow researchers to carry out synoptic studies of the ocean on time and space scales appropriate to the phenomena under study. Furthermore, these vessels will play a major role in enabling scientists to actually program and follow the execution of missions at sea from the safety and comfort of their laboratories.

17.3 Vehicle control

This section provides a brief summary of challenging problems in the area of marine vehicle control and guides the reader through some of the techniques used for solving them. The presentation is naturally biased towards the research work done at IST/ISR in the process of developing control algorithms for the vehicles described in the previous section. However, the types of problems addressed and the references cited are believed to be sufficiently broad and contain enough information to give the reader a balanced vision of the main trends in the field. See also Reference 9 and the references therein for background material.

17.3.1 Control problems: motivation

There is considerable interest in the development of advanced methods for motion control of marine vehicles (including surface and underwater robots) in the presence of unknown ocean currents, wave action and vehicle modelling uncertainty. Among the problems studied, the following categories are especially relevant and will be briefly described below:

(i) Vertical and horizontal plane control.
(ii) Pose (position and attitude) control.
(iii) Trajectory tracking and path following control.
(iv) Cooperative motion control of multiple marine vehicles.

Vertical and horizontal plane control – In a vast number of mission scenarios, underwater vehicles are required to manoeuvre in the vertical and horizontal planes while tracking a desired speed profile bounded away from zero. Examples include heading control in the horizontal plane and depth or altitude control (above the seabed) in the vertical plane. See References 4 and 9–11 and the references therein. More challenging applications require depth control close to the sea surface, in the presence of strong wave action [12]. This type of control is required for both streamlined and bluff bodies of which the *Infante* AUV and the *Sirene* AUV, respectively are representative examples. The first class of bodies have a preferred direction of motion and control is usually accomplished by resorting to simplified dynamic models of motion obtained by linearising their nonlinear dynamics about trimming conditions. The second class of bodies, however, do not have a preferred direction of motion. This makes the task of controlling them harder, for one must resort to more complex nonlinear dynamic models of motion. The problem of control in the horizontal plane is also relevant in the case of autonomous surface craft such as the *Delfim* or *Caravela* vessels.

Pose control – A completely different class of problems arises when an underwater vehicle must be steered to a final target point with a desired orientation. This situation calls for the development of controllers to manoeuvre the vehicle at speeds around zero. The problem is especially challenging when the number of actuators of the vehicle is fewer than its degrees of freedom, as in the case of the *Sirene* AUV [13]. In this situation, theoretical limitations arising from the fact that the vehicles are nonholonomic [14] dictate that discontinuous, hybrid, or even time-varying feedback control laws be used. See References 13 and 15–18, and the references therein for discussions on this subject.

Trajectory tracking and path following – Trajectory tracking refers to the problem of making a marine vehicle track a time-parameterised reference curve in two- or three-dimensional space [4,19]. Stated in simple terms, one requests that the vehicle be at assigned spatial coordinates at assigned instants of time. This requires that the velocity of the vehicle be controlled with respect to an inertial frame. As is well-known, this may lead, in the case of an AUV faced with strong currents, to a situation where the vehicle surfaces stall and control authority is drastically reduced. Furthermore, trajectory tracking control often leads to jerky motions of the vehicle (in its attempt to meet stringent spatial requirements) and to considerable actuator activity. These problems are somehow attenuated when the temporal constraints are lifted, which brings us to the problem of path following. By this we mean the problem of forcing a vehicle to converge to and follow a desired spatial path, without any temporal specifications [4,9,20–24]. However, we will still require that the vehicle track a desired temporal speed profile. The latter objectives occur for example when an autonomous surface vessel must cover a certain area by performing a 'lawn mowing' manoeuvre along desired tracks with great accuracy, at speeds determined by a scientific end-user. The underlying assumption in path following control is that the vehicle's forward speed tracks the desired speed profile, while the controller acts on the vehicle's orientation to drive it to the path. Typically, smoother convergence to the path is achieved when path following strategies are used instead of trajectory tracking control laws, and the control signals are less likely to be pushed to saturation. This interesting circle of ideas opens the door to more sophisticated strategies that naturally combine some of the attributes of trajectory tracking and path following, as first suggested in the pioneering work of Hauser and Hindman [25] and recently pursued in References 9 and 26–28.

Cooperative motion control. Coordinated path following – In a great number of mission scenarios multiple autonomous marine vehicles must work in cooperation. The rationale for this problem can be best understood by referring to a number of practical examples:

(i) *Combined autonomous surface craft/autonomous underwater vehicle control.* In this scenario an ASC must follow a desired path accurately and an AUV operating at a fixed depth must follow exactly the same horizontal path (shifted in the vertical), while tracking the ASC motion along the upper path. The AUV serves as a mobile sensor suite to acquire scientific data while the ASC plays the role of a fast communication relay between the AUV and a support ship. Thus, the ASC effectively explores the fact that high data rate underwater communications can best be achieved if the emitter and the receiver are aligned along the same vertical line in order to avoid multipath effects. Notice how both vehicles must follow exactly the same type of path, which is imposed by the scientific missions at hand. This operational scenario was first advanced in the scope of the Asimov project of the European Union (EU) [3] and is depicted in Figure 17.6 (left), which illustrates coordinated operation of the *Infante* AUV and *Delfim* ASC. The same figure on the right shows also the type of experiments that were carried out at sea in the Azores, with the *Aries* AUV of the NPS, Monterey and the *Delfim* ASC communicating with each other using an acoustic modem.

(ii) *Combined autonomous underwater vehicle control: image acquisition.* This scenario occurs when an underwater vehicle carries a strong light source and illuminates the scenery around a second underwater vehicle that must follow a predetermined path and acquire images for scientific purposes.

(iii) *Combined autonomous underwater vehicle control: fast acoustic coverage of the seabed.* In this important case, two vehicles are required to manoeuvre above the seabed at identical or different depths, along parallel paths, and map the sea bottom using two copies of the same suite of acoustic sensors (e.g., sidescan,

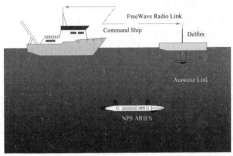

Figure 17.6   Coordinated motion control. (Left) The Infante *AUV and the* Delfim *ASC. (Right) The* ARIES *AUV and the* Delfim *ASC – planning of the 2001 Azores mission (Courtesy of Prof. Anthony Healey, NPS, Monterey, CA, USA)*

mechanically scanned pencil beam, and sub-bottom profiler). By requesting the vehicles to traverse identical paths so as to make the acoustic beam coverage overlap on the seabed, large areas can be covered quickly. One can also envision a scenario where the vehicles use a set of vision sensors to inspect the same scenery from two different viewpoints to try and acquire three-dimensional images of the seabed.

In the above cases, one of the vehicles (leader) follows a path and the second vehicle (follower) is required to track the first one along a path that is related to that of the leader. A cursory analysis of the problem seems to indicate that a solution is at hand once a path following and a trajectory controller have been found for the leader and the follower vehicle, respectively. However, the problem is far more complex than a simple analysis suggests. Consider for example the first mission scenario, where the (leader) surface vehicle may exhibit relatively large path following errors due to wind, currents and wave action. It would be a bad strategy for the underwater vehicle to track the (possibly 'jerky') trajectory of the ASC closely. In fact, it is far better for the AUV (that is subject to far less external disturbances) to remain on its nominal spatial path and to manoeuvre along that path so as to 'stay in the vicinity' of the leader. This will enable each vehicle to remain inside the projected area of the cone of communications of the other.

The problems described will henceforth be referred to as 'coordinated path following,' a name that was chosen to stress the fact that the vehicles follow assigned paths but adjust their speeds to coordinate themselves in time as the mission unfolds. See References 9, 23 and 29–33 and the references therein for an introduction to and an historical perspective of this vibrant topic of research. See also References 34 and 35 for a very interesting type of cooperative motion control problems with applications in ship rendezvous manoeuvres. Coordinated path following falls in the scope of the general problem of cooperative control, which has received considerable attention in the fields of air, space and ground robotics [36–43] and, to a less extent, in the field of marine robots [44–46]. The work reported in the literature addresses a large class of topics that include, among others, formation flying, coordination of groups of mobile autonomous agents, control of the 'centre of mass' and radius of dispersion of swarms of vehicles, and uniform coverage of an area by a group of ground robots, to name but a few. We chose to focus on coordinated path following because this topic is well rooted in solid practical applications and also because its mathematical formulation is closely related to that of path following, which is also covered in this chapter. At this point, it is important to stress that the type of problems tackled in the field of marine robotics are far more difficult than the corresponding ones in air or on land, because underwater navigation and communications are exceedingly difficult. Even at a theoretical level, these limitations pose formidable challenges to system designers because coordination must be achieved in the presence of time-dependent, low bandwidth communication links that are often plagued with temporary failures.

17.3.2 Control problems: design techniques

This section describes a number of control techniques that can be used to solve the control problems introduced above. Space limitations preclude us from presenting complete details of the mathematical machinery needed. Instead, we cite relevant publications and present the key ideas involved. The presentation is naturally biased towards the research work done at IST/ISR.

For the sake of completeness we start by describing the general form of the equations of motion of marine vehicles, with a bias towards AUVs. The basic notation will however apply to all kinds of marine vehicles. See Reference 9 and the references therein for a lucid presentation of this subject and for an extension to the modelling of surface craft. The interested reader will find complete dynamic models for the *Sirene*, *Delfim* and *Infante* marine vehicles in studies by Aguiar [15], Prado [47] and Silvestre [4], respectively, including details for their implementation in Matlab. The equations are developed using an inertial frame $\{I\}$ and body-fixed frame $\{B\}$ that moves with the vehicle. The following notation is required:

$\mathbf{p} = [x, y, z]^T$ – position of the origin of $\{B\}$ in $\{I\}$;
$\mathbf{v} = [u, v, w]^T$ – linear velocity of the origin of $\{B\}$ in $\{I\}$, expressed in $\{B\}$ (u, v, w denote surge, sway and heave speed, respectively);
$\boldsymbol{\lambda} = [\phi, \theta, \psi]^T$ – vector of Euler angles (roll, pitch and yaw) that describes the orientation of frame $\{B\}$ with respect to $\{I\}$;
$\boldsymbol{\omega} = [p, q, r]^T$ – angular velocity of $\{B\}$ relative to $\{I\}$, expressed in $\{B\}$;
$R := {}_I^B R(\lambda)$ – rotation matrix from $\{B\}$ to $\{I\}$, parameterised locally by λ; R is orthonormal and $R = I$ for $\lambda = 0$.
$Q := Q(\lambda)$ – matrix that relates body-fixed angular velocity ω to Euler angles rates. Matrix Q satisfies $d\lambda/dt = Q\omega$ and equals the identity for $\lambda = 0$.

Let

$$\mathbf{x}_{\text{dyn}} = \begin{bmatrix} \mathbf{v} \\ \boldsymbol{\omega} \end{bmatrix}; \quad \mathbf{x}_{\text{kin}} = \begin{bmatrix} \mathbf{p} \\ \boldsymbol{\lambda} \end{bmatrix}$$

where $\mathbf{x}_{dyn}$ and $\mathbf{x}_{kin}$ denote the dynamic and kinematic variables, respectively, that are used to describe the motion of the vehicle. Further, let

$$L(\lambda) = \begin{bmatrix} R(\lambda) & 0 \\ 0 & Q(\lambda) \end{bmatrix}$$

Using the above notation, the vehicle dynamics and kinematics can be described by [4,19]

$$M_{RB}\frac{d}{dt}\mathbf{x}_{dyn} + C_{RB}(\mathbf{x}_{dyn})\mathbf{x}_{dyn} = \tau\left(\frac{d}{dt}\mathbf{x}_{dyn}, \mathbf{x}_{dyn}, \lambda, \delta, \mathbf{n}\right)$$

$$\frac{d}{dt}\mathbf{x}_{kin} = L(\lambda)\mathbf{x}_{dyn}$$

where M_{RB} and C_{RB} denote the rigid body inertia matrix and the matrix of Corriolis and centripetal terms, respectively, and τ is the vector of external forces and torques applied to the rigid body. Vector τ can be further decomposed as

$$\tau := \tau\left(\frac{d}{dt}\mathbf{x}_{dyn}, \mathbf{x}_{dyn}, \lambda, \delta, \mathbf{n}\right) = \tau_{rest}(\lambda) + \tau_{add}\left(\frac{d}{dt}\mathbf{x}_{dyn}, \mathbf{x}_{dyn}\right)$$

$$+ \tau_{surf}(\mathbf{x}_{dyn}, \delta) + \tau_{visx}(\mathbf{x}_{dyn}, \delta) + \tau_{prop}(\mathbf{x}_{dyn}, \mathbf{n})$$

where τ_{rest} is the vector of (restoring) forces and moments caused by the interplay between gravity and buoyancy and τ_{add} captures the so-called added mass terms. Vector τ_{surf} contains the forces and moments generated by the deflecting surfaces, τ_{visc} consists of the hydrodynamic forces and moments exerted on the vehicle's body (including damping terms), and τ_{prop} is the vector of forces and moments generated by the propellers. In the case of the *Infante* AUV, the input vector $\delta = [\delta_b, \delta_s, \delta_r]^T$ consists of: δ_b – common bow plane deflection, δ_s – common stern plane deflection and δ_r – common rudder deflection. Vector $\mathbf{n}$ contains the speeds of rotation of the two stern propellers. It is now routine to rewrite the above equations in standard state-space form as [19]

$$P := \begin{cases} \dfrac{d}{dt}\mathbf{x}_{dyn} = F_{dyn}(\mathbf{x}_{dyn}, \mathbf{x}_{kin}) + B(\mathbf{x}_{dyn}) + H(\mathbf{x}_{dyn})\mathbf{u} \\ \dfrac{d}{dt}\mathbf{x}_{kin} = F_{kin}(\mathbf{x}_{dyn}, \mathbf{x}_{kin}) = L(\lambda)\mathbf{x}_{dyn} \end{cases} \quad (17.1)$$

by making $\mathbf{u} = [\delta^T, n^T]^T$. The total speed of the vehicle will be denoted $v_t = \|\mathbf{v}\|_2$. At this point we also recall the classical definitions of angle of attack $\alpha = \sin^{-1}(w/(u^2 + w^2)^{1/2})$, sideslip angle $\beta = \sin^{-1}(v/v_t)$ and flight path angle (angle that the total velocity vector makes with the horizontal and equals $\gamma = \theta - \alpha$ when vehicle motion is restricted to the vertical plane).

17.3.2.1 Vertical and horizontal plane control

There are a number of techniques available for the control of AUVs in the vertical and horizontal planes. See, for example, the seminal work of Healey and Lienard [48] and the techniques described in Reference 9 for an introduction to control design using linear state-space feedback, sliding mode control theory and adaptive control. For our purposes, we assume that the general AUV model presented above can be divided into two sub-models for the vertical and horizontal planes. This procedure is fully justified for the case where the vehicle executes manoeuvres that require light interaction between steering in the horizontal plane and diving in the vertical plane. We further assume that the vehicle under consideration has a preferred direction of motion that corresponds to the situation where the vehicle is levelled and 'flies straight' at a constant speed $v_t > 0$. Mathematically, this corresponds to an equilibrium or trimming condition at which the dynamic state $\mathbf{x}_{\text{dyn}}$ is defined by $u = v_t, v = w = p = q = r = 0$, the roll and pitch angles are set to zero, and the input vector $\mathbf{u}$ is defined accordingly. Assuming the vehicle motion does not deviate too much from this equilibrium condition, simple dynamic models can be obtained by linearising the nonlinear dynamics about trimming. Naturally, the simplified models can be parameterised by the total speed v_t. This motivates the approach taken at IST/ISR towards the development of AUV control laws that are well rooted in gain-scheduling control theory [4,10,11,49]. With the set-up adopted, the design of a controller to achieve stabilisation and adequate performance of a given nonlinear plant (system to be controlled) involves the following steps [50,51]:

(i) Linearising the plant about a finite number of representative trimming conditions (also called equilibrium or trimming points).
(ii) Designing linear controllers for the plant linearisations at each trimming point.
(iii) Interpolating the parameters of the linear controllers of Step (ii) to achieve adequate performance of the linearised closed-loop systems at all points where the plant is expected to operate; the interpolation is performed according to the vehicle's forward speed, and the resulting family of linear controllers is referred to as a gain scheduled controller.
(iv) Implementing the gain scheduled controller on the original nonlinear plant.

The strategy described effectively reduces the problem of nonlinear control system design to that of designing a finite number of linear controllers, as described in Step (ii). This allows the system designer to use techniques that explicitly address the issues of robust stability and performance in the presence of plant uncertainty [52]. In our work, the methodology selected for linear control system design relies on the reduction of an H_∞ (H-infinity) performance criterion [53]. The starting point in this design technique is the standard linear feedback system of Figure 17.7 (left) where $\mathbf{w}$ is the input vector of exogenous signals (commands, disturbances and sensor noise), $\mathbf{z}$ is an output vector that includes the signals (tracking errors, actuation signals, etc.) to be reduced, $\mathbf{y}$ is the vector of measurements that are available for feedback and $\mathbf{u}$ is the vector of actuator signals (inputs to the plant).

The generalised plant G consists of the linearised model of the plant together with appended dynamic weights that shape the exogenous and internal signals in the frequency domain (Figure 17.7, right). For example, in the design of a depth controller using stated feedback, detailed in Reference 10, $\mathbf{w}_1$ is the depth command z_{cmd} that must be tracked. Vector $\mathbf{w}_2$ includes the input noise to each of the sensors that provide measurements of the state variables, as well as disturbance inputs to the states w and q

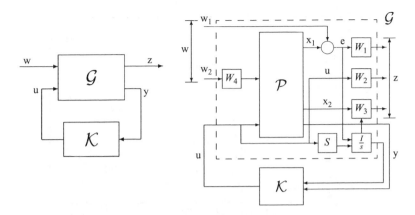

Figure 17.7   Control design. (Left) Plant/controller in a feedback configuration. (Right) Design model with appended weights

of the plant. Vector **u** consists of the actuation signals for the bow and stern plane deflections, and x_1 is the depth variable z. Vector x_2 includes the variables α, q, θ that are also penalised in the design process. Notice the existence of a block of integrators I/s that operates on the tracking error e and on the entries of the control input vector **u** that are selected by the matrix S. Integral action on the error is required to ensure zero steady state in response to step commands in w_1. Integral action on the entries of **u** introduces a 'washout' on the particular control inputs selected. In the present case, the 'washout' ensures zero bow plane deflection at trimming conditions. With the above choices, $\mathbf{y} = [\alpha, q, \theta, z, (z_{cm} - z)/s, \delta_b/s]^T$. The dynamic or scalar weights W_1 through W_4 are introduced to achieve command and input–output requirements.

Suppose the feedback system is well-posed, and let T_{zw} denote the closed-loop transfer matrix from **w** to **z**. The H_∞ control problem can now be briefly described as follows: given a number $\gamma > 0$ find, if possible, a controller K that yields closed-loop stability and makes the infinity norm $\|T_{zw}\|_\infty$ (i.e., maximum input–output 'energy amplification' of T_{zw}) smaller than γ. The positive number γ and the weights appended in $\mathcal{G}$ play the role of tuning knobs to try and meet adequate closed-loop performance specifications in the frequency domain.

To solve this problem, one can resort to linear matrix inequalities (LMIs) [54], which are steadily becoming the tool par excellence for advanced control system design. In fact, many control problems can be cast as LMI problems that can be solved efficiently using convex programming techniques. The case of AUV control using state feedback is studied in Reference 10. The far more complex and realistic cases of static output feedback and reduced order static output feedback are reported in References 11 and 49, respectively. These references include also details on how to solve the practical problem of gain scheduled controller implementation mentioned in Step (iv) by using a dedicated velocity algorithm that is also referred to as the 'δ-implementation' [55]. Using this implementation, the trimming values for the plant inputs and for the state variables that are not explicitly required to track kinematic

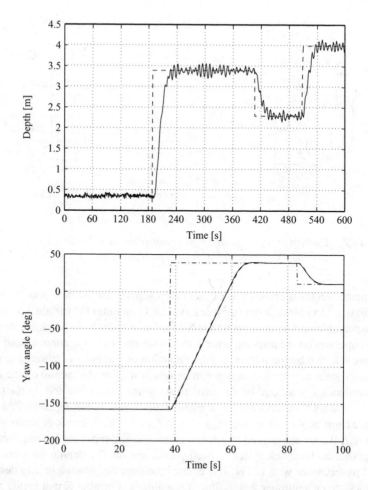

Figure 17.8   Depth and heading control of the Infante *AUV*

reference inputs are automatically 'learned' during operation and do not need to be computed off-line. Figure 17.8 shows the type of performance obtained with a static output feedback controller during tests with the *Infante* AUV at sea.

17.3.2.2   Pose control

The problem of pose control is clearly of a different category and will only be briefly touched upon in this section. The vehicles involved are usually bluff bodies that must manoeuvre at low speed during their final approach to a target position. As discussed before, the problem is especially hard to solve when the vehicles are underactuated because there is no smooth (or even continuous), time-invariant state-feedback control law that will yield asymptotic stability of the desired pose (position and orientation) [14].

At IST/ISR, work on this subject was motivated by the participation in the Desibel project [5]. Work evolved at both a theoretical and 'practical level'. From a theoretical standpoint, two methodologies were developed for pose control of the *Sirene* AUV. The first method sought inspiration from previous related work in the field of wheeled robots [56] where the kinematics of a robot are rewritten in polar coordinates, thereby introducing a discontinuity in the control law as a form of obviating some of the limitations imposed by Brockett's result. See References 13 and 15 and the references therein. The second method used a totally different approach that borrowed from logic-based hybrid control theory [18]. The transition from theory to practice, done in the scope of the Desibel project, witnessed the development of a set of control laws for vehicle manoeuvring that were tested off the coast of Toulon, France, down to depths of 2000 m [6,57]. Figure 17.9 shows practical results of heading and depth control of the vehicle.

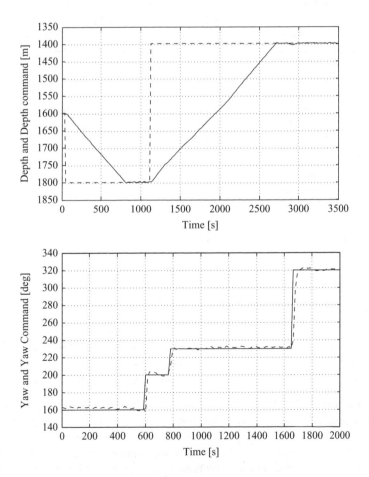

Figure 17.9 Depth and heading control of the Sirene AUV

17.3.2.3 Trajectory tracking and path following control

This section offers a short summary of some design techniques that can be used for trajectory tracking and path following of marine vehicles. Once again, we skip the mathematical details. However, we guide the reader to the appropriate references.

17.3.2.3.1 Trajectory tracking

In a number of aeronautical applications, trajectory tracking controllers for autonomous vehicles have traditionally been designed using the following methodology. First, an inner loop is designed to stabilise the vehicle dynamics. Then, using timescale separation criteria, an outer loop is designed that relies essentially on the vehicle's kinematic model and converts trajectory tracking errors into inner loop commands. In classical missile control literature this outer loop is usually referred to as a guidance loop. Following this classical approach, the inner control loop is designed based on vehicle dynamics, whereas the outer guidance law is essentially based on kinematic relationships only. During the design phase, a common rule of thumb is adopted whereby the inner control system is designed with sufficiently large bandwidth to track the commands that are expected from the guidance system (the so-called timescale separation principle). However, since the two systems are effectively coupled, stability and adequate performance of the combined systems are not guaranteed. This potential problem is particularly serious in the case of marine vehicles, which lack the agility of fast aircraft and thus impose tight restrictions on the closed-loop bandwidths that can be achieved with any dynamic control law. Motivated by the above considerations, a new methodology was introduced in References 4 and 19 for the design of guidance and control systems for marine vehicles whereby the guidance and control are designed simultaneously. Before we proceed, we introduce the following notation and concepts. See Reference 19 for a rigorous exposition.

We start by noticing that the kinematic variables x_{kin} in (17.1) can be split as

$$\mathbf{x}_{kin} = \left[\mathbf{x}_{kin,i}^T, \mathbf{x}_{kin,o}^T\right]^T$$

where $\mathbf{x}_{kin,i}$ denotes the kinematic variables that appear explicitly in the top equations of (17.1) and $\mathbf{x}_{kin,o}$ are the remaining variables (the yaw variable and the position vector $\mathbf{p}$ do not appear explicitly in the dynamic equations). A generalised trimming trajectory Y_C^g for the set of Equations (17.1) can be defined as

$$Y_C^g := \left\{ (\mathbf{x}_{dyn_C}, \mathbf{x}_{kin_C}(.), \mathbf{u}_C) : \begin{bmatrix} F_{dyn}(\mathbf{x}_{dyn_C}, \mathbf{x}_{kin,i_C}) + B(\mathbf{x}_{dyn_C}) \\ +H(\mathbf{x}_{dyn_C})\mathbf{u}_C = 0 \\ F_{kin,i}(\mathbf{x}_{dyn_C}, \mathbf{x}_{kin,i_C}) = 0 \end{bmatrix} \right\}$$

where it is assumed that the kinematic equations for $\mathbf{x}_{kin,o}$ do not depend on $\mathbf{x}_{kin,i}$. Stated in simple terms, a generalised trimming trajectory is obtained by freezing the input $\mathbf{u}$ at some value $\mathbf{u}_C$. A trimming trajectory of (17.1), denoted Y_c, is now simply obtained from Y_C^g by extracting the kinematic components $\mathbf{x}_{kin_C}$, that is, a trimming trajectory is determined by the evolution of the linear position and orientation of the vehicle when the input vector is frozen. Often, by a trimming trajectory we also mean the evolution of the position coordinates only. The meaning will be clear from the context. Associated with (17.1) we can of course consider other (not necessarily trimming) trajectories that are obtained by letting the input vector evolve according to an arbitrary time profile. In this setting, the problem of trajectory tracking can be defined as that of making the state space of a vehicle tend asymptotically to a desired generalised trajectory, by proper choice of the input **u**. To do this, an adequate generalised tracking error vector must be defined. Instead of considering the general case, we now focus on the case of trimming trajectories. A possible choice for the error space is given through the nonlinear transformation

$$NLT := \begin{cases} \mathbf{v}_E = \mathbf{v} - \mathbf{v}_C \\ \omega_E = \omega - \omega_C \\ \mathbf{p}_E = R^{-1}(\mathbf{p} - \mathbf{p}_C) \\ \lambda_E = \arg(R_E) \end{cases}$$

where R is the rotation matrix from body to inertial frame, R_E denotes the rotation matrix from vehicle body-axis to the 'desired' target orientation along the trajectory, and $\arg(\cdot)$ is the operator that extracts the arguments (Euler angles) of R_E.

The new design method proposed builds on the following results: (i) the trimming trajectories of autonomous vehicles correspond to helices parameterised by the vehicle's linear speed, yaw rate, and flight path angle (in the case of ocean surface vehicles, the trimming parameters are simply linear speed and yaw rate), (ii) tracking of a trimming trajectory by a vehicle is equivalent to driving a conveniently defined generalised tracking error (NLT above) to zero and (iii) the linearisation of the generalized error dynamics about any trimming trajectory is time invariant (this fact is not obvious).

Based on the above results, the problem of integrated design of guidance and control systems for accurate tracking of trajectories that consist of the juxtaposition of trimming trajectories can be cast in the framework of gain-scheduled control theory. In this context, the vehicle's linear speed, yaw rate and flight path angle play the role of scheduling variables that interpolate the parameters of linear controllers designed for a finite number of representative trimming trajectories. This leads to a new class of trajectory tracking controllers that exhibit two major advantages over classical ones: (i) stability of the combined guidance and control system is guaranteed, and (ii) zero steady state error is achieved about any trimming trajectory. As in the previous section, controller scheduling and implementation is achieved by using a generalisation of the δ-implementation strategy derived by Kaminer et al. [55], see the details in Reference 4. Interestingly, with this strategy the structure of the final tracking control law is such that the trimming values for the plant inputs and for the states variables that are not explicitly required to track kinematic reference inputs are automatically 'learned' during operation. The importance of this property cannot be overemphasised, for it is in striking contrast with most known methods for trajectory tracking which build on the unrealistic assumption that all input and state variables along the trajectory to be followed are computed in advance.

17.3.2.3.2 Path following

As explained previously, path following is the problem of making a vehicle converge to and follow a desired spatial path, denoted Γ, while tracking a desired speed profile. The temporal and spatial goals are therefore separated. Often, it is simply required that the speed of the vehicle remain constant. In what follows, it is generally assumed that the path is parameterised in terms of its length. A point on the path is therefore specified in terms of its curvilinear abscissa, denoted $s \geq 0$. However, the path can also be parameterised in terms of any other convenient parameter ζ given by $\zeta = g(s); g(0) = 0$, where $g(\cdot)$ is an invertible function.

The solutions to the problem of path following described below are rooted in the work described in References 20 and 21 for wheeled robots. When extended to marine robots, the key ideas explored can be briefly explained by considering Figure 17.10 (left), which depicts the situation where a vehicle follows a two-dimensional path denoted Γ. A path following controller should compute: (i) the distance y_e between the vehicle's centre of mass and the closest point P on the path (if this distance is well defined) and (ii) the angle between the vehicle's total velocity vector v_t and the tangent t to the path at P, and reduce both to zero [58]. Stated equivalently, the objective is to align the total velocity vector with t. At this point, it is important to recall the definition of flow frame $\{W\}$ of a vehicle: $\{W\}$ is obtained from the body frame by rotating it through the angle $\psi + \beta$, thus leaving the x-axis of the flow frame aligned with the total velocity vector v_t. Recall also the definition of the Serret-Frenet $\{F\} = (t, n)$ along a path, consisting of the tangent and normal to that path. Clearly, $\{F\}$ plays the role of the flow frame $\{W_v\}$ of a 'virtual target vehicle' that should be tracked by the flow frame $\{W\}$ of the actual vehicle. The mismatch between the two frames (as measured by linear distance y_e and angle $\psi_e = \psi + \beta - \psi_F$) plays a key role in the definition of the error space where the path following control problem can be formulated and solved. These concepts can of course be generalised to the three-dimensional case [23,24]. Notice that in the case of a wheeled robot the current frame is simply replaced by its body frame because the robot does not exhibit sideslip; consequently, the total velocity vector is aligned with the x-body-axis.

At this point, different solutions to the problem of path following can be proposed. A solution that relies on gain-scheduling control theory and on the linearisation of a generalised error vector about trimming paths, akin to that previously described for trajectory tracking, is reported in Reference 4. See also References 59 and 60 for an application of the same techniques to aircraft control. Formally, to define a trimming path we let $Y_c = x_{\text{kin}_C}(\cdot)$ be a trimming trajectory of a vehicle and let $s(t) = \|v_C\| t; \|v_C\| \neq 0, t \geq 0$ where $\|v_C\|$ denotes the trimming speed. Given an arbitrary invertible function $\zeta = g(s); s \geq 0, g(0) = 0$, then

$$\Gamma_C = \Gamma_C(\zeta) := \left\{ \Pi_\mathbf{p} x_{\text{kin}_C} \left(\frac{g^{-1}(\zeta)}{\|v_C\|} \right); \ \zeta \geq 0 \right\}$$

where $\Pi_\mathbf{p} x_{\text{kin}_C}$ is the projection of x_{kin_C} onto its first three components $\mathbf{p}_C$ is a trimming path of the vehicle parameterised by ζ. It is now possible to define a generalised path following error (about a trimming path) that includes y_e and ψ_e referred to above in the two-dimensional case and to compute the time-invariant linearisation of the generalised error dynamics. The procedure to develop a gain-scheduled path

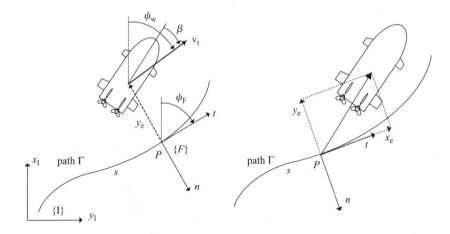

Figure 17.10 Path following. (Left) Closest point strategy. (Right) An extra degree of freedom (virtual target strategy)

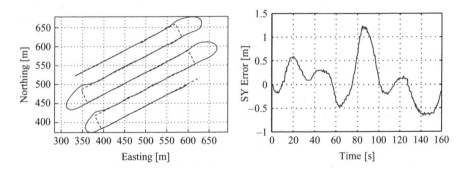

Figure 17.11 Path following at sea (Delfim ASC) in the presence of ocean currents, waves and wind. (Left) Lawn mowing manoeuvre. (Right) Deviation from the path (in metres)

following controller now follows closely the procedure adopted for trajectory tracking system design. This methodology is at the core of the path following controllers that were successfully implemented and run on the *Delfim* ASC and the *Bluebird* aircraft, property of the Naval Postgraduate School, Monterey, California, USA [59,60]. Figure 17.11 shows the results obtained with the *Delfim* ASC doing a 'lawn mowing' manoeuvre over a seamount, off the coast of Terceira Island, in the Azores. In this mission, the ASC ran a path following algorithm along the longer transects, in the presence of a strong ocean current.

With the approach described stability and performance properties can only be guaranteed locally. To obtain global stability results, nonlinear control design methods must be brought to the fore. This was the approach taken in Reference 20 where an elegant and fruitful technique for path following was first proposed for wheeled robots. The new methodology is applicable to a very general set of paths, builds on solid results in nonlinear control theory, and allows for the design of stabilising feedback controllers by resorting to Lyapunov-like arguments. However, controller design was based on the vehicle kinematics only. This is clearly insufficient for marine robots, because their equations of motion exhibit dynamic terms with parameter uncertainty that must be taken into account directly in the control design process. Furthermore, the motion of marine craft may be subjected to the influence of wind, ocean currents and wave action, which poses additional challenges to control system design.

Motivated by these considerations, the work by Encarnação and co-workers [23,24,61] and later refined by Lapierre et al. [62], generalised the results derived for wheeled robots to ocean surface and underwater vehicles by deriving control laws to steer marine robots along desired paths. The key ideas behind the development of the nonlinear algorithms proposed can be simply explained for the two-dimensional case as follows (see also the discussion at the beginning of this section). Assume without loss of generality that the total speed of the vehicle is held constant and compute the evolution of the path following error vector consisting of variables y_e and ψ_e, as functions of yaw rate r. Define a candidate Lypapunov function that is quadratic in the error variables, and use it to find a 'kinematic', nonlinear, feedback control law for r (as if it were a true input) to reduce the error vector to zero. Finally, go from the virtual control law for r to the actual physical input of the vehicle (torque N) using backstepping techniques [63]. This procedure was extended to the three-dimensional case and also to deal explicitly with unknown sea currents by Encarnação et al. [61] and Encarnação et al. [24], respectively. The latter result requires that a nonlinear controller and a current observer be put together. Proving that the ensemble works correctly and biases the heading of the vehicle to counteract the current is not trivial, because of the lack of a separation principle for nonlinear systems.

At this point it is important to remark that the results obtained above inherit the limitation that is present in the path following control strategy for wheeled robots described, for example, in Reference 21: to prove convergence of the error vector, the initial position of the vehicle is restricted to lie inside a tube around the path, the radius of which must be smaller than the smallest radius of curvature that is present in that path. This restriction was entirely removed in Reference 62 by controlling explicitly the rate of progression of the virtual target to be tracked along the path, thus bypassing the problems that arise when the position of that target is simply defined by the projection of the actual vehicle on that path (Figure 17.11, right). See also Reference 64 where a similar technique was first proposed for wheeled robots and Reference 65 for an extension of the same technique to deal with the problem of nonlinear path following for marine vehicles in the presence of parameter uncertainty. The design methodology proposed effectively creates an extra degree of freedom that can then be exploited to avoid the singularities that occur when the distance to path is not well defined (this occurs, e.g., when the vehicle is located exactly at the centre of curvature of a circular path). Interestingly, related strategies were explored in the work of Skjetne et al. [27,28] on output manoeuvring and also in the work of del Rio et al. [66].

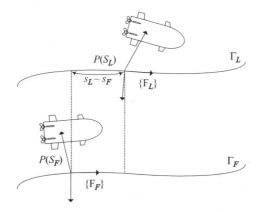

Figure 17.12 Coordinated path following (parallel paths)

17.3.2.4 Coordinated motion control

For reasons that have been explained before, we restrict ourselves to the problem of coordinated path following. An in-depth exposition of this challenging topic of research is outside the scope of this chapter. Instead, we give a fast paced presentation of the subject by keeping the mathematical formalism to a minimum. The key ideas explored in the sequence are easy to grasp and can be explained by referring to Figure 17.12, which depicts a 'leader' AUV L and a 'follower' AUV F and two parallel paths Γ_L and Γ_F parameterised by their 'along-path' curvilinear abscissas s_L and s_F, respectively. Path Γ_F is obtained from Γ_L by shifting it down, vertically. The problem of coordinated path following can now be posed: given paths Γ_L and Γ_F and a desired total speed profile v_t^d for vehicle L, derive control laws for vehicles L and F so that they: (i) converge to their respective paths while tracking the desired speed profile (spatial assignment), and (ii) synchronise their motions along the paths so that the line connecting their centres of mass remains vertical (temporal assignment). Using the terminology of coordinated motion control, we require that the two vehicles reach an 'in-line' 'formation pattern' while manoeuvring along the paths. Another underlying requirement is that the amount of information exchanged between the two vehicles should be kept to a minimum. Ideally, only position information should be exchanged.

A solution to this problem was advanced by Lapierre *et al.* [32] by resorting to a technique that 'almost-decouples' the spatial and temporal assignments referred to above: both the leader and follower execute path following algorithms, the leader travelling along its path at the desired speed profile. It is the task of the follower to adjust its total speed based on the measurement of a generalised 'along-path distance' between the two vehicles. In the simple case illustrated in Figure 17.12, this distance is denoted $s_{L,F}$ and is simply the difference between the along-path coordinates s_F and s_L of P_L and P_F, respectively. Intuitively, the follower speeds up or slows down in reaction to the distance between the 'virtual target vehicles' involved in the path following algorithms. This strategy drastically reduces the amount of information that must be

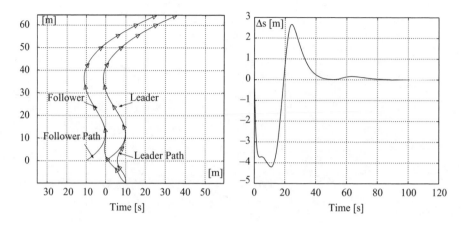

Figure 17.13   Coordinated path following. (Left) Paths. (Right) Along-path distance between leader and follower

exchanged between the two vehicles. Controller design builds on Lyapunov theory and backstepping techniques. The resulting nonlinear feedback control law yields convergence of the two vehicles to the respective paths and forces the follower to accurately track the leader asymptotically. Thus, the mathematical machinery supports the intuition behind the spatial/temporal almost decoupling assumption. Figure 17.13 shows the results of simulations with two underwater vehicles. See Reference 32 for details. The right part of the figure shows clearly how the along-path distance between the two vehicles tends asymptotically to zero.

It is interesting to remark that the rationale for this strategy is already implicit in the work of Encarnação and Pascoal [29] for coordinated path following of an ASC and an AUV. However, the strategy adopted is not easily generalised to more than two vehicles and requires that a large amount of information be exchanged between them. The solution described in Reference 32 overcomes the latter constraint for the case of two vehicles. To overcome the first constraint, a different strategy must be adopted. This brings us to the body of the work initiated by Ghabcheloo *et al.* [67,68] for wheeled robots and extended in Reference 33 for fully actuated underwater vehicles. The main results obtained show how to design coordinated path following controllers for multiple vehicles, arbitrary paths (not necessarily obtained through parallel displacements of a template path), and very general coordination patterns that are compatible with the paths to be followed.

Dealing with general paths and formation patterns is done by re-parameterising the paths in terms of variables, say $\zeta_i : i = 1, 2, \ldots, N$ (where N is the number of vehicles) that are not necessarily their curvilinear abcissae s_i. For example, suppose one wishes to make N marine vehicles coordinate their motions along N concentric circumferences with radii $R_i : i = 1, 2, \ldots, N$ so that their centres of mass are aligned radially. Further assume we parameterise the circumferences in terms of parameters $\zeta_i = s_i/(2\pi R_i)$ (this is equivalent to normalising the total lengths of the circumferences to unity). Clearly, the vehicles are aligned as desired when $\zeta_1 = \zeta_2 = \cdots = \zeta_N$. Having thus solved the problem of defining when coordination is achieved, one is now left with that of coordinating multiple vehicles in the presence of communication constraints. In particular, one wishes to specify the structure of the communication network. Namely, the communications lattice (what vehicle talks to what vehicle) and the type of information that is transmitted among the vehicles (ideally, only the positions along their paths should be transmitted). The pioneering work in References 69–71 showed how these issues can be addressed in the scope of graph theory [72]. Possible assumptions are: (i) the communications are bidirectional, that is, if vehicle i sends information to j, then j also sends information to i, and (ii) the communications graph is connected (a communication graph is said to be connected if two arbitrary vertices, representing vehicles, can be joined by a communication path of arbitrary length). Notice that if assumption (ii) is not verified, then there are two or more clusters of vehicles and no information is exchanged among the clusters. Under these assumptions, using again an almost decoupling type of approach, it is possible to show that there exists a decentralised control law that will drive the vehicles to their paths and achieve coordination. See work by Ghabcheloo *et al.* [33,67,68,73,74] for background material and for proofs of this and other related results. The tools used rely heavily on Lyapunov stability theory [75].

The methodologies required to deal with the general problem of coordinated motion control of marine robots (of which that of coordinated path following is an important example) are still at their infancy. Challenging issues that warrant further research include the study of guaranteed stability and performance of coordinated path following systems when the communications network changes in time and/or fails temporarily. See, for example, Reference 76 and the references therein for a discussion of topics that may have some bearing on these issues.

17.4 Mission control and operations at sea

The previous section described some of the techniques used for single and multiple marine vehicle control. In what follows we describe briefly how to transition from theory to practice. To do this, two key ingredients are needed: (i) a distributed computer architecture, and (ii) a software architecture for system implementation and human–machine interfacing. When implemented in a fully operational vehicle (equipped with the systems for navigation, guidance and control, together with the remaining enabling systems for energy and scientific payload management, actuator control and communications), the latter is often referred to as a mission control system. The literature on mission control is vast and lacks a unified treatment. In fact, the development of a mission control system for single or multiple vehicles reflects the background of the developing team, the applications envisioned, and the hardware available for mission control system implementation. Space limitations prevent us from giving an overview of the main trends in the important area of mission control. The interested reader is referred to References 2 and 77–81 and the references therein for some background material and an historical perspective.

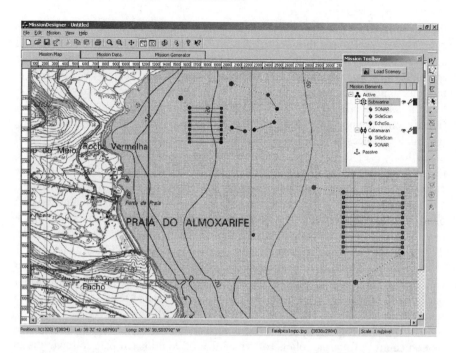

Figure 17.14   Multiple vehicle mission design: graphical interface

17.4.1   The CORAL mission control system

For our purposes, a mission control system is simply viewed as a tool allowing a scientific end-user not necessarily familiarised with the details of marine robotics to program, execute and follow the progress of single or multiple vehicles at sea. With the set-up adopted at IST/ISR, mission design and mission execution are done seamlessly by resorting to simple, intuitive human–machine interfaces. Missions are simply designed in an interactive manner by clicking and dragging over the desired target area maps and selecting items out of menus that contain a list of possible vehicle actions. See Figure 17.14, which is a printout of a graphical interface for mission design.

Notice the presence of a mission map (map of the area to be covered, possibly including the localisation of the obstacles to be avoided), together with a menu of the vehicles available to execute the mission that is being designed. Available to a mission designer are the functionalities of each vehicle (including the types of scientific sensors available), a set of mechanisms enforcing spatial/temporal multi-vehicle synchronisation, and a path planning application to help in the mission design process (so as to meet adequate spatial /temporal/energy requirements). The figure shows the situation where an AUV and an ASC must perform 'lawn mowing' manoeuvres in different regions of the map.

The process of mission design and mission execution unfolds into four basic steps, see Figure 17.15. First, the mission is designed using the graphical interface described

Vehicle and mission control of single and multiple autonomous marine robots 377

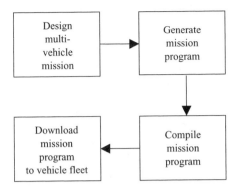

Figure 17.15   Mission design: automatic program generation

above. A mission program is automatically generated in Step 2 and compiled in Step 3. Finally, the mission program is sent to the vehicle or fleet of vehicles in Step 4 and run in real-time. During program execution, the human operator follows the progress of the mission using a similar graphical interface, which now shows the trajectories of the vehicles as they become available via the inter-vehicle communications network.

The methodology adopted for mission control system design and implementation can be best explained for the case of a single vehicle [80]. The methodology builds on the key concept of 'vehicle primitive', which is a parameterised specification of an elementary operation performed by a marine vehicle (e.g., keeping a constant vehicle speed, maintaining a desired heading, holding a fixed altitude over the seabed, or taking images of the seabed at pre-assigned time instants). Vehicle primitives are obtained by coordinating the execution of a number of concurrent (vehicle) system tasks, which are parameterised specifications of classes of algorithms or procedures that implement basic functionalities in an underwater robotic system (e.g., the vehicle primitive in charge of maintaining a desired heading will require the concerted action of system tasks devoted to motion sensor data acquisition, navigation and vehicle control algorithm implementation, and actuator control). Vehicle primitives can in turn be logically and temporally chained to form mission procedures, aimed at specifying parameterised robot actions at desired abstraction levels. For example, it is possible to recruit the concerted operation of a set of vehicle primitives to obtain a parameterised Mission Procedure that will instruct a vehicle to follow an horizontal path at a constant speed, depth and heading, for a requested period of time. Mission procedures allow for modular mission program generation, and simplify the task of defining new mission plans by modifying/expanding existing ones.

With the methodology adopted, system task design is carried out using well established tools from continuous/discrete-time dynamic system theory while finite state automata are used to describe the logical interaction between system tasks and vehicle primitives. The design and analysis of vehicle primitives, mission procedures and mission programs, build on the theory of Petri nets, which are naturally oriented towards the modelling and analysis of asynchronous, discrete event systems with

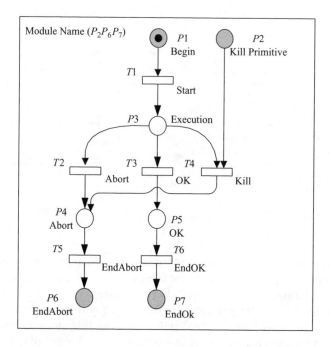

Figure 17.16   Vehicle primitive structure

concurrency [82,83]. This approach leads naturally to a unifying framework for the analysis of the logical behaviour of the discrete event systems that occur at all levels of a mission control system and guarantee basic properties such as the absence of deadlocks. A mission program is thus effectively embodied into a – higher level – Petri net description that supervises the scheduling of mission procedures (and thus indirectly of vehicle primitives) concurring to the execution of a particular mission. Actual implementation of the building blocks referred to above is done by resorting to a powerful Petri net description language named CORAL (proprietary of IST/ISR) that makes the process depicted in Figure 17.15 automatic. The extension of these concepts and tools to deal with multi-vehicle operations is described by Oliveira [81].

For the sake of clarity, the basic structure of a vehicle primitive, embodied in a Petri net description, is shown in Figure 17.16. The name of the primitive is written in 'ModuleName'. Notice the presence of places $P1$, $P2$, $P6$ and $P7$ that play a key role in integrating a particular vehicle primitive in the overall mission program. Placing a mark in $P1$ enables the execution of the primitive. A mark in $P2$ will force the primitive to abort. A mark in $P6$ represents the successful execution of the primitive, whereas a mark in $P7$ means that there was a failure in the execution. The firing of transition $T3$ is the event required to actually start the execution of the system tasks that concur to the execution of the vehicle primitive.

The mission control system developed and tested by IST/ISR using the marine vehicles described in Section 17.1 is supported by a distributed computer architecture.

Vehicle and mission control of single and multiple autonomous marine robots 379

Distributed processes (both inside a single vehicle or across several vehicles) are coordinated using inter-process/inter-computer communication and synchronisation mechanisms implemented over CAN Bus and Ethernet, using internet protocol (IP) and other proprietary communication protocols. This distributed computer architecture is designed around PCs (PC104) running the Windows Embedded NT operating system, and around 8 and 16 bit microcontrollers (such as the Siemens C509L and the Philips XAS3) that communicate using a standard Intel 82527 Controller Area Network controller (CAN 2B protocol). All microcontroller boards were developed at IST/ISR with the purpose of meeting stringent requirements on power consumption, reliability and cost.

17.4.2 Missions at sea

A large series of missions with different types of marine vehicles have consistently shown the reliability of the mission control system developed at IST/ISR. The figures that follow illustrate the types of missions and scientific data acquired with the *Delfim* ASC and the *Infante* AUV in the Azores, Portugal. Figure 17.17 (left) shows a bathymetric map of the D. João de Castro Bank seamount (sunken volcano), off the coast of Terceira island, obtained with an echosounder that is part of the scientific equipment of *Delfim*. The vehicle ran transects over the seamount in a purely autonomous mode. Figure 17.17 (right) shows echosounder data obtained when moving from the outside to the inside of the crater. The figure captures the contour of the seamount. Notice also the presence of acoustic reflections off the bubbles that occur near the hydrothermal vents located around the rim of the crater. Figure 17.18 (left) shows a bathymetric map of a scenario of operation for the *Infante* AUV, near Faial island, in the Pico canal. A geological fault is clearly seen protruding from the island (not represented, but located at the top of the figure) in the direction of the canal. Figure 17.18 (right) shows sidescan data obtained with *Infante* AUV while manoeuvring at a fixed depth

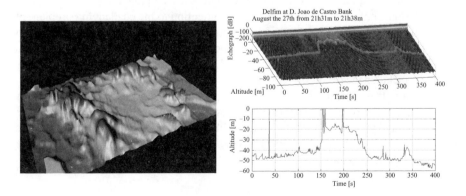

Figure 17.17 Mission with the Delfim *ASC over the D. Joao de Castro seamount. (Left) Bathymetric map of the area. (Right) Echosounder data over the seamount*

Figure 17.18 Mission with the Infante *AUV over the Espalamaca ridge, Azores. (Left) Bathymetric map of the area. (Right) Sidescan data over the ridge*

above the fault, crossing it from left to right. The fault is easily identified in the sidescan sonar image.

17.5 Conclusions

This chapter provided an overview of theoretical and practical problems in the field of marine robotics with a focus on the areas of vehicle and mission control. At the vehicle control level, four categories of problems were introduced: vertical and horizontal plane control, pose control, trajectory tracking and path following, and coordinated motion control of multiple marine robots. Recent advances in linear and nonlinear control theory were shown to provide solid bases for their solution. The technical machinery needed borrows from gain scheduling control theory, linear matrix inequalities, Lyapunov based controller design, backstepping and graph theory. At the Mission Control level, the chapter called attention to the challenging problem of bringing together time- and event-driven systems under a unifying framework. Petri nets were presented as the tool par excellence to tackle this problem, from both an analysis and synthesis viewpoint. To ground the presentation on practical issues, the chapter included the results of tests carried out at sea with prototype AUVs and ASCs. The picture that emerges is that theory and practice must go hand in hand if one is to develop a future breed of marine vehicles capable of operating reliably at sea in a cooperative manner. The challenging problems of cooperative motion control and navigation under severe communications constraints will certainly guide much of the research in the years to come.

Acknowledgements

The authors would you like to thank R. Ghabcheloo for his very constructive review of the section on vehicle control and also J. Alves, R. Oliveira, L. Sebastião and

M. Rufino for their valuable assistance in documenting the tests at sea and preparing the section on mission control.

References

1 Egeskov, P., A. Bjerrum, A. Pascoal, C. Silvestre, C. Aage, and L. Smitt (1994). Design, construction and hydrodynamic tests of the AUV MARIUS. In Proceedings of the Symposium on Autonomous Underwater Vehicle Technology, Cambridge, Massachusetts, USA.
2 Pascoal, A., C. Silvestre, P. Oliveira, A. Bjerrum, G. Ayela, J. Paul-Pignon, S. Bruun, and C. Petzelt (1997). MARIUS: an autonomous underwater vehicle for coastal oceanography. *IEEE Robotics and Automation Magazine*, Special Issue on Robotics and Automation in Europe: Projects funded by the Commission of the European Union, 4(4), pp. 46–59.
3 Asimov Team (2000). Robotic ocean vehicles for marine science applications: the European ASIMOV project. In Proceedings of the OCEANS 2000 MTS/IEEE, Rhode Island, USA.
4 Silvestre, C. (2000). Multi-objective optimization theory with applications to the integrated design of controllers/plants for underwater vehicles. Ph.D. thesis, Instituto Superior Técnico, Lisbon, Portugal.
5 Brisset, L., M. Nokin, D. Semac, H. Amann, W. Schneider, and A. Pascoal (1995). New methods for deep sea intervention on future benthic laboratories: analysis, development, and testing. In Proceedings of the Second Mast Days and Euromar Market Conference, Sorrento, Italy.
6 Aguiar, A. and A. Pascoal (1997). Modelling and control of an autonomous underwater shuttle for the transport of Benthic Laboratories. In Proceedings of the MTS/IEEE OCEANS'97 Conference, Halifax, NS, Canada.
7 Oliveira, P., C. Silvestre, A. Pascoal, and A. Aguiar (1998). Vehicle and mission control of the SIRENE underwater shuttle. In Proceedings of the IFAC Conference on Control Applications in Marine Systems, Fukuoka, Japan.
8 Caravela (2001). Caravela Final Report (in Portuguese). Project L-104, PRAXIS XXI, IMAR-Rinave-IST-Conafi, Lisbon, Portugal.
9 Fossen, T. (2002). *Marine Control Systems: Guidance, Navigation and Control of Ships, Rigs and Underwater Vehicles*. Marine Cybernetics AS, Trondheim, Norway.
10 Silvestre, C. and A. Pascoal (1997). Control of an AUV in the vertical and horizontal planes: system design and tests at sea. *IEE Transactions of the Institute of Measurement and Control*, 19(3), pp. 126–138.
11 Silvestre, C. and A. Pascoal (2004). Control of the Infante AUV using gain-scheduled static output feedback. *IFAC Journal Control Engineering Practice*, 12(12), pp. 1501–1509.
12 Silvestre, C. and A. Pascoal (1997). AUV control under wave disturbances. In Proceedings of the 10th International Symposium on Unmanned Untethered Submersible Technology, New Hampshire, USA.

13 Aguiar, A. and A. Pascoal (2001). Regulation of a nonholonomic autonomous underwater vehicle with parametric modelling uncertainty using Lyapunov functions. In Proceedings of the 40th IEEE Conference on Decision and Control, Orlando, Florida, USA.
14 Brockett, R. (1983). Asymptotic stability and feedback stabilization. In R. Brockett, R. Millman, and H. Sussman, Eds, *Differential Geometric Control Theory*, Birkhauser, pp. 181–191.
15 Aguiar, A. (2002). Nonlinear motion control of nonholonomic and underactuated systems. Ph.D. thesis, Instituto Superior Técnico, Lisbon, Portugal.
16 Aguiar, A. and A. Pascoal (2002). Dynamic positioning of an underactuated AUV in the presence of a constant unknown current disturbance. In Proceedings of the 15th IFAC World Congress, Barcelona, Spain.
17 Aguiar, A. and A. Pascoal (2002). Dynamic positioning and way-point tracking of underactuated AUVs in the presence of ocean currents. In Proceedings of the 41st IEEE Conference on Decision and Control, Las Vegas, Nevada, USA.
18 Aguiar, A. and A. Pascoal (2002). Stabilization of an underactuated autonomous underwater vehicle via logic-based hybrid control. In Proceedings of the 10th Mediterranean Conference on Control and Automation, Lisbon, Portugal.
19 Silvestre, C., A. Pascoal, and I. Kaminer (2002). On the design of gain-scheduled trajectory tracking controllers. *International Journal of Robust and Nonlinear Control*, 12(9), pp. 797–839.
20 Samson, C. (1992). Path following and time-varying feedback stabilization of a wheeled mobile robot. In Proceedings of the ICAARCV 92 Conference, Singapore.
21 Micaelli, A. and C. Samson (1993). Trajectory-tracking for unicycle–type and two-steering-wheels mobile robots. Technical Report No. 2097, Project Icare, INRIA, Sophia-Antipolis, France.
22 Aicardi, M., G. Casalino, G. Indiveri, A. Aguiar, P. Encarnação, and A. Pascoal (2001). A planar path following controller for underactuated marine vehicles. In Proceedings of the 9th Mediterranean Conference on Control and Automation, Dubrovnik, Croatia.
23 Encarnação, P. (2002). Nonlinear path following control systems for ocean vehicles. Ph.D. thesis, Instituto Superior Técnico, Lisbon, Portugal.
24 Encarnação, P. and A. Pascoal (2000). 3-D path following for autonomous underwater vehicles. In Proceedings of the 39th IEEE Conference on Decision and Control, Sydney, Australia.
25 Hauser, J. and R. Hindman (1995). Manoeuvre regulation from trajectory tracking: feedback linearizable systems. In Proceedings of the IFAC Symposium on Nonlinear Control Systems Design, Lake Tahoe, CA, USA.
26 Encarnação, P. and A. Pascoal (2001). Combined trajectory tracking and path following for marine vehicles. In Proceedings of the 9th Mediterranean Conference on Control and Automation, Dubrovnik, Croatia.
27 Skjetne, R., A. Teel, and P. Kokotovic (2002). Stabilization of sets parametrized by a single variable: application to ship manoeuvring. In Proceedings of the 15th

International Symposium on Mathematical Theory of Network Science, Notre Dame, IN, USA.
28 Skjetne, R., T. Fossen, and P. Kokotovic (2004). Robust output manoeuvring for a class of nonlinear systems. *Automatica*, 40, pp. 373–383.
29 Encarnação, P. and A. Pascoal (2001). Combined trajectory tracking and path following: an application to the coordinated control of autonomous marine craft. In Proceedings of the 40th IEEE Conference on Decision and Control, Orlando, Florida, USA.
30 Skjetne, R., S. Moi, and T. Fossen (2002). Nonlinear formation control of marine craft. In Proceedings of the 41st IEEE Conference on Decision and Control, Las Vegas, NV, USA.
31 Skjetne, R., I. Flakstad, and T. Fossen (2003). Formation control by synchronizing multiple manoeuvring systems. In Proceedings of the 6th IFAC Conference on Manoeuvring and Control of Marine Craft, Girona, Spain.
32 Lapierre, L., D. Soetanto, and A. Pascoal (2003). Coordinated motion control of marine robots. In Proceedings of the 6th IFAC Conference on Manoeuvering and Control of Marine Craft, Girona, Spain.
33 Ghabcheloo, R., D. Carvalho, and A. Pascoal (2005). Coordinated motion control of fully actuated underwater vehicles. Internal Report, Code CPFUV01-2005, Institute for Systems and Robotics, Lisbon, Portugal.
34 Kyrkjebo, M. and K. Pettersen (2003). Ship replenishment using synchronization control. In Proceedings of the 6th IFAC Conference on Manoeuvring and Control of Marine Craft, Girona, Spain.
35 Kyrkjebo, M. Wondergem, K. Pettersen, and H. Nijmeijer (2004). Experimental results on synchronization control of ship rendezvous operations. In Proceedings of the IFAC Conference on Control Applications in Marine Systems, Ancona, Italy.
36 Desai, J., J. Otrowski, and V. Kumar (1998). Controlling formations of multiple robots. In Proceedings of the IEEE International Conference on Robotics and Automation, Leuven, Belgium.
37 Beard, R., J. Lawton, and F. Hadaegh (1999). A coordination architecture for spacecraft formation control. *IEEE Transactions on Control System Technology*, 9, pp. 777–790.
38 Giuletti, F., L. Pollini, and M. Innocenti (2000). Autonomous formation flight. *IEEE Control Systems Magazine*, 20(6), pp. 34–44.
39 Queiroz, M., V. Kapila, and Q. Yan (2000). Adaptive nonlinear control of multiple spacecraft formation flying. *Journal of Guidance, Control and Dynamics*, 23(3), pp. 385–390.
40 Mesbahi, M. and F. Hadaegh (2001). Formation flying control of multiple spacecraft via graphs, matrix inequalities, and switching. *Journal of Guidance, Control and Dynamics*, 24(2), pp. 369–377.
41 Pratcher, M., J. D'Azzo, and A. Proud (2001). Tight formation control. *Journal of Guidance, Control and Dynamics*, 24(2), pp. 246–254.

384 *Advances in unmanned marine vehicles*

42 Ogren, P., M. Egerstedt, and X. Hu (2002). A control Lyapunov function approach to multiagent coordination. *IEEE Transactions on Robotics and Automation*, 18(5), pp. 847–851.
43 Jadbabaie, A., J. Lin, and A.S. Morse (2003). Coordination of groups of mobile autonomous agents using nearest neighbor rules. *IEEE Transactions on Automatic Control*, 48(6), pp. 988–1001.
44 Stilwell, D. and B. Bishop (2000). Platoons of underwater vehicles. *IEEE Control Systems Magazine*, 20(6), pp. 45–52.
45 Bachmayer, R. and N. Leonard (2002). Vehicle networks for gradient descent in a sampled environment. In Proceedings of the 41st IEEE Conference Decision and Control, Las Vegas, Nevada, USA.
46 Bhatta, P. and N. Leonard (2002). Stabilization and coordination of underwater gliders. In Proceedings of the 41st IEEE Conference Decision and Control, Las Vegas, Nevada, USA.
47 Prado, M. (2005). Modelling and control of an autonomous oceanographic vessel. M.Sc. thesis (in Portuguese), Instituto Superior Técnico, Lisbon, Portugal.
48 Healey, A. and D. Lienard (1993). Multivariable sliding mode control for autonomous diving and steering of unmanned underwater vehicles. *IEEE Journal of Ocean Engineering*, 18(3), pp. 327–339.
49 Silvestre, C. and A. Pascoal (2005). Depth control of the Infante AUV using gain-scheduled reduced-order output feedback. In Proceedings of the 16th IFAC World Congress, Prague, Czech Republic, 2005.
50 Khalil, H.K. (2000). *Nonlinear Systems*. 3rd edition, Pearson Education International, Inc. New Jersey.
51 Rugh, W. and J.S. Shamma (2000). A survey of research on gain-scheduling. *Automatica*, September, pp. 1401–1425.
52 Athans, M., S. Fekri, and A. Pascoal (2005). Issues on robust adaptive feedback control. Plenary Talk. In Proceedings of the 16th IFAC World Congress, Prague, Czech Republic.
53 Doyle, J., K. Glover, P. Khargonekar, and B. Francis (1989). State space solutions to standard H_2 and H_∞ control problems. *IEEE Transactions on Automatic Control*, 34(8), pp. 831–847.
54 Boyd, S., El Gahoui, E. Feron, and V. Balakrishnan (1994). *Linear Matrix Inequalities in Systems and Control Theory*. Society for Industrial and Applied Mathematics, SIAM, Philadelphia, USA.
55 Kaminer, I., A. Pascoal, P. Khargonekar, and E. Coleman (1995). A velocity algorithm for the implementation of gain-scheduled controllers. *Automatica*, 31(8), pp. 1185–1191.
56 Aicardi, M., G. Casalino, A. Bicchi, and A. Balestrino (1995). Closed loop steering of unicycle-like vehicles via Lyapunov techniques, *IEEE Robotics and Automation Magazine*, March, pp. 27–35.
57 Oliveira, P., C. Silvestre, P. Aguiar, and A. Pascoal (1998). Guidance and control of the SIRENE underwater shuttle: from system design to tests at sea. In Proceedings of the OCEANS'98 Conference, Toulon, France.

58 Encarnação, P., A. Pascoal, and A. Arcak (2000). Path following for autonomous marine craft. In Proceedings of the 5th IFAC Conference on Manoeuvring and Control of Marine Crafts (MCMC 2000), Aalborg, Denmark.
59 Kaminer, I., A. Pascoal, E. Hallberg, and C. Silvestre (1998). Trajectory tracking for autonomous vehicles: an integrated approach to guidance and control. *Journal of Guidance, Control, and Dynamics*, 21(1), pp. 29–38.
60 Hallberg, E., I. Kaminer, and A. Pascoal (1999). Development of a rapid flight test prototyping system for unmanned air vehicles. *IEEE Control Systems Magazine*, 19(1), pp. 55–65.
61 Encarnação, P., A. Pascoal, and A. Arcak (2000). Path following for marine vehicles in the presence of unknown currents. In Proceedings of the SYROCO 2000, Vienna, Austria.
62 Lapierre, L., D.Soetanto, and A. Pascoal (2003). Nonlinear path following control of autonomous underwater vehicles. In Proceedings of the IFAC Workshop on Guidance and Control of Underwater Vehicles, Newport, South Wales, UK.
63 Krstic, M., I. Kanellakopoulos, and P. Kokotovic (1995). *Nonlinear and Adaptive Control Design*. John Willey & Sons, New York.
64 Soetanto, D., L. Lapierre, and A. Pascoal (2003). Adaptive nonsingular path following control of wheeled robots. In Proceedings of the 42nd IEEE Conference on Decision and Control, Hawaii, USA.
65 Kaminer, I., A. Pascoal, and O. Yakimenko (2005). Nonlinear path following control of fully actuated marine vehicles with parameter uncertainty. In Proceedings of the 16th IFAC World Congress, Prague, Czech Republic.
66 Del Rio, F., G. Jimènez, J. Sevillano, C. Amaya, and A. Balcella (2002). A new method for tracking memorized paths: applications to unicycle robots. In Proceedings of the MED2002, Lisbon, Portugal.
67 Ghabcheloo, R., A. Pascoal, C. Silvestre, and I. Kaminer (2004). Coordinated path following control of multiple wheeled robots. In Proceedings of the 5th IFAC Symposium on Intelligent Autonomous Vehicles, Lisbon, Portugal. To appear in International Journal of Systems Science: Special Issue on Cooperative Control Approaches for Multiple Mobile Robots, 2005.
68 Ghabcheloo, R., A. Pascoal, and C. Silvestre (2005). Nonlinear coordinated path following control of multiple wheeled robots with communication constraints. In Proceedings of the ICAR 2005, Seattle, Washington, USA.
69 Fax, A. and R. Murray (2002). Information flow and cooperative control of vehicle formations. In Proceedings of the 15th IFAC World Congress, Barcelona, Spain.
70 Fax, A. and R. Murray (2002). Graph Laplacians and stabilization of vehicle formations. In Proceedings of the 15th IFAC World Congress, Barcelona, Spain.
71 Olfati Saber, R. and R. Murray (2003). Agreement problems in networks with directed graphs and switching topology. In Proceedings of the 42nd IEEE Conference on Decision and Control, Hawaii, USA.
72 Biggs, N. (1993). *Algebraic Graph Theory*. 2nd edition, Cambridge University Press, Cambridge, U.K.

73 Ghabcheloo, R., A. Pascoal, C. Silvestre, and I. Kaminer (2004). Coordinated path following control using linearization techniques. Internal Report, Code CPF01-2004, Institute for Systems and Robotics, Lisbon, Portugal.
74 Ghabcheloo, R., A. Pascoal, C. Silvestre, and I. Kaminer (2004). Coordinated path following control using nonlinear techniques. Internal Report, Code CPF02-2004, Institute for Systems and Robotics, Lisbon, Portugal.
75 Rouche, N., P. Habets, and M. Laloy (1993). *Stability Theory by Liapunov's Direct Method*. Springer-Verlag, New York.
76 Mureau, L. (2005). Stability of multiagent systems with time-dependent communication links. *IEEE Transactions on Automatic Control*, 50(2), pp. 169–182.
77 Healey, A., D. Marco, P. Oliveira, A. Pascoal, C. Silvestre, and V. Silva (1996). Strategic level mission control and evaluation of CORAL and PROLOG implementations for mission control specifications. In Proceedings of the IEEE Autonomous Underwater Vehicle (AUV) 96 Conference, Monterey, CA, USA.
78 Healey, A., D. Marco, and R. McGhee (1996). Autonomous underwater vehicle control coordination using a tri-level hybrid software architecture. In Proceedings of the IEEE Robotics and Automation Conference, Minneapolis, USA.
79 Oliveira P., A. Pascoal, V. Silva, and C. Silvestre (1996). Design, development, and testing of a mission control system for the MARIUS AUV. In Proceedings of the 6th International Advanced Robotics Program IARP-96, Toulon, France.
80 Oliveira, P., A. Pascoal, V. Silva, and C. Silvestre (1998). Mission control of the MARIUS autonomous underwater vehicle: system design, implementation, and sea trials. *International Journal of Systems Science, Special Issue on Underwater Robotics*, 29(10), pp. 1065–1080.
81 Oliveira, R. (2003). Supervision and mission control of autonomous vehicles. M.Sc. thesis (in Portuguese), Instituto Superior Técnico, Lisbon, Portugal.
82 Cassandras, C. (1993). *Discrete Event Systems, Modelling and Performance Analysis*. Aksen Associates Incorporated Publishers.
83 Moody, J. and Antsaklis, P. (1998). *Supervisory Control of Discrete Event Systems Using Petri Nets*. Kluwer Academic Publishers, Dordrecht.

Chapter 18

Wave-piercing autonomous vehicles

H. Young, J. Ferguson, S. Phillips and D. Hook

18.1 Introduction

Autonomous surface and underwater vehicles have been widely developed, but those operating at the interface, submerged just below the surface yet penetrating to the air, have not received the same attention. Yet they combine many of the advantages of both surface and underwater vehicles, but with some less capability than either. This chapter discusses unmanned wave-piercing vehicles, which operate at the interface between sea and air. It will outline their development history, and the two main design approaches evolved so far, together with some detail on the vehicles which have resulted.

18.1.1 Abbreviations and definitions

In this chapter, the term 'unmanned underwater vehicles', abbreviated to UUVs, covers all types of underwater vehicles operating without a human being onboard, and 'unmanned surface vehicles' (USVs), are the equivalent for surface vehicles. These abbreviations are generally accepted throughout the maritime community. However, 'unmanned wave-piercing vehicles' lie between the surface and underwater, and as they are discussed throughout the chapter they will be abbreviated to UWVs. These may be 'wave-piercing surface vehicles' (WSVs) or 'wave-piercing underwater vehicles' (WUVs).

Unmanned vehicles are also categorised as either 'remotely operated vehicles', abbreviated to ROVs or 'autonomous', depending on whether they have an umbilical attaching them to the control or are free swimming. So far as is known, ROVs are always underwater vehicles, but autonomous vehicles can operate on or below the surface, giving rise to AUVs for autonomous underwater vehicles and ASVs for autonomous surface vehicles.

The term 'autonomous' can also imply either close control by an operator of each manoeuvre, or the additional ability to programme a complete mission to be conducted without intervention unless the vehicle fails. In this chapter the latter, fuller use of the term will be assumed.

18.1.2 Concepts

Wave-piercing vehicles, in common with other surface vehicles, have the advantage of communicating and sensing over long ranges with high bandwidths by radio, accurate positioning by GPS, and reliable, well-developed propulsion systems breathing air. Unlike other surface vehicles, however, wave-piercing vehicles have improved stability, efficiency and reduced vessel motion due to the waves. This is of particular importance for unmanned wave-piercing vehicles, which are usually rather small in size. It permits sophisticated instrumentation for both above water and underwater sensing to be fitted to small autonomous craft, and it improves speed, efficiency and sea-keeping without losing the advantages entirely.

Fully submerged unmanned underwater vehicles on the other hand have limited communication and sensor range and bandwidth, have difficulty in accurate positioning and are generally short of power. The former can be overcome to some extent by interrupting the mission and surfacing, but the latter remains a serious disadvantage. UWVs bridge the gap between fully submerged UUVs and unmanned surface vehicles (USVs). They can also be regarded as an economical substitute for surface ships for data gathering, carrying their sensors to considerable ranges and transmitting the data back in real time.

An autonomous wave-piercing vehicle (AWV) has the attraction of being able to use low-cost propulsion, control and positioning equipment although it is of course restricted to operations close to the sea surface with all that implies. However, excepting the cases when totally covert operation, high data resolution or submergence under obstacles is a performance requirement, many survey tasks can be performed from near the surface. Thus, the main reasons for developing UWVs are:

(a) to replicate the data gathering capability of the ship to reduce ship's time required for a specified task,
(b) to remove sensors from noise, bubbles and pollution round the ship's hull,
(c) to remove personnel from areas of potential danger,
(d) to provide a link between surface control and submerged AUVs.

The next section explores the historical development of UWVs, which is comparatively recent. Later sections discuss the technologies employed, their applications and potential for further development.

18.1.3 Historical development

UWVs were developed to overcome the disadvantages of unmanned surface vehicles. USVs were first tried for hydrographic survey in Canada and Japan in the 1970s. The Canadians concluded that in even quite low sea states the vessel movement was

Figure 18.1   Hydrographic 'Dolphins' on board ISE research vessel

too severe to carry hydrographic instrumentation. An unmanned surface research vehicle has been built with Eureka funding called 'Caravela'. This is a self-righting platform for long-range data gathering, and does not carry instrumentation requiring stability. Recently, a large number of unmanned rigid inflatable boat (RIB) vehicles and motor launches have been developed, mainly for surveillance and military use. These vehicles are comparatively inexpensive, and behave adequately in calm water, but are of limited value in a seaway due to their unstable motion at survey sensor speeds.

The Canadian response to the need for a stable hydrographic survey platform was to commission the design of an unmanned wave-piercing vehicle by International Submarine Engineering Ltd (ISE), of Port Coquitlam, near Vancouver. The result of this development was a high-speed diesel driven survey vehicle, 'Dolphin', built in 1984. Dolphin has a torpedo-shaped hull containing the propulsion and control systems, together with the payload. Air for propulsion is brought in by a schnorkel with a patent valve on top to prevent ingestion of sea water. The schnorkel also carries communications and GPS equipment at radio frequencies. This vehicle was in effect an autonomous schnorkelling submarine and was the first wave-piercing autonomous underwater vehicle (WAUV) to be developed. The vehicle is driven to operating depth just below the waves and controlled there by hydroplanes; see Figure 18.1. Dolphin carried out successful surveys for the Canadian government, and proved the advantages of autonomous wave piercing vehicles. The design has evolved over time and there is now a group of vehicles from Canada, France and the USA based on the same principles, which can be thought of as 'Dolphin derivatives'.

An experimental military UWV vehicle which used electrical propulsion and so did not exploit the advantages of an air-breathing engine was prototyped by the US Navy in the mid-1990s but does not appear to have been developed further.

In the early 1990s, a consortium of Norwegian and UK companies with ISE planned to fit a fish stock assessment sonar to Dolphin to extend its field of application and avoid the disturbance to the fish caused by the noise of the research vessel. The plan foundered when funds could not be found to modify the vehicle for this new application. Subsequently, a UK consortium, led by Seespeed Technology Ltd and Hugh Young & Associates, decided to develop a vehicle to meet the specific needs of fish stock assessment. This resulted in a UWV which floats at operating depth instead of being driven there by hydroplanes. The techniques employed can be called wave-piercing autonomous surface vehicle (WASV) technology, and the resulting vehicles are known as survey autonomous semi-submersible (SASS) vehicles.

While the Dolphin derivatives are fully mature designs, successfully marketed and operating in several countries, SASS vehicles are in their infancy. The SASS-6M, the first 5.5 m vehicle, has recently been launched and has started sea trials. Both approaches to UWVs are thought to have potential for further development, see Section 18.6.

18.2 Wave-piercing autonomous underwater vehicles

In 1980, ISE Ltd. saw both a commercial and a military need for unmanned submersibles. At the time, the Canadian Hydrographic Service (CHS) was looking for a small vessel that was more stable than the manned survey launches used at the time from larger survey ships. These small launches were unsafe to operate in sea state 3 or above and because this is the prevalent sea condition in the offshore areas of Canada, the hydrographic survey was not cost-effective.

Previous experiments with an unmanned planing hull showed that although the hydrographic data could be collected remotely, a more stable platform was required. Investigation with scale models in test tanks indicated that decreasing the water-plane of the vehicle would improve stability in high sea states as well as reduce drag and the subsequent powering requirement. A design evolved around a small water-plane area single hull vehicle (SWASH) in which the water-plane was reduced solely to the mast of the vehicle. The result was an unmanned schnorkelling submarine, or wave-piecing autonomous underwater vehicle (WAUV)

In 1981, ISE was contracted to develop a prototype for evaluation. Following studies, the 'Dolphin' vehicle was built in 1983. Tests of the prototype demonstrated the ability to conduct sustained stable operations in sea state 5 conditions at speeds of up to 15 knots and to be controllable over distances of up to 10 km; see Figures 18.2 and 18.3.

Following acceptance testing in September 1983, a production prototype vehicle was delivered to the CHS. The prototype was trialled and then two production vehicles

Wave-piercing autonomous vehicles 391

Figure 18.2   Dolphin hydrographic vehicle at sea

were extensively sea trialled over the next 2 years. The vehicle has an overall length of 7.5 m, a diameter of 1.0 m and a weight of 3.6 tonnes. It was powered by a 110 kW h diesel to give a range (at 12 knots) of 620 km, limited to sea state 6.

These vehicles were integrated with a Simrad EM 1000 multibeam echo sounder and carried out hydrographic survey operations on the east coast of Canada until the CHS lost funding to continue offshore mapping in the early 1990s. In the same timeframe, however, interest arose from the military in using the system for mine-hunting.

18.2.1 Robotic mine-hunting concept

The concept has been to outfit the 'Dolphin' vehicle with mine-hunting sensors, a high bandwidth sonar data link and a GPS positional receiver, and to operate it either from a craft of opportunity or from a dedicated naval ship. This approach allows the operator to work from a safe vantage point and if desired, from a point which is over the horizon.

Configured in this manner, the 'Dolphin' can operate up to 15 km ahead of the controlling vessel. As part of the overall concept, both hull mount and variable depth sonars can be provided so that simultaneous detection can be made on the seabed and in the water column. A towfish attached to the 'Dolphin' with a tow cable, 'flies' at constant depth close to the seabed – up to 200 m deep. The towfish is equipped with both a multibeam sidescan sonar and positioning equipment.

Figure 18.3 'Dolphin' multi-vehicle operation

On the mother ship, a mine-hunting control system is fitted. In the remote mine-hunting concept, four consoles are provided for control, mission planning and execution, and sonar control and analysis. Important to the concept is containerised packaging. In this case, the vehicle and all of its support equipment can be shipped in a 13 m ISO container.

Because of the long stand-off range which the remote mine-hunting concept offers, no special magnetic, or acoustic requirements are imposed upon vessels of opportunity, and modifications are modest. A 300–500 tonne vessel of 30–40 m in length, with a beam of 10–12 m is an excellent candidate.

The principal advantages of the concept over other approaches are:

(a) Data are available in real time rather than following the mission as is the case with AUVs.
(b) Search rate is three times that of AUVs.
(c) Standoff range can be up to 15 km.
(d) The system can be used for real-time mine avoidance as well as mine reconnaissance.
(e) Unlike USV concepts, wave-making drag is non-existent; thus, the power requirement and size of the vehicle is reduced.
(f) The platform is more stable than USV concept platforms. This results in better towfish stability and improved GPS and radio data antenna stability.

18.2.2 Early tests

The first integration of mine-hunting sensors occurred in 1987 under the sponsorship of the US Defence Advanced Research Projects Agency (ARPA). ISE integrated a single beam Klein 595 sidescan sonar and a Collins AN/ARC 182 data radio into the Dolphin and demonstrated that the system was capable of detecting moored mines remotely. Following this demonstration, the US Navy acquired two systems for development.

Drawbacks to the 1987 integration were the limiting speed of 5 knots imposed by single beam sidescan sonar technology and the lack of a cable winch for the sonar. Subsequently, both the Canadian and the US Navies started to develop the capability of operating faster multibeam sidescan sonars from the Dolphin. In 1989 and 1990, the Canadian Defence Research Establishment sponsored the development of a winch capable of towfish handling to depths of 100 m.

The next step was to validate the system as a stable sonar platform and to compare its motion with that of other potential mine-hunting vessels. This was done in 1991 off the West Coast of Vancouver Island and the results clearly demonstrated the superiority stability and sea-keeping abilities of the semi-submersible drone over a surface drone or a small ship of mine-hunter size. Perhaps the most striking indication of the relative sea-keeping capabilities of the vessels occurred on a particularly rough day with significant wave heights of 3.5 m and an average period of only 6.7 s. The 16 m, 65 tonne surface 'drone' could not operate safely at all and did not leave the harbour, and the 500 tonne ship also ran for shelter after a few circuits. The 3.6 tonne Dolphin operated throughout.

18.2.3 US Navy RMOP

In late 1993, a US government–industry team was assembled to integrate a multibeam sonar system in a Dolphin-based platform to be known as the Remote Minehunting Operational Prototype (RMOP). The integration involved the following equipment:

(a) Westinghouse AQS 14 sidescan sonar and towfish.
(b) Reson 6012 SeaBat hull-mount sonar.
(c) Rockwell-Collins PLGR GPS.
(d) TAC 3 display console.
(e) Broadcast MicroWave System high bandwidth (10 Mb/s) radio.

The successful demonstration of this vehicle led to the inclusion of the RMOP in exercise 'Kernal Blitz' in 1995. Operations were conducted day and night over a 4-day period in sea states 3 and 4 conditions and currents of 1 knot. The performance of the system was universally described as excellent; see Figure 18.4.

Following this, the RMOP continued service with the US Navy as the RMS V1 and RMS V2 system until 1999. Further successful deployments were conducted including a deployment to the Persian Gulf on the USS *Cushing* (DD985). At this point, the US focus shifted to the RMS V3 development with Lockheed Martin. This development continues with US Navy in-service operations planned for 2006.

Figure 18.4 US Navy RMOP vehicle

18.2.4 The Canadian 'Dorado' and development of the French 'SeaKeeper'

The RMS V2 provided a basic remote mine-hunting capability to the naval force commander. Its small size enabled it to be rapidly shipped to forward areas and operated from either naval ships or craft of opportunity. The Canadian interest was in extending the towing capability to 200 m at 10 knots. A number of modifications were necessary to achieve this capability including a more powerful engine of 315 kW h, the development of a contra-rotating propulsor and an enlarged vehicle hull. The development of the larger Dorado hull was commenced in the summer of 1998; see Figure 18.5.

In 1997, ISE teamed with the French defence company DCN to develop the system for the international market. Much of this work has been done using the Canadian Defence Research vehicle and considerable cooperation between French and Canadian development agencies has occurred.

Operations in Canada have included testing and trials to demonstrate the 200 m towing capability as well as a 36-month Technical Demonstration of the system ability in a remote mine-hunting capacity. In France, two major sensor integrations have been sponsored in 1999 and 2003, and a tactical control system has been developed. In both countries, a number of trials have been undertaken and three demonstrations to foreign navies have been conducted. In 2003, the system was successfully evaluated by the French Navy for survey operations in the approaches to Brest Harbour. The 1999 French integrations have entailed a Klein Associates 5500 Multibeam Sidescan

Figure 18.5 'Dorado' vehicle with towfish at Victoria, British Columbia

Sonar, a Triton Elics ISIS Sonar Processor and a C-Spec 2.4 GHz 10 Mb/s Dataradio. This system was named 'SeaKeeper'.

The system completed sea trials in the Vancouver area. The integrated system was shipped from Vancouver to Brest in France for trials. Following these tests, a demonstration was given to 16 foreign navies. Operations were supported from the French research agency, Groupe d'Etudes Sous-Marine de l'Atlantique (GESMA) trials vessel '*L'Áventuriere*'.

In 2000, the Canadian DREA (Defence Research Establishment (Atlantic)) sponsored a trial at the Canadian Navy range in Nanoose to demonstrate the ability to tow the mine-hunting towfish at depths of 200 m. Actual operations to 215 m were carried out. This led to the start of the Technical Demonstrator project in June of 2000. Over the next 36 months, a series of improvements to the system were undertaken.

Many of these improvements lie in the area of positioning. The goal was to establish the ability to locate mine-like targets to within ±10 m. In August 2001, a series of trials with the system were undertaken and a final set of trials was conducted in Esquimalt in 2002. Following these trials, a demonstration was provided to the French Navy as well as to seven other navies. The findings of the Technical Demonstration were that all of the objectives had been met, and in some cases, exceeded.

Finally, a scientific trial was carried out by the Canadian DREA and the French GESMA, and the French navy evaluation was conducted. The following were the principal findings for the evaluation team.

18.2.4.1 Operational availability

During 13, 8-h days of trials and demonstrations, the SeaKeeper achieved more than 100 h of sonar operation, showing a remarkable reliability for a prototype. A few failures occurred, which did not impact on the availability or overall trial programme.

18.2.4.2 Mine search performance

All mines laid in the area were detected with a classification as a mine-like object. The automatic track-following on search patterns and classification 'flower' patterns demonstrated manoeuvrability and flexibility of use of the system, with a short turning radius and a capability of fast reconfiguration of the vehicle tracks.

18.2.4.3 Positioning precision

Measurement of the positioning error of each sonar contact on the targets gave a detailed knowledge of the performance of each of the subsystems used in SeaKeeper for MLO localisation. The average positioning error on sonar contacts during the demonstrations was 2.5 m.

18.2.4.4 System efficiency

The system proved itself very fast and easy to deploy, the average time for handling the vehicle in the water was 5 min, and the average speed of transit was 9.5 knots, with a maximum of 12 knots, which allowed operations in 5-knot currents; providing permanent availability of the system whatever the tidal conditions. The area coverage rate was three times faster than with traditional means. The use of a multibeam echo sounder on the SeaKeeper vehicle for piloting the towfish altitude on a harsh bottom profile proved very successful. A total of 100% sonar coverage was achieved even in areas with slopes of 20%.

18.2.4.5 Other applications

While the major focus has been to develop a mine-hunting system, considerable effort has also been undertaken to establish the design for both a seismic survey platform towing up to 10 km seismic cables, and a ROV support platform operating from an offshore oil platform. The ROV support platform is designed to operate a 100 HP work class ROV.

18.3 Wave-piercing autonomous surface vehicles

Wave-piercing autonomous surface vehicles (WASVs) are semi-submersibles, that is to say they are surface vehicles with a high waterline. The essential feature of any wave-piercing vessel is to have a small uniform water-plane area over the vertical section of hull affected by waves, so that waves can move up and down the hull in that area without causing an increase in buoyancy and lift. Above and below this uniform section the hull can assume any required shape for propulsion, services, payload etc. Typical shapes for manned vehicles involve a torpedo-like section below the waves,

Wave-piercing autonomous vehicles 397

narrowing to a smaller uniform section and increasing to a larger area above the waves for accommodation and payload.

A WASV is also designed so that the ratio of the water-plane area to the buoyancy causes it to float with the waterline roughly halfway up the uniform small waterplane area section, even when the vehicle is stationary. The WAUV, because it is a schnorkelling submarine, on the other hand, floats with the hull broaching the surface. While the small uniform cross section of a submarine's schnorkel gives it a wave piercing capability, it has to be driven to depth by the lift of the hydroplanes, and hence must either have hovering thrusters or have way upon it.

The WASV technology has been implemented in the survey autonomous semi-submersible (SASS) vehicle. It involves a submerged torpedo-like body running just below wave depth with a ballasted keel for stability, and a strong upright fared fin or spar of small uniform section protruding through the water surface. The spar provides the buoyancy to float at the correct depth with the mean waterline half way up the spar. It allows the use of air for conventional prime movers and radio frequency coverage for communications as well as GPS or DGPS positioning. It also provides a stable structural component above water essential for launch and recovery. A schematic drawing is shown in Figure 18.6.

A large detachable instrumentation compartment occupies the forward section of the vehicle for operators to install their own underwater equipment, and the top of the strut or spar provides a site both for mounting the main air-breathing engine and for above water sensors such as aerials or surveillance cameras. Fitting the engine above the waterline has the advantage of easy access for maintenance and saves having to make a watertight seal round a hatch. The engine is splash proof, and can have automatic restart in case of more severe splash over.

One of the main deterrents to the use of autonomous vehicles at sea is the difficulty of launch and recovery, especially in worsening weather. The stable spar penetrating the water gives a convenient attachment for recovery and also enables refuelling to

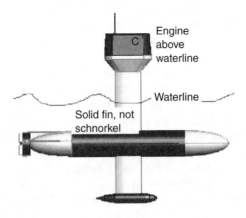

Figure 18.6   Schematic diagram of an SASS vehicle

be carried out alongside, thus avoiding the most frequent reason for return on board. This is discussed later in the chapter.

Although the initial incentives for developing the WASV technology were the same as those for WAUVs, that is, stability in a small unmanned vehicle, with good propulsion, communication and positioning characteristics, it became clear that the ability to operate at slow and zero speed conferred the additional capabilities of acting as a mother-ship for daughter ROVs and AUVs, and also the ability to loiter and act as a mobile buoy. These attributes extend the applications for SASS vehicles.

18.3.1 Development programme

The WASV technology is capable of realisation in a wide range of sizes. Computer simulations were carried out over the range 1–20 m hull length, and showed no anomalies over this range.

After theoretical studies, tank testing was carried out with a 0.75 m model to check drag and stability characteristics. The model was made in sections so that the ratio of hull diameter to length could be varied, and it was towed at different submergence depths. These tests supported theoretical calculations and showed that hull shapes could be varied over a wide range without changing the characteristics unduly.

Following these tests a self-propelled SASS vehicle demonstrator model 1.75 m in length was built and tested without an active control system in order to ensure (and to provide evidence) that the vehicle design was inherently stable. Both calm water and rough water tests were undertaken on this demonstrator model with the help of a NERC/DTI Seasence award.

The unstabilised demonstrator was been shown to be directionally and vertically stable in slow and high speed runs in calm water; it also performed well in rough water, particularly in head sea conditions. Later a two-fin stabilisation system was added and improved the stability to the required standard. Both forms were trialled at several speeds at all headings in the rough conditions. In particular the ability of a semi-submersible vehicle to perform at low speeds, which constitutes much of the novel technology, was proven without doubt. A video loop of these trials is available, and also see Figure 18.7.

The SASS demonstrator is not an operational vehicle; it is small, with only simple controls, and lacks sophistication in design; its sole purpose was to demonstrate the SASS concept as workable, and this has been achieved. To develop a marketable product required the building of a prototype vehicle, command and control system, launch and recovery system, and four-fin stabilisation. The first prototype, SASS-6M, 5.5 m in length and weighing 1.3 tonnes has been designed and built, and is undergoing sea trials. The vehicle is shown in Figures 18.8 and 18.9.

The propulsion system for this prototype consists of a 25 HP diesel engine driving the propeller via a gearbox. An electric drive alternative via a clutch will be incorporated into the first production vehicles for ultra quiet running, for repeated slow manoeuvring, or in case of failure of the prime mover.

Wave-piercing autonomous vehicles 399

Figure 18.7   SASS demonstrator vehicle on trials

Figure 18.8   SASS-6M vehicle ready to launch

Figure 18.9   SASS-6M vehicle at sea

A fin keel has been selected in preference to bilge keels for efficiency, although bilge keels would be useful for stability when ashore. Four-fin stabilisation, depth keeping and trim control have been incorporated. These interact with the autopilot, and hence interface with the command and control system, and have been designed as a single system.

For stowage, the spar is capable of folding forward when recovered as shown in Figure 18.10. A trolley stand has also been designed for movement and stowage on shore. The spar can also be removed from the hull completely, and in this configuration the vehicle and all its components required for operation will fit into an ISO container.

18.3.2   Command and control

The command and control system for the prototype has a display showing local charts in either raster or vector format, with the ability to show tracks of mother-ship and several daughters, and to do way-point prediction and planning The system then sends commands to the SASS in the following five modes:

1. *Direct control.* This mode gives direct control of helm and throttle, usually for the final stages of manoeuvre.
2. *Set course and speed.* This uses the autopilot to relieve the controller of the need to make incremental adjustments.

Figure 18.10   SASS-6M in stowage position

3. *Way-point to way-point.* This mode enables the SASS to carry out a pattern or grid automatically, using GPS or DGPS positioning.
4. *Take station.* This mode enables the SASS to maintain a relative position from the mother-ship, either ahead (typically for fisheries and MCM applications), or abeam (for area surveys), or astern (to save recovery on completion of a mission).
5. *Hover.* This enables the SASS to hold a fixed position by DGPS, which is useful to deploy sensors below the vehicle or to act as a stationary buoy without moorings. Because the SASS floats at depth this can be achieved with minimal or zero forward way to remain stable

Customising and interfacing to telemetry and autopilot has been carried out, involving the design of an integrated system. These features are part of the integrated command and control system. A vehicle instrumentation recording facility is included in this system. A collision avoidance system will probably be fitted in later vehicles.

18.3.3   Launch and recovery

One of the main criticisms of autonomous vehicles generally from operators is the difficulty of launching and recovering them. The SASS vehicle approaches this problem from two directions. The first is to avoid recovery whenever possible. This can be achieved by use of the 'Take Station' mode and by bringing the vehicle alongside for refuelling and simple diagnostics using the spar, instead of recovery. The second is to use the stable, rigid spar which is an integral part of the structure and which penetrates the water surface, to ease the recovery task when this is unavoidable. A method of locking on to the spar for recovery and releasing it for launch which permits the use of a standard RIB davit is being designed.

After recovery the spar can be folded forward for stowage and maintenance (see Figure 18.10), and it is possible to remove it completely for repair or replacement. In this configuration the SASS and all its equipment will fit into an ISO container for transport over greater distances.

18.3.4 Applications

The SASS vehicles can be regarded as replacements for ships or launches for data gathering at lower cost and in some cases with improved stability. As the technology is still being developed, no SASS vehicles are in use, but discussions with potential users indicate that the ability to fit above water and underwater sensors gives rise to a large number of applications.

18.3.4.1 Civil applications

Civil applications include the following:

- Bathymetric and hydrographic survey, using swath bathymetric sonar or echo sounder mounted in the nose cone.
- Oceanographic and meteorological survey, using a range of techniques, both in water and above water, including themistor chains, cameras, sampling and acoustic devices.
- Fish stock assessment and fish behaviour studies, as a quiet substitute for fishery research ships with various acoustic sensors.
- Archaeological survey, with acoustic sensors.
- Anti-terrorist or smuggling surveillance, employing visual, infrared (IR) and radio sensors, and using the SASS vehicle's ability to loiter while conserving fuel.

18.3.4.2 Military applications

Military applications include the following:

- Mine countermeasures, with sidescan, swath or synthetic aperture sonar in water down to 50 m, possibly allowing ships their own off board mine countermeasures capability in these waters.
- Electronic warfare platforms, fitted with radio or acoustic equipment.
- Metoc and hydrography, with sensors similar to those used in civil surveys.
- Beach survey, using acoustic, video and IR sensors.
- Battle-space control, as an acoustic/radio relay.
- Above water and underwater surveillance with acoustic, video and IR sensors.

18.4 Daughter vehicles

Although streamers and towfish have been towed by 'Dolphin', and a winch has been fitted in a 'Dolphin' keel, so far ROVs and fully submerged AUVs have not been carried on board any UWV as daughter vehicles. However, the concept of using a UWV as a mother-ship is consistent with the view that these vehicles are ship substitutes, provided the UWV can remain stable at zero speed in the case of an ROV or the very low speeds usual for AUVs.

The ability of the SASS vehicles to remain stable at zero speed makes the concept of carrying a daughter ROV credible, and plans to fit a small video ROV in the nose cone of SASS-6M prototype vehicle are well advanced. The video signal from the ROV will be passed up the umbilical to the SASS and transmitted over a telemetry link to the operator, who will also be able to control the SASS vehicle. This arrangement will give the operator the ability to investigate the sea bed with a video camera at a range of several kilometres, set by the capability of the telemetry, from an office ashore or afloat.

It is planned to fit a Seabotix LBV 300 ROV to the SASS-6M for trials. This ROV carries lights and video camera at the end of an umbilical 300 m long, and is usually controlled from the dockside or a survey launch. It also has a self-spooling reel system, which will fit the dimensions of the SASS-6M nose cone, leaving room for the vehicle itself forward of the reel behind opening doors.

If these trials are successful, the concept could be extended to military mine disposal ROVs such as Archerfish and Seafox, using a fibre umbilical.

Extending the idea of UWVs as mother-ship substitutes leads to the concept of a daughter AUV. All AUVs suffer from the disadvantage of limited speed, and for smaller vehicles endurance and range also impose severe limitations, and they are usually carried to an operating area by survey launch or ship. A UWV can in concept carry, launch and recover a small AUV, provided it can remain stable at slow speeds for recovery. The SASS-6M prototype vehicle is not itself capable of carrying AUVs, but the basic SASS technology is flexible and concept studies are planned to develop suitable vehicles for deploying *Remus* and *Gavia* sized AUVs, possibly in pairs so that one can be on station while the other is recharged.

However, the UWV need not carry the AUV. It can carry out another function of a mother-ship by escorting the AUV, updating its inertial navigation system and checking that the sensor instrumentation is working without requiring it to interrupt its task and come to the surface. It is planned to fit ultra short base-line (USBL) equipment to the SASS-6M to track a submerged AUV, and remain vertically above it so as to obtain optimum accuracy and the best acoustic path for communications. This also enables an operator to know exactly where the AUV is, as well as having assurance that sensors and positioning systems are correct. Thus, an AUV and a UWV can work more usefully as a pair, giving information from the seabed at high resolution and long radio ranges.

If a UWV can remain stable at slow or zero speed it gives the potential to extend their application to carrying and escorting UUVs, extending the range and endurance of the latter. Plans to detail and exploit this potential are in hand.

18.4.1 Applications

The use of daughter vehicles greatly increases the potential applications of the UWV technology. It overcomes their main disadvantage of being unable to obtain data from close to the seabed at any depth. Provided the UWV can remain stable at very low speeds it can control a daughter carrying the sensors needed to explore and investigate deep areas of interest at close range, and it allows data obtained from close to the seabed to be transmitted to the operator by the umbilical and radio instead of the much more limited ranges and bandwidths available to acoustic transmissions.

In the case of a daughter ROV the full bandwidth of the sensors can reach the UWV on the surface via the umbilical, but a daughter AUV is limited by the bandwidth of the best acoustic channel between it and the surface. In many applications this is adequate for passing information on critical events or for quality assurance, but the full information can only be obtained by recovering or surfacing the daughter.

18.4.1.1 Civil applications

Bathymetric, hydrographic, oceanographic and archaeological survey can all benefit from information near the seabed if the improvement in quality can justify the increased cost of using a daughter or escorted vehicle.

Cable route survey in the deep ocean uses either towfish or AUVs. In either case their position needs to be accurately known throughout the survey. This is currently done using a trail boat positioned over the towfish and passing accurate positioning information to the survey ship. A UWV can replace the trail boat at less cost.

Fish stock assessment and fish behaviour studies can benefit from information close to the nets and trawls. The fully submerged AUV Autosub has been successfully used in conjunction with a fisheries research ship, and it may in future be found that an AUV/SASS combination is useful. However the increased cost is unlikely to make the use of daughter vehicles worthwhile.

18.4.1.2 Military applications

The main military application of daughter AUVs is for mine countermeasures, where it is important to get as much information about the mine as is safely possible. Side-scan and synthetic aperture sonars need to be carried below the surface in water depths greater than 50 metres, and AUVs with acoustic links to a UWV can allow control and communication between operator in a safe place and sensor. Specialised vehicles may need to be developed for these applications.

Daughter ROVs are already used with the PAP (Poisson Auto Propulsé) system (a mine hunting and disposal ROV developed for the French navy) and ROVs for visual identification using video and lights give good images, but at a limited range from a ship. UWVs with daughter ROVs can increase this distance as well as allowing a sacrificial ROV to destroy the mine safely. A further development is to use the UWV as a discrete, if not totally covert, carrier for a mine countermeasures AUV to close the range to the area to be investigated so as to give the AUV longer useful time on station. In effect this would be a two-stage system.

The ability to control AUVs can also make a useful contribution to battle space control, while a UWV could be developed as a weapon launch platform.

18.5 Mobile buoys

The WASV technology permits a UWV to remain stable while remaining in one position determined by GPS. It can therefore act as a mobile buoy, able to take or change station on command and without ground tackle. This permits positioning in deep water where ground tackle would be impossible.

GPS position can be maintained to a specified accuracy either by repeated drifting with the environment and recovering to the desired position, or by executing a slow circular or figure-of-eight manoeuvre depending or whether a line which must not be twisted is deployed from the vehicle. If the drifting method is used, electric propulsion will be engaged to avoid repeated starting and stopping of the diesel, recharging the batteries as required. Concept studies for a mobile buoy are planned. While opinions are divided on the market for such a vehicle, applications to fisheries research show promising results.

18.5.1 Applications

The idea of UWVs as controllable mobile buoys is novel, raising applications which so far as is known have not been possible heretofore. Drifting and moored buoys, however, have been employed in a number of data gathering experiments.

18.5.1.1 Civil applications

Fish behaviour studies at present use moored buoys to gather data, and this could be replaced by mobile buoys able to avoid being run down, and able to deploy, relocate and return without the need of a servicing vessel.

Oceanographic surveys use drifting buoys, and these experiments could be given more controlled direction by using mobile buoys. Air/sea interface studies could also employ mobile buoys in a similar manner.

Emergency positioning requirements currently use moored buoys. These could be positioned more quickly using mobile buoys.

18.5.1.2 Military applications

The chief military application is as a gateway buoy for battle space control, but port security and surveillance – acoustic, electronic and visual – are potential applications yet to be explored.

18.6 Future development of unmanned wave-piercing vehicles

The technique of UWVs is in its infancy. New applications for both the WASV and WAUV approaches are being investigated, and further new ideas are in mind and can be developed when funding allows. As far as immediate applications are concerned the four major development areas can be foreseen: these are as robotic mine countermeasure vehicles, as a substitute for survey ships of all kinds, as a replacement for mother ships tending AUVs and ROVs, and as mobile buoys taking on some of the functions of moored buoys and drifters.

The amalgamation of features of both UUVs and USVs gives UWV technology a capability which has not yet been fully grasped. Investment is needed to engage with potential users and develop further applications and research. The increased availability of GPS equipments and satellite communications has greatly enhanced the opportunities for UWV applications, and an expanding future is seen for UWV technology. The day can be foreseen when UWVs will be commonplace.

Chapter 19

Dynamics, control and coordination of underwater gliders

R. Bachmayer, N.E. Leonard, P. Bhatta, E. Fiorelli and J.G. Graver

19.1 Introduction

Sampling the oceans has traditionally been conducted from ships, with the first global oceanographic research cruise by Sir Wyville Thomson on the HMS *Challenger* from 1872 to 1876, which led to numerous discoveries such as the mid-Atlantic ridge and the *Challenger Deep* in the Mariana Trench to name only two. It took over 23 years to compile the results from this cruise.

Today, with increasing use of remote sensing techniques from satellites and airplanes, more and more data are becoming available. With modern computers, these data can be analysed in a relatively short period of time. However, current remote sensing technologies, airborne or from space, do not penetrate very far below the ocean's surface. In order to gain more insight into the temporal and spatial processes below the surface, we were until recently still depending on ship based measurements and moorings. Over the last couple of decades alternative technologies such as profiling floats, remotely operated vehicles (ROVs) and autonomous underwater vehicles (AUVs) have emerged to complement the existing sensing techniques [1]. Visions of autonomous platforms roaming the oceans as described in References 2 and 3 have not been fully realised yet, but technological advances driven by these visions have brought us a long way from the HMS *Challenger* cruise. In this chapter, we address issues related to the design and operation of a class of AUVs referred to as underwater gliders.

An underwater glider is a winged, buoyancy-driven AUV developed especially to autonomously collect oceanographic data over the course of weeks or months at a time. The standard, on-board, oceanographic measurements include conductivity, temperature and pressure, but other sensors such as bathy-photometers, optical backscatter sensors, fluorometers, photosynthetically active radiation (PAR) sensors and various acoustic sensors have been successfully tested and deployed on gliders. Henry Stommel and Douglas C. Webb [2], initially conceived the glider concept envisioning an endurance vehicle that could sample the ocean while circumnavigating the globe.

Unlike propeller-driven AUVs, developed in parallel with ROV technology, underwater gliders were developed primarily by oceanographers and engineers who were building and using autonomous profiling floats [4]. Large numbers of floats were initially deployed in the 1990s for the World Ocean Circulation Experiment (WOCE). Deployment of profilers for the Argo global sensor array, a part of the integrated global observation strategy, began in 2000. During the 30 days prior to 18 November 2004, as many as 1478 active floats successfully transmitted data [5]. The data sent from those profilers complement satellite-based remote sensor measurements of the sea surface.

The main component of a float is a buoyancy engine that enables the float to rise, sink or equilibrate to a pre-programmed density. The float travels with a parcel of water matching the programmed density, only rising to the surface at pre-programmed intervals, usually every 10 days, recording a temperature and salinity profile during the ascent. Once at the surface the float reports its collected data via satellite to shore. The floats are designed to operate outside the 2000 m isobath and to cycle at least 150 times to a maximum depth of 2000 m. Because of these design choices and the inability of the floats to actively control their horizontal movements, these profiling floats have only limited use in coastal areas.

Underwater gliders, also propelled by a buoyancy engine, provide additional capabilities for more versatile autonomous ocean sampling. Gliders have lifting surfaces that add manoeuvrability, and they have active attitude control which enables control of their motion in the horizontal plane. Underwater gliders use a buoyancy engine similar to the ones used in the profiling floats described above. A change in weight allows the glider to move up and down in the water-column. The combination of upward/downward force with changes in attitude (i.e., pitch and roll) allow the wings and body to generate the hydrodynamic lift and drag forces which propel the glider horizontally and vertically through the water. Current operational gliders, such as *Seaglider* [6] developed by the University of Washington's Applied Physics Lab and the School of Oceanography, *Spray* [7] developed by Scripps Institution of Oceanography and the electric *Slocum* glider [8] developed by Webb Research Corporation use an electromechanical displacement actuator, pump or piston, to change their weight. A prototype glider using an alternative thermally driven buoyancy engine is currently under development [8]. A laboratory-scale glider is described in Reference 9.

19.2 A mathematical model for underwater gliders

In this section, we present a mathematical model that describes the longitudinal dynamics of underwater gliders. This model is the restriction of a more general three-dimensional (3D) model to the longitudinal plane. The 3D model and the planar model presented in this section were derived in References 10 and 11. The planar model has been further studied in the context of stability analysis and control design in References 10–13 and parameter identification in Reference 14.

We model the underwater glider as a rigid body with fixed wings and tail. The glider is equipped with buoyancy control and controlled internal moving mass. We take the hull of mass m_h to be ellipsoidal with wings and tail attached so that the centre of buoyancy (CB) is at the centre of the ellipsoid. Buoyancy control is modelled with a mass m_b, located at the CB, that can be changed with a control input u_4 so that $\dot{m}_b = u_4$ (where the dot indicates rate of change with respect to time). Control of mass distribution is modelled with a controlled moving mass $\bar{m}$. The total mass of the vehicle is $m_v = m_h + m_b + \bar{m}$. Let the mass of the displaced fluid be m; then, $m_0 g$ is the weight of the vehicle where $m_0 = m_v - m$.

We assign a coordinate frame fixed on the vehicle body, with the origin at the CB. The axes of the body frame are aligned with the principal axes of the ellipsoid as shown in Figure 19.1. The longitudinal plane is formed by the e_1 and e_3 axes. We fix the inertial frame such that k points in the direction of gravity. In the planar model, we assume the glider has zero roll so that we can choose the j axis to be parallel to the e_2 axis. Then, the i–k plane is parallel to the longitudinal plane as shown in the right side of Figure 19.1.

We neglect the contribution of the wings and the tail to the added mass of the glider so that the added mass and inertia matrices of the glider are diagonal. We denote by m_1 (m_3) the sum of the added mass associated with the e_1 (e_3) body axis plus the stationary mass m_s of the glider where $m_s = m_h + m_b$. Similarly, J_2 is

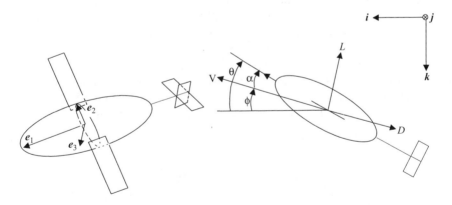

Figure 19.1 (Left side) Body frame assignment. The body axis e_1 lies along the long axis of the vehicle (positive in the direction of the nose of the glider), the body axis e_2 lies in the plane of the wings and the body axis e_3 points toward the bottom of the glider, so that the body frame is right handed. (Right side) External forces and moment in the longitudinal plane. L and D are the hydrodynamic lift and drag forces, respectively, and M is the hydrodynamic moment. θ is the pitch angle, ϕ the flight path angle, α the angle of attack and V the glider speed

the sum of added inertia and stationary rigid body moment of inertia about the CB in the longitudinal plane.

The glider in the longitudinal plane is shown on the right side of Figure 19.1. Denote by v_1 and v_3 the components of glider velocity along the e_1 and e_3 directions, respectively. The glider speed V is then given by $V = \sqrt{v_1^2 + v_3^2}$. The pitch angle of the glider is θ and the angular velocity in the plane is $\Omega_2 = \dot{\theta}$. The glide path angle is ϕ, the angle of attack is $\alpha = \arctan(v_3/v_1)$ and $\theta = \phi + \alpha$. Let x and z represent the position of the glider in the i and k directions, respectively. Further, let r_{P1} and r_{P3} be the coordinates of the moving mass $\bar{m}$ with respect to the body frame.

The external forces acting on the glider include the weight of the glider $m_0 g$ and the hydrodynamic lift L and drag D, shown in Figure 19.1. Lift and drag forces are assumed to act at the glider CB. The moment due to lift is denoted as M. The moment due to gravity is determined by the force $\bar{m}g$ which acts at the position (r_{P1}, r_{P3}). The hydrodynamic forces and moment are modelled as

$$L = (K_{L0} + K_L \alpha)V^2 \tag{19.1}$$

$$D = (K_{D0} + K_D \alpha^2)V^2 \tag{19.2}$$

$$M = (K_{M0} + K_M \alpha + K_q \Omega_2)V^2 \tag{19.3}$$

The hydrodynamic coefficients $K_{L0}, K_L, K_{D0}, K_D, K_{M0}, K_M$ and K_q can be determined using reference data for generic aerodynamic bodies, wind tunnel tests, static and dynamic parameter identification, detailed computational fluid dynamics (CFD) analysis or a combination of these methods. Identification of hydrodynamic coefficients for the *Slocum* glider using data from sea trials is summarised in Section 19.4.2 and described in more detail in References 14 and 15.

Let u_1 and u_3 be the components of the total force acting on the moving mass $\bar{m}$ in the body frame, along the e_1 and e_3 directions. Let P_{P1} and P_{P3} be the (body frame) components of linear momentum of the moving mass. The equations describing the longitudinal dynamics of the glider first derived in Reference 10 are

$$\dot{x} = v_1 \cos\theta + v_3 \sin\theta \tag{19.4}$$

$$\dot{z} = -v_1 \sin\theta + v_3 \cos\theta \tag{19.5}$$

$$\dot{\theta} = \Omega_2 \tag{19.6}$$

$$\dot{v}_1 = \frac{1}{m_1}\{-m_3 v_3 \Omega_2 - m_0 g \sin\theta + L \sin\alpha - D \cos\alpha - u_1\} \tag{19.7}$$

$$\dot{v}_3 = \frac{1}{m_3}\{m_1 v_1 \Omega_2 + m_0 g \cos\theta - L \cos\alpha - D \sin\alpha - u_3\} \tag{19.8}$$

$$\dot{\Omega}_2 = \frac{1}{J_2}\{(m_3 - m_1)v_1 v_3 - \bar{m}g(r_{P1}\cos\theta + r_{P3}\sin\theta) + M + r_{P1}u_3 - r_{P3}u_1$$
$$- \Omega_2(r_{P1}P_{P1} + r_{P3}P_{P3})\} \tag{19.9}$$

$$\dot{r}_{P1} = \frac{1}{m}P_{P1} - v_1 - r_{P3}\Omega_2 \tag{19.10}$$

$$\dot{r}_{P3} = \frac{1}{m}P_{P3} - v_3 + r_{P1}\Omega_2 \tag{19.11}$$

$$\dot{P}_{P1} = u_1 \tag{19.12}$$

$$\dot{P}_{P3} = u_3 \tag{19.13}$$

$$\dot{m}_0 = u_4 \tag{19.14}$$

Steady, uncontrolled motions of the glider in the longitudinal plane correspond to upward or downward glide paths at constant speed, glide angle and pitch. These steady motions are the equilibria of the dynamical system represented by Eqs. (19.6)–(19.14). Note that horizontal motion (purely in the i direction) with a non-zero speed is not a steady motion of the glider.

The equations of motion for steady glides in the longitudinal plane modified to match the configuration of a *Slocum* glider are presented in Section 19.4.2. More detailed studies can be found in References 10 and 12–14.

The hydrodynamic lift and drag coefficients, K_{L0}, K_L, K_{D0} and K_D, determine the permissible values of the steady flight path angle, ϕ_e, in the range $(-(\pi/2), (\pi/2))$. The magnitude of the shallowest steady glide angle for the glider model is smaller for lower values of $K_D > 0$. The largest permissible equilibrium speed V_e is determined by the largest permissible value of the equilibrium weight $m_{0_e}g$. Given a permissible value of V_e and of ϕ_e, we can compute the corresponding m_{0_e}, the equilibrium angle of attack α_e and the equilibrium position of the internal mass (r_{P1_e}, r_{P3_e}). In fact, there is a one-parameter family of solutions for (r_{P1_e}, r_{P3_e}) given a permissible steady glide (V_e, ϕ_e). For any steady glide, a large, positive r_{P3_e} contributes to stability since a large r_{P3_e} increases the vehicle's 'bottom heaviness' [10].

We note that the above equations of motion (19.7)–(19.14) were derived with the internal mass assumed to be able to move freely within the glider body. This leads to stability problems much like the fuel slosh problem in a spacecraft. Further, the moving mass in operational gliders is not allowed to move freely, but rather is constrained to move along a suspension system. In Section 19.3, we describe a formulation of the forces u_1 and u_3 that realises these constraints [12]. The new control inputs w_1 and w_3 correspond to the components of acceleration of the moving mass. Stability is discussed further in Section 19.3.

19.3 Glider stability and control

In this section, we describe linear and nonlinear stability analysis of steady glides, which are equilibria of the longitudinal dynamics of the glider given by (19.6)–(19.14). We discuss controllability and observability of linear models, and control laws for stabilising glider motion to a desired steady glide. We also discuss the relationship between the glider model and Lanchester's phugoid-mode model for an aircraft [16, 17]. We describe how we are using this connection to further help with nonlinear analysis and control design of gliders. Derivations and further discussion of the developments described in this section can be found in References 10 and 12–14.

19.3.1 Linear analysis

The equations of motion (19.6)–(19.14) can be linearised about any steady glide path of the glider. The linearised dynamics can then be used to analyse a given vehicle design, for example, to test for stability and controllability of a desired glide path given a choice of vehicle design parameters. Control laws can also be designed using these linearised dynamics. The linearisation of the glider model is carried out in References 10 and 11, with Eqs. (19.4)–(19.5) for x and z replaced by a single equation for z', where $z' = x \sin \phi_e + z \cos \phi_e$ is the relative position of the glider with respect to a given (desired) glide path, in the direction perpendicular to the glide path, as shown in Figure 19.2. The dynamics for z' are

$$\dot{z}' = \sin \phi_e (v_1 \cos \theta + v_3 \sin \theta) + \cos \phi_e (-v_1 \sin \theta + v_3 \cos \theta). \tag{19.15}$$

The linear stability of four example glide paths of an experimental glider called ROGUE, developed at Princeton University, is studied in References 10 and 11. Using the model (19.6)–(19.14), all four glide paths exhibit a relatively slow unstable mode. However, this instability is not representative of the real operational dynamics; in

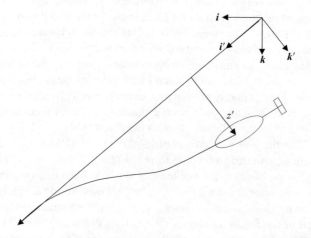

Figure 19.2 Planar gliding controlled to a line, $z' = x \sin \phi_e + z \cos \phi_e$

this analysis it is due to the coupling between the unconstrained moving mass and the vehicle, as discussed in Section 19.2. Since the moving mass is in practice constrained to move along a suspension system in all operational gliders (including ROGUE), we set the force on the moving mass (u_1, u_3) to be equal to the force that constrains the moving mass to move along a suspension system plus a control term parameterised by (w_1, w_3), the acceleration of the moving mass $\bar{m}$. The components of acceleration (w_1, w_3) can then be freely designed as control inputs.

First, we change coordinates so that instead of linear momentum of the moving mass, we use velocity of the moving mass as states. The coordinate transformation, derived from Eqs. (19.10) and (19.11), is

$$\begin{pmatrix} P_{P1} \\ P_{P3} \end{pmatrix} \mapsto \begin{pmatrix} \dot{r}_{P1} \\ \dot{r}_{P3} \end{pmatrix} = \begin{pmatrix} \frac{1}{\bar{m}} P_{P1} - v_1 - r_{P3}\Omega_2 \\ \frac{1}{\bar{m}} P_{P3} - v_3 + r_{P1}\Omega_2 \end{pmatrix} \tag{19.16}$$

Next, we assign the force (u_1, u_3) to be

$$\begin{bmatrix} u_1 \\ u_3 \end{bmatrix} = \begin{bmatrix} \left(\frac{1}{\bar{m}} + \frac{1}{m_1} + \frac{r_{P3}^2}{J_2}\right) & -\frac{r_{P1}r_{P3}}{J_2} \\ -\frac{r_{P1}r_{P3}}{J_2} & \left(\frac{1}{\bar{m}} + \frac{1}{m_3} + \frac{r_{P1}^2}{J_2}\right) \end{bmatrix}^{-1}$$

$$\times \left(\begin{bmatrix} \frac{1}{m_1} X_1 + \dot{r}_{P3}\Omega_2 + \frac{r_{P3}Y}{J_2} \\ \frac{1}{m_3} X_3 - \dot{r}_{P1}\Omega_2 - \frac{r_{P1}Y}{J_2} \end{bmatrix} + \begin{bmatrix} w_1 \\ w_3 \end{bmatrix} \right) \tag{19.17}$$

where

$$X_1 = -m_3 v_3 \Omega_2 - \bar{m}(v_3 + \dot{r}_{P3} - r_{P1}\Omega_2)\Omega_2 - m_0 g \sin\theta + L\sin\alpha - D\cos\alpha,$$
$$X_3 = m_1 v_1 \Omega_2 + \bar{m}(v_1 + \dot{r}_{P1} + r_{P3}\Omega_2)\Omega_2 + m_0 g \cos\theta - L\cos\alpha - D\sin\alpha,$$
$$Y = (m_3 - m_1)v_1 v_3 - \bar{m}g(r_{P1}\cos\theta + r_{P3}\sin\theta) + M$$
$$\quad - \Omega_2(r_{P1}P_{P1} + r_{P3}P_{P3})$$

For notational purposes, we also define

$$u_4 = w_4 \tag{19.18}$$

One can interpret (19.16) as a nonlinear coordinate transformation and (19.17) as a nonlinear feedback control law.

If we define $\eta = (x, z, \theta, \Omega_2, v_1, v_3)^T$, $\zeta = (r_{P1}, \dot{r}_{P1}, r_{P3}, \dot{r}_{P3}, m_0)^T$, $w = (w_1, w_3, w_4)^T$, then the resulting equations of motion of the glider are

$$\dot{\eta} = q(\eta, \zeta, w) \tag{19.19}$$

$$\dot{\zeta} = A\zeta + Bw \tag{19.20}$$

where,

$$A = \begin{bmatrix} 0 & 1 & 0 & 0 & 0 \\ 0 & 0 & 0 & 0 & 0 \\ 0 & 0 & 0 & 1 & 0 \\ 0 & 0 & 0 & 0 & 0 \\ 0 & 0 & 0 & 0 & 0 \end{bmatrix}, \quad B = \begin{bmatrix} 0 & 0 & 0 \\ 1 & 0 & 0 \\ 0 & 0 & 0 \\ 0 & 1 & 0 \\ 0 & 0 & 1 \end{bmatrix}$$

and the vector field q is obtained by substituting the mapping (19.16) and Eq. (19.17) into Eqs. (19.4)–(19.9).

The steady glide paths for the system described by (19.19) and (19.20) are the same as those for the original set of equations of motion. Linearisation shows that the same steady glide equilibria that were unstable before the feedback (19.17) and (19.18) are now stable for the feedback controlled system. That is, the steady glides are stable in the practical case that the internal moving mass is constrained to move along a suspension system inside the vehicle; this is consistent with operational experience. These stable glide paths become unstable if there is insufficient bottom heaviness, that is, it is a requirement for stability that the centre of gravity at the equilibrium be sufficiently far below the centre of buoyancy.

For the linearisation examples computed in References 10 and 11, all states including z' are locally controllable. This implies that we can design a linear controller that can (at least locally) stabilise gliding to a prescribed line. In fact, it is shown, that these glides are fully controllable even when the internal mass is restricted to move only along the e_1 (or e_3) direction. We note, however, that all of this linear analysis holds only for small deviations from the equilibrium glide path. Control of large motions, such as switching from an upward glide to a downward glide path, might not be well handled with an arbitrary choice of control authority. For instance, while the motion of the moving mass restricted to the e_1 direction would be sufficient for sawtooth manoeuvres, motion restricted to the e_3 direction would not allow for both upward and downward steady glide motions.

All states of the linearised models computed in References 10 and 11, except z', are observable with measurements restricted to moving mass position (r_{P1}, r_{P3}) and the vehicle weight $m_0 g$. All states, except z', are also observable with measurements limited to the pitch angle θ, r_{P1} (or r_{P3}) and $m_0 g$, and these are all typically measured in the operational gliders. In order to control only the speed and direction of the glider path we need not measure z'. This is because the speed of the glider depends only on v_1 and v_3, the direction of the glider path depends only on v_1, v_3 and θ, and the glider dynamics do not depend on z'. So we can stabilise the glider about a certain desired V_e and ϕ_e, by measuring m_0 and any two of the three variables r_{P1}, r_{P3}, θ, and using a linear dynamic observer such as the one designed in Reference 11 to estimate the variables not directly measured. Since battery power in a glider is limited, it is of great value to be able to estimate accurately dynamic states of the glider without having to introduce additional sensors. In the current operational gliders, linear velocity is estimated using a somewhat cruder method that assumes angle of attack is constant.

If we want to control the glider to a prescribed line in the plane, then we need a measurement of z'. This requires vertical position z, which is easily measured, and horizontal position x, which is not so easily measured. However, with an initial measurement of x from a GPS fix taken when the glider is at the surface, the horizontal motion of the glider can be estimated using integration, and the glider can be controlled to a prescribed line on the plane.

If we use a linear control law

$$w = K(\zeta - \zeta_e), \tag{19.21}$$

where ζ_e is a desired equilibrium value of ζ, and K is a control gain matrix such that $A+BK$ is Hurwitz, the closed-loop system (19.19) and (19.20) is locally exponentially stable in θ, Ω_2, v_1, v_3 and ζ if the equilibrium is sufficiently bottom heavy [12]. Recall from Section 19.2 that we arrive at a one-parameter family of solutions for (r_{P1_e}, r_{P3_e}) for a permissible (V_e, ϕ_e). Thus, for a given desired equilibrium glider speed and direction, we can choose a sufficiently large r_{P3_e} such that the closed-loop system is locally exponentially stable in θ, Ω_2, v_1, v_3 and ζ.

Simulations suggest a very large region of attraction for control laws such as (19.21). This is illustrated by a stable switch from a downward 45° glide to an upward 45° glide demonstrated in Reference 12.

19.3.2 Phugoid-mode model

In this subsection, we describe the phugoid-mode model, which is the integrable system of equations describing the longitudinal dynamics of an aircraft as given by Lanchester [16, 17]. We discuss the utility of this model in our derivation of a Lyapunov function, an energy-like function, used to prove stability and design stabilising controllers for the nonlinear glider dynamics. Although the linear analysis described in Section 19.3.1 is quite powerful, a nonlinear tool allows for a more systematic means to analyse and design controllers. For instance, in the nonlinear context, analysis and design does not have to be redone for each glide path of interest. Furthermore, nonlinear tools provide the means to determine global behaviour and one is not restricted to studying and controlling steady glide paths.

Lanchester derived an integrable model of the longitudinal dynamics of an aircraft by making some simplifying assumptions. He considered the situation when the thrust of the aircraft exactly balances the drag. He also assumed a fixed angle of attack, arguing that the relatively small moment of inertia and the large angle of attack moment coefficient allows a fast convergence of the angle of attack to its equilibrium value. With these assumptions the equations of motion of the aircraft are

$$\dot{V} = -g \sin \phi \tag{19.22}$$

$$\dot{\phi} = \frac{1}{mV}(KV^2 - mg \cos \phi) \tag{19.23}$$

where V is the aircraft speed, ϕ is the flight path angle, m is the mass of the aircraft and K is the constant lift coefficient. The equilibrium of this phugoid-mode model corresponds to a steady, level flight motion of the aircraft. The equilibrium speed is $V_e = \sqrt{mg/K}$ and the equilibrium flight path angle is $\phi_e = 0$.

Lanchester showed that the phugoid-mode model is integrable by deriving two conservation laws, the energy E and a quantity C:

$$E = \frac{1}{2}mV^2 + mg \int_0^t V(\tau) \sin(\phi(\tau))\, d\tau$$

$$C = \frac{V}{V_e} \cos\phi - \frac{1}{3}\left(\frac{V}{V_e}\right)^3$$

The quantity C parameterises the classes of trajectories followed by the aircraft: steady level flight, wavy flight paths, loops and the singular trajectory [17].

In the glider model, suppose for illustration that the added masses are equal, that is, $m_1 = m_3$, the buoyancy mass m_b is constant so that weight m_0 is constant and the internal moving mass $\bar{m}$ is fixed at the CB. In this case, the equations of motion of the glider in the variables V, ϕ, α and Ω_2 are

$$\dot{V} = -\frac{1}{m_1}\{m_0 g \sin\phi + D\} \tag{19.24}$$

$$\dot{\phi} = \frac{1}{m_1 V}\{L - m_0 g \cos\phi\} \tag{19.25}$$

$$\dot{\alpha} = \Omega_2 - \frac{1}{m_1 V}\{L - m_0 g \cos\phi\} \tag{19.26}$$

$$\dot{\Omega}_2 = \frac{M}{J_2} \tag{19.27}$$

In this glider model, if we neglect the drag and assume a constant $\alpha = \alpha_e$, the equations for V and ϕ can be rewritten as follows:

$$\dot{V} = \frac{-m_0 g \sin\phi}{m_1} \tag{19.28}$$

$$\dot{\phi} = \frac{1}{m_1 V}\{(K_{L0} + K_L \alpha_e)V^2 - m_0 g \cos\phi\} \tag{19.29}$$

Equations (19.28) and (19.29) can also be obtained from the Lanchester phugoid-mode model (19.22) and (19.23) by rescaling t by a factor of m_0/m_1, replacing m by m_0 and K by $(K_{L0} + K_L \alpha_e)$.

Using a Lyapunov function derived from the conserved quantity C, we make rigorous Lanchester's simplifying assumptions and the conditions under which the phugoid-mode model yields dynamics that are representative of the more complete glider model dynamics [13]. Further, we use C to derive a Lyapunov function for proving asymptotic stability of any stable steady glide for the glider model described by Eqs. (19.24)–(19.27). We show how we can use this Lyapunov function to estimate the region of attraction of the steady glide, that is, the set of initial conditions which converge to the glide path of interest. We also use this Lyapunov function to design stabilizing controllers that have provably good properties.

19.4 *Slocum* glider model

Having derived the equations of motion for the longitudinal plane dynamics of a general underwater glider and discussed stability and control in the previous sections, we describe next a representative realisation of the glider concept using the example of the commercially available *Slocum* glider. In Section 19.4.1, we describe the *Slocum* glider. In Section 19.4.2, we adapt the general dynamic glider model to match the *Slocum* glider and summarise a model identification approach and results for the steady glide case.

19.4.1 The Slocum *glider*

The *Slocum* glider [8] is designed and manufactured by Webb Research Corporation, Falmouth, MA, USA. The operational envelop of the glider includes a 200 m depth capability and a projected 30-day endurance, which translates into approximately 1000 km operational range with a 0.4 m/s fixed horizontal and 0.2 m/s vertical speed. The glider has an overall length of 1.5 m and a mass of 50 kg. The buoyancy engine is an electrically powered piston drive, located in the nose section of the glider; Figure 19.3. The mechanism allows a close to neutrally buoyant trimmed glider to change its displacement in water by $\pm 250\,\text{cm}^3$, which corresponds to approximately ± 0.5 per cent of the total volume displaced. This change in buoyancy generates a vertical force which is translated through two swept wings into a combined forward and up/downward motion. Due to the location of the buoyancy engine, the change in direction of the buoyant force also creates the main pitching moment for the glider. In addition to the buoyancy engine, the glider possesses two more control actuators: a 9.1 kg battery pack, the internal moving mass referred to as 'sliding mass', that can be linearly translated along the main axis of the glider and a rudder attached to the vehicle tail fin structure. The sliding mass is used for fine tuning the pitch angle.

The glider has two onboard computers, a control computer and a science computer. Sensors on the glider report the following dynamic vehicle states: heading, pitch,

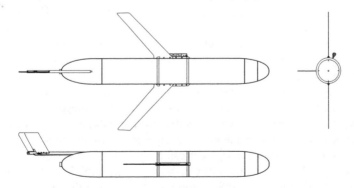

Figure 19.3 Slocum *electric glider layout (Drawing courtesy of Webb Research Corporation, Falmouth MA, USA)*

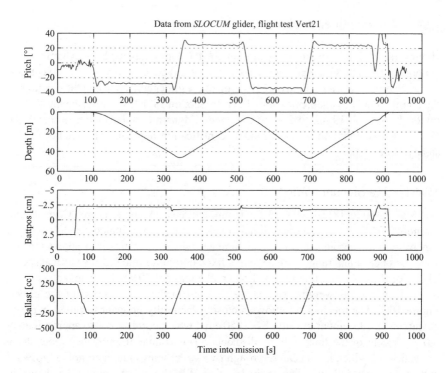

Figure 19.4 Pitch, absolute depth, battery position and buoyancy engine data as a function of mission time recorded from a Slocum *glide segment, January 2003, Chub Key, Bahamas (Printed with permission of D. Fratantoni, WHOI)*

roll, depth, sliding mass position and the state of the buoyancy engine; Figures 19.4 and 19.5. These readings are recorded and processed by the control computer. Vehicle position at the surface is determined by a GPS receiver, with the antenna located on the rear fin. While submerged, the glider velocity and horizontal position are not directly measured. Instead, the glider navigates underwater using a deduced reckoning algorithm. At present, the pitch angle, the rate of change in depth and an assumed constant angle of attack are used by the onboard computer to estimate the horizontal speed of the glider.

The *Slocum* glider can be programmed to navigate in various ways. For a typical mission scenario, the glider navigates to a set of preprogrammed way-points downloaded prior to execution in a mission specification file and operates under closed-loop pitch and heading control. A mission is composed of yos and segments. A yo is a single down/up cycle, while a segment can be composed of several yos and starts with a dive from the surface and ends with a surfacing, Figure 19.4. At all surfacings the glider tries to acquire its GPS location. On the surface the glider compares its desired way-point to its actual GPS position and determines a heading correction for the next way-point before it dives again for the next segment of the mission. Other

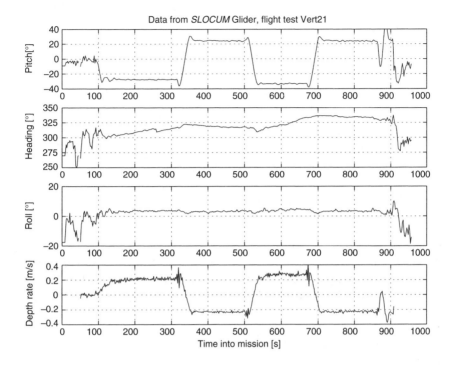

Figure 19.5 Pitch, heading, roll and rate of depth change as a function of mission time recorded from a Slocum glide segment, January 2003, Chub Key, Bahamas (Printed with permission of D. Fratantoni, WHOI)

modes of operation include gliding at a given compass heading, fixed rudder angle or fixed battery position.

19.4.2 Glider identification

The performance of the glider's navigation and control may improve with better estimates of model parameters. Hydrodynamic coefficients are particularly difficult to determine analytically. In this section we summarise a useful data oriented approach to determine the steady state coefficients of a glider model derived from the equations of motion as presented in Section 19.2. We illustrate the approach using data from sea trials conducted with a *Slocum* glider. In addition to providing hydrodynamic coefficients, the approach provides an estimate of static buoyancy trim offset. This suggests that the approach could possibly be adapted to trim the glider at the beginning of a deployment and to detect systems changes or faults in the glider during deployment. For a more detailed description of the methodology the reader is referred to References 14 and 15.

We begin by adapting the glider equations from Section 19.2 to more closely represent the *Slocum* glider dynamics. We add terms to the model to account for the *Slocum* ballast system location and the sliding mass range of travel. For the former,

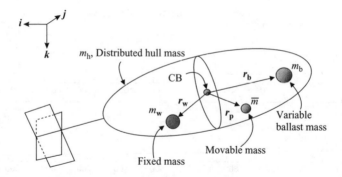

Figure 19.6 Glider mass definitions

we add a point mass m_w that may be offset from the CB by the vector r_w. This makes the total stationary mass of the glider m_s the sum of three terms rather than two: $m_s = m_h + m_w + m_b$. Further, the mass m_b, representing the variable ballast point mass is now also offset from the CB by the vector r_b. The total mass of the *Slocum* vehicle becomes

$$m_v = m_h + m_w + m_b + \bar{m} = m_s + \bar{m}$$

The different masses and position vectors are illustrated in Figure 19.6.

We compute the *Slocum* vertical plane equilibrium steady glides from the modification to the model equations (19.6)–(19.14) as follows:

$$\dot{x} = v_1 \cos\theta + v_3 \sin\theta \tag{19.30}$$

$$\dot{z} = -v_1 \sin\theta + v_3 \cos\theta \tag{19.31}$$

$$0 = (m_3 - m_1)v_{1_{eq}}v_{3_{eq}} - \bar{m}g(r_{P1_{eq}}\cos\theta_{eq} + r_{P3_{eq}}\sin\theta_{eq})$$
$$- m_{b_{eq}}g(r_{b1}\cos\theta_{eq} + r_{b3}\sin\theta_{eq})$$
$$- m_w g(r_{w1}\cos\theta_{eq} + r_{w3}\sin\theta_{eq}) + M_{eq} \tag{19.32}$$

$$0 = L_{eq}\sin\alpha_{eq} - D_{eq}\cos\alpha_{eq} - m_{0_{eq}}g\sin\theta_{eq} \tag{19.33}$$

$$0 = L_{eq}\cos\alpha_{eq} + D_{eq}\sin\alpha_{eq} - m_{0_{eq}}g\cos\theta_{eq} \tag{19.34}$$

where the subscript 'eq' denotes the state at equilibrium steady glide. The equilibrium terms corresponding to the offset mass m_w and the location r_b of the ballast mass m_b do not appear in our earlier model [10].

Equations (19.33) and (19.34) can be rearranged into a more compact form:

$$0 = L_{eq} - m_{0_{eq}}g\cos(\theta_{eq} - \alpha_{eq}) \tag{19.35}$$

$$0 = D_{eq} + m_{0_{eq}}g\sin(\theta_{eq} - \alpha_{eq}) \tag{19.36}$$

The hydrodynamic forces and moment are generally modelled as

$$D_{eq} = \tfrac{1}{2}\rho C_D(\alpha_{eq})AV^2 \qquad (19.37)$$

$$L_{eq} = \tfrac{1}{2}\rho C_L(\alpha_{eq})AV^2 \qquad (19.38)$$

$$M_{eq} = \tfrac{1}{2}\rho C_M(\alpha_{eq})AV^2 \qquad (19.39)$$

where C_D, C_L and C_M are the standard aerodynamic drag, lift and moment coefficients, A is the maximum glider cross-sectional area, and ρ is the fluid density. For the longitudinal quasi-steady fluid model, C_D, C_L and C_M are functions of α.

Using these equations the equilibrium glide-path angle can be computed as

$$\tan\phi_{eq} = -\frac{D_{eq}}{L_{eq}} = -\frac{C_D(\alpha_{eq})}{C_L(\alpha_{eq})} \qquad (19.40)$$

The data shown in Figures 19.4 and 19.5 are taken from a series of glides specifically conducted for the purpose of model identification. The experiments took place during January 2003 near Chub Key, Bahamas using a *Slocum* glider owned by Dr D. Fratantoni of Woods Hole Oceanographic Institution, Woods Hole, MA, USA. For more details on the experiments the reader is referred to References 14 and 15.

Experimental data together with reference data, such as found in References 18–20, data from wind tunnel tests [21] as well as from a CFD analysis [22] were collected to determine a set of model parameters matching the *Slocum* model equilibria equations (19.30)–(19.34). To illustrate our approach, we rewrite the equilibrium conditions in a more convenient form as follows.

First, we note that parameters corresponding to the mass of the glider as well as the position of the CG can be measured directly, while the position of the CB can be computed from the glider geometry. Since the vertical velocity (i.e., change of depth) $\dot{z}$ and pitch angle θ are measured, the total velocity V can be expressed as a function of the angle of attack α

$$V = \left|\frac{\dot{z}}{\sin(\theta - \alpha)}\right| \qquad (19.41)$$

Substituting (19.41) and the hydrodynamic forces (19.37), (19.38), and moment (19.39) into Eqs. (19.35) and (19.36) gives

$$0 = \frac{1}{2}\rho C_L(\alpha_{eq})A\left(\frac{\dot{z}_{eq}}{\sin(\theta_{eq} - \alpha_{eq})}\right)^2 - m_{0_{eq}}g\cos(\theta_{eq} - \alpha_{eq}) \qquad (19.42)$$

$$0 = \frac{1}{2}\rho C_D(\alpha_{eq})A\left(\frac{\dot{z}_{eq}}{\sin(\theta_{eq} - \alpha_{eq})}\right)^2 + m_{0_{eq}}g\sin(\theta_{eq} - \alpha_{eq}) \qquad (19.43)$$

These equations for steady glides are now in the convenient form for identification in which they include only measured quantities such as $\dot{z}$, θ, m_0 and depend only on functions of α. Hydrodynamic coefficients $C_L(\alpha)$, $C_D(\alpha)$ and $C_M(\alpha)$ can be represented as polynomials in α such as in Eqs. (19.1)–(19.3).

19.4.2.1 Hydrodynamic lift and drag coefficient

We focus here on a demonstration of our methodology to identify the hydrodynamic lift and drag coefficients. The *Slocum* glider is assumed to be top/down symmetric with flat plate wings. Accordingly, the lift coefficient is an odd function with respect to α, $C_L(\alpha_{eq}) = -C_L(-\alpha_{eq})$, while the drag coefficient is an even function with respect to α, $C_D(\alpha_{eq}) = C_D(-\alpha_{eq})$. The drag coefficient has its minimum at $\alpha_{eq} = 0$. The hydrodynamic moment is an odd function with respect to α, such that $C_M(\alpha_{eq}) = -C_M(-\alpha_{eq})$. For a description of the identification of hydrodynamic pitch moment the reader is referred to References 14 and 15.

The difficulty in estimating hydrodynamic coefficients purely from experimental data comes from the limited instrumentation on the operational gliders and from the limitations on experiments conducted in an uncontrolled environment. To overcome these limitations, we relied in part on reference data from other sources. For instance, we obtained an estimate of the lift coefficient by comparing data from three different sources: aerodynamic reference data, CFD analysis [22] and wind tunnel data [21]. The comparison revealed good agreement between the different sources and the following lift coefficient $C_L(\alpha)$ from Reference 22 was used in the model identification procedure:

$$C_L(\alpha) = 11.76\alpha + 4.6\,\alpha|\alpha| \tag{19.44}$$

where α is in radians.

Assuming the lift described by (19.44) is given, the identification problem is then to determine the drag coefficient so that steady glides computed from the equilibrium equations are consistent with flight test data. Looking at the depth rate in Figure 19.5 we observe that the glider is descending at a slower rate than it is ascending. Under the assumption of a symmetric glider, the difference between descent and ascent speed for identical pitch angles can only be due to different buoyant forces. The trimming of a glider is usually done in a freshwater tank. The difference in density of freshwater versus the water of the deployment area is corrected as best as possible, but a degree of uncertainty remains. Additionally, an operator often trims a glider slightly light for safety reasons. These factors can contribute to a trim offset.

In order to accurately determine the drag coefficient we calculate the trim offset in the buoyant force Δm_0. First we substitute $m_{0_{eq}} = m_{b_{eq}} + \Delta m_0$ into Eq. (19.42):

$$\frac{1}{2}\rho C_L(\alpha_{up})A\frac{\dot{z}_{up}^2}{\sin(\theta_{up} - \alpha_{up})^2} = (m_{b_{up}} + \Delta m_0)g\cos(\theta_{up} - \alpha_{up}) \tag{19.45}$$

$$\frac{1}{2}\rho C_L(\alpha_{down})A\frac{\dot{z}_{down}^2}{\sin(\theta_{down} - \alpha_{down})^2} = (m_{b_{down}} + \Delta m_0)g\cos(\theta_{down} - \alpha_{down}) \tag{19.46}$$

We assume that identical magnitude of pitch angle implies identical magnitude of glide path angle, that is, $\theta_{up} = -\theta_{down}$ implies $\phi_{up} = -\phi_{down}$. We further assume a symmetric buoyancy engine, that is, $m_{b_{down}} = -m_{b_{up}}$. Adding Eqs. (19.45) and (19.46) we obtain

$$\frac{1}{4}\rho C_L(\alpha_{up})A\frac{1}{\sin(\theta_{up}-\alpha_{up})^2\cos(\theta_{up}-\alpha_{up})g}(\dot{z}_{up}^2 - \dot{z}_{down}^2) = \Delta m_0 \quad (19.47)$$

Using the available steady glide data, we now can estimate the buoyancy trim offset Δm_0 by using data from several up and down glides. We estimate the buoyancy trim offset to be $\Delta m_0 = -73$ g. This means that, for the water density and the weight of the glider during these tests, the glider is 73 g light (positively buoyant) when the buoyancy engine is set to $m_b = 0$, that is, the 'zero buoyancy' point.

We next substitute $m_{0_{eq}} = m_{b_{eq}} + \Delta m_0$ into Eq. (19.43). The resulting equation can be solved numerically to determine the angle of attack for each individual steady glide data point. Assuming a second-order polynomial for $C_D(\alpha_{eq})$, we compute the following coefficient using a least-squares fit:

$$C_D(\alpha) = 0.214 + 32.3\alpha^2 \quad (19.48)$$

with α in radians.

Figure 19.7 shows the drag coefficients with error bars determined using experimental data superimposed on the line that represents C_D as a smooth function of α given in Eq. (19.48).

Substituting the hydrodynamic coefficients as identified above into Eq. (19.40) allows us to plot pitch angle or glide path angle as a function of α under steady

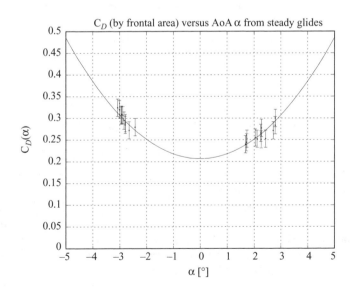

Figure 19.7 C_D computed from equilibrium glide data using a buoyancy trim offset of -73 g. The solid line is the fitted curve for C_D given in Eq. (19.48)

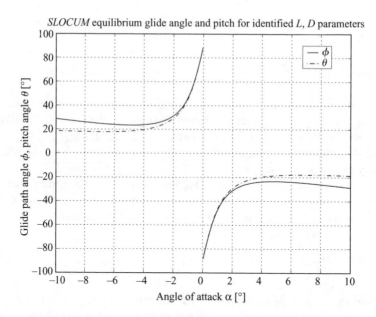

Figure 19.8   Equilibrium glides using lift, drag fit to data

glide conditions as shown in Figure 19.8. Substituting the angles into Eq. (19.42) allows us to compute the total, horizontal and vertical glide velocity as a function of α. From Figure 19.9 it can be seen that for a 25° glide angle our steady glide model yields a depth rate of 20 cm/s and a horizontal speed of 42 cm/s. This is consistent with estimates from glider operations conducted by Webb Research Corporation and WHOI.

19.5   Coordinated glider control and operations

Having looked at the issues of modelling and control of individual gliders, we now focus on multiple glider operations. In Section 19.5.1, we describe the virtual bodies and artificial potentials (VBAP) methodology for cooperative control of multiple autonomous vehicles. VBAP was implemented on a group of autonomous underwater gliders during the Autonomous Ocean Sampling Network (AOSN) II field experiment conducted in Monterey Bay, during summer 2003. In Section 19.5.2, we discuss issues and constraints specific to the coordination of underwater gliders and describe our VBAP implementation for AOSN II. In Section 19.5.3, we present an overview and brief results from the sea trials performed during AOSN II. Two sea-trials demonstrated our ability to coordinate a group of *Slocum* underwater gliders into dynamic formations. A third trial demonstrated how a single *Slocum* glider could track the path of a Lagrangian drifter in real time.

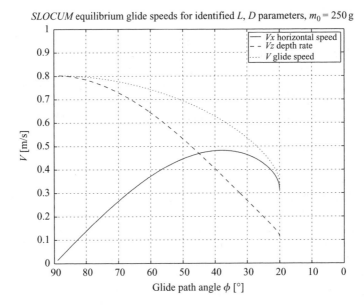

Figure 19.9   Equilibrium speed using lift, drag fit to data

19.5.1   Coordinating gliders with virtual bodies and artificial potentials

The VBAP multi-vehicle control methodology relies on artificial potentials and virtual bodies to coordinate a group of vehicles modelled as point masses into regular formations in a provably stable manner. A virtual body consists of linked, moving reference points called virtual leaders. The artificial potentials couple the dynamics of vehicles and virtual leaders and enable us to prescribe a desired spacing between vehicles and between vehicles and virtual leaders. The control law for each vehicle is derived from the gradient of the sum of the artificial potentials and drives the vehicles toward a minimum of the potential. The formation design problem then becomes one of designing the potentials so that the desired formation is a minimum. The combination of virtual leaders and artificial potentials allows for a wide range of vehicle group shapes [23].

In a separate design step, the dynamics of the virtual body are prescribed to drive the virtual body, and thus the vehicle group, to perform manoeuvres that include translation, rotation and contraction/expansion, all while ensuring that the formation error remains bounded. In the case that the vehicles are equipped with sensors to measure the environment, the manoeuvres can be driven by measurement-based estimates of the environment; for example, the formation can be controlled to climb gradients and track boundaries of oceanographic features. This is an important benefit of VBAP and permits the vehicle group to perform as an adaptable sensor array, see References 24 and 25.

19.5.2 VBAP glider implementation issues

A number of practical considerations and constraints make the implementation of coordinated control strategies for gliders challenging. These include limitations on inter-vehicle communication, latencies and asynchronicities in the feedback loop as well as strong ocean currents which produce large external forces on the gliders. For reasons related to the sensing of the ocean environment, forward glider speed (of approximately 40 cm/s on average) was fixed during AOSN II. Other issues include surface traffic and shallow-water limits.

Coordinated and cooperative control requires measurement of the relative states of the gliders. For example, in the VBAP methodology, relative glider positions are needed. However, in the AOSN II experiment, the gliders were not equipped with the means to directly measure the relative position of other gliders in the fleet. Indeed, it is often a design choice with gliders to do without acoustic modems although modems could, in principle, be used to communicate positions and headings among gliders. This choice stems from the trade-off between good control and long endurance. In AOSN II, the priority was to reserve onboard power for vehicle endurance and for science sensors.

Instead, relative position measurements were computed on a shore computer when gliders came to the surface. In AOSN II, gliders came to the surface every 2 h to get a new GPS fix, to transmit data to the shore station and, if appropriate, to collect new mission directives. Accordingly, the VBAP control law was computed at these surfacings and a new mission was input to the gliders. Because the gliders did not surface at the same time, VBAP was run at each cycle only after all the gliders in the coordinated group had surfaced. The output of the VBAP run was a set of trajectories for each of the gliders over the next few hours, in the form of an ordered way-point list for each glider. Because of the asynchronous glider surfacings and latencies in the processing and communicating of glider measurements, the implementation made use of a number of estimates. Furthermore, the feedback loop could not be closed at a given surfacing, indeed an updated mission input to a glider was necessarily based on data measured a cycle or two earlier. The implementation strategy and detailed simulations, made in advance of the field experiment, to tune and test performance for different design alternatives are described in Reference 25.

The VBAP methodology was additionally constrained by the constant glider speed assumption. This constraint restricts what formations are feasible using VBAP. Furthermore, when external currents vary across the formation, there is, in general, no configuration of vehicles, each moving at the same constant speed, such that all relative velocities between vehicles remains zero. Procedures to circumvent these and related problems are also described in Reference 25.

19.5.3 AOSN II sea trials

Here we present a short description and overview of results from the coordinated glider sea trials performed during AOSN II. On 6 and 16 August 2003, we used our VBAP methodology with a single virtual leader serving as the virtual body to coordinate three *Slocum* underwater gliders. We explored various schemes to control the motion and orientation of the formation as well as the inter-vehicle spacing of the formation, as it made its way in and around the bay. On 23 August 2003 we demonstrated how a single *Slocum* glider could track a Lagrangian drifter in real time. More thorough quantitative analysis of our coordination performance is presented in Reference 26. All gliders used during these trials are the property of David Fratantoni of the Woods Hole Oceanographic Institution (WHOI).

On 6 August 2003 three *Slocum* gliders were coordinated into an equilateral triangle formation and directed towards the northwest part of Monterey Bay in response to the onset of an upwelling event. The desired inter-vehicle spacing was fixed at 3 km throughout but orientation of the formation was regulated. During the first 8 h of the trial the triangle formation was free to rotate about the virtual leader. Only during the last 8 h was the orientation of the group about the virtual leader regulated so that an edge of the triangle formation would be perpendicular to the group's path.

Snapshots of the glider formation, shown in Figure 19.10(a), indicate that VBAP successfully coordinated the gliders into the desired formation and this formation tracked the desired reference trajectory. Temperature gradients were estimated using glider measurements. These are shown in Figure 19.10(a) as a series of vectors. The smoothness of the gradient estimates demonstrates the promise of glider formations serving as ocean sensor arrays. In Reference 26, we corroborate these gradients with satellite-acquired sea surface temperature data, and we provide a quantitative study, using objective analysis techniques [27, 28], of the glider formation's ability to resolve length and time scales.

On 16 August 2003 a sea trial with three gliders was performed to demonstrate our ability to coordinate gliders into arrays that can be dynamically made to expand and contract while translating. The mission plan was to direct the gliders to criss-cross a region while in a equilateral triangle formation. Orientation of the formation about the virtual leader (triangle centroid) was also regulated (again so that an edge of the triangle formation would be perpendicular to the group's path). The initial desired inter-vehicle spacing was 6 km, and then, several hours into the trial, it was reduced to 3 km.

Figure 19.10(b) presents the instantaneous glider formations for this trial. Starting from their initial distribution, the gliders expand to the desired 6 km inter-vehicle spacing while the group centroid tracks the desired reference trajectory. During the first few dive cycles it can be seen that the group centroid does not closely follow the reference trajectory. The relatively large excursions are due to high currents in excess of 36 cm/s at the surface (recall that a *Slocum* travels at approximately 40 cm/s). During the latter stages of the trial, currents were less aggressive and performance was significantly improved. The formation snapshots clearly show that the formation contracts and that an edge of the triangle is regulated to be perpendicular to the path.

On 23 August 2003 a *Slocum* glider was directed to follow a Lagrangian drifter in real time. This sea trial was meant to demonstrate the utility of the glider to track Lagrangian features. To achieve good tracking, it was necessary to predict the future trajectory of the drifter. This prediction was based on a persistence rule, using a quadratic curve fit of measured drifter positions and corresponding time stamps. The strategy employed during this demonstration directed the glider to travel back and

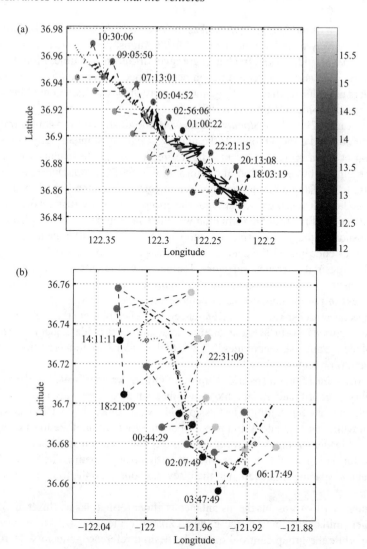

Figure 19.10 Multi-glider sea trials at AOSN II. (a) Snapshots of glider formations during the August 6 sea trial along with estimated gradient. Dots portraying gliders are shaded to indicate temperature at the 10 m isobath. Dashed lines illustrate instantaneous formations at 2-h intervals. The dotted line is the path of the formation centroid. The black arrows correspond to the negative gradient of temperature (and thus point in the direction of the colder water). Time is UTC starting on 6 August 2003. (b) Glider formation snapshots during the August 16 sea trial. Dashed lines illustrate instantaneous formations. The dotted line is the path of the formation centroid. Dash–dot line is the virtual leader path, that is, the desired centroid trajectory. Time is UTC on 16 August 2003

forth along a chord of a circle (of specified radius) with respect to the drifter, as shown in Figure 19.11(a). The resulting trajectories of glider and drifter are shown in Figure 19.11(b).

As seen in Figure 19.11(b), for the most part the glider followed the drifter. The average speed of the drifter during this demonstration was approximately 7 cm/s and we see the cross-track excursions which result from the glider moving faster than the drifter. Some of the gross tracking error can be attributed to estimated currents onboard the glider varying significantly from currents experienced by the drifter. Furthermore, nearing the end of the trial the drifter appeared stationary on the GPS scale. Our implementation was particularly prone to error in this situation, due to unknown velocity fields and time delays in the control implementation. Further details and suggestions to improve performance can be found in Reference 26. Although we demonstrated drifter tracking with a single glider, we could in principle enable a group of gliders to track a drifter and serve as a Lagrangian sensor array, thus providing a new means of collecting physical and biological oceanographic data.

19.6 Final remarks

Autonomous underwater gliders have come of age. Since their initial development in the early 1990s, underwater gliders have become a recognised tool in physical oceanography, and their use as an observation tool has started to spill over into other areas, such as marine chemistry and marine biology. One of the main reasons for the growing popularity of gliders is their low initial and operational costs. Further, ease of deployment, long endurance and quietness make gliders an attractive tool for applications such as surveyance and acoustic tracking of marine mammals. Yet the current glider designs are still fairly uniform in their speed, navigation and communication capabilities and their full potential may well be still to come. Future designs with capabilities such as increased speed and underwater navigation would expand the current glider operational envelope to include high current regimes and long-term sub-surface missions, for example, under-ice operations.

At various locations, projects are underway to improve upon long- and mid-range underwater navigation capabilities using acoustic navigation. Other projects are addressing the speed constraint imposed by the size of the glider, its buoyancy engine capacity and the general glider shape. The research presented in this chapter provides a tool not only to analyse existing glider performance but also to predict glider behaviour for future designs. For the interested reader, glider design and performance prediction is described in greater detail in Reference 15.

With the increase of available autonomous platforms, the use of gliders in multi-vehicle scenarios has become a reality. Fleets of gliders can be used for a variety of applications including cooperative, adaptive ocean sampling as demonstrated in the methodology and sea trials of the previous section. We are currently developing new algorithms for coordinated control of gliders where the goal is to optimise the data collected by the gliders so that they are of greatest oceanographic value. Our metric is error in the estimation of oceanographic fields, such as temperature or salinity, given

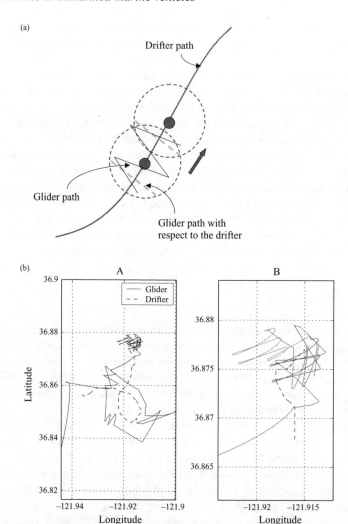

Figure 19.11 Drifter tracking (a) Drifter tracking plan: The solid circles indicate drifter positions at two time instants, and the line connecting the solid circles is the drifter path. The solid line crossing the drifter path is the desired glider path. The glider path with respect to the drifter is a chord of a circle of specified radius about the drifter. (b) Tracks followed by the glider and the drifter during the 23 August 2003 demonstration. (A) The complete demonstration. (B) The last four glider dive cycles. The dashed line is the drifter track and the solid line is the glider track. The shading of the solid line changes at the start of every new dive (approximately every 2 h)

the data from the gliders. The objective is then to design the glider array so that it minimises this error, or equivalently, maximises the information content in the data it collects. Sea trials are already scheduled as part of a collaborative effort, and we plan to test our algorithms on a large group of gliders in the ocean.

Gliders present a mature technology that provides new opportunities for exploring the oceans and collecting data over long periods of time at reasonable cost. Experiments with operational gliders have proved the merit of using gliders individually and collectively. Future work will focus on vehicle, navigation, communication, control and coordination design and application that realises the full potential of autonomous underwater gliders.

References

1 Bachmayer, R., S. Humphris, D. Fornari, C. Van Dover, J. Howland, A. Bowen, R. Elder, T. Crook, D. Gleason, W. Sellers, and S. Lerner (1998). Oceanographic research using remotely operated underwater robotic vehicles: exploration of hydrothermal vent sites on the mid-Atlantic ridge at 37° North 32° West. *Marine Technology Society Journal*, 32(3), pp. 37–47.
2 Stommel, H. (1989). The Slocum mission. *Oceanography*, 2, pp. 22–25.
3 Curtin, T., J.G. Bellingham, J. Catipovic, and D. Webb (1993). Autonomous ocean sampling networks. *Oceanography*, 6(3), pp. 86–94.
4 Gould, W.J. (2003). A brief history of float developments. WOCE IPO, http://www.soc.soton.ac.uk/JRD/HYDRO/argo/history.php [online].
5 Webpage, Welcome to the Argo Home page [online]. Available at: http://www-argo.ucsd.edu/ [Accessed: 18 November 2004].
6 Eriksen, C.C., T.J. Osse, R.D. Light, T. Wen, T.W. Lehman, P. L. Sabin, J.W. Ballard, and A.M. Chiodi (2001). Seaglider: a long-range autonomous underwater vehicle for oceanographic research. *IEEE Journal of Oceanic Engineering*, 26(4), pp. 424–436.
7 Sherman, J., R.E. Davis, W.B. Owens, and J. Valdes (2001). The autonomous underwater glider 'Spray'. *IEEE Journal of Oceanic Engineering*, 26(4), pp. 437–446.
8 Webb, D.C., P.J. Simonetti, and C.P. Jones (2001). SLOCUM: an underwater glider propelled by environmental energy. *IEEE Journal of Oceanic Engineering*, 26(4), pp. 447–452.
9 Graver, J.G., J. Liu, C. Woolsey, and N.E. Leonard (1998). Design and analysis of an underwater vehicle for controlled gliding. In Proceedings of the 32nd Conference on Information Sciences and Systems, Princeton, NJ, pp. 801–806.
10 Leonard, N.E. and J.G. Graver (2001). Model-based feedback control of autonomous underwater gliders. *IEEE Journal of Oceanic Engineering*, 26(4), pp. 633–645.
11 Graver, J.G. and N.E. Leonard (2001). Underwater glider dynamics and control. In Proceedings of the 12th International Symposium on Unmanned Untethered Submersible Technology, Durham, NH, USA.

12 Bhatta, P. and N.E. Leonard (2002). Stabilization and coordination of underwater gliders. In Proceedings of the 41st IEEE Conference on Decision and Control, Las Vegas, NV, pp. 2081–2086.
13 Bhatta, P. and N.E. Leonard (2004). A Lyapunov function for vehicles with lift and drag: stability of gliding. In Proceedings of the 43rd IEEE Conference on Decision and Control, Paradise Island, Bahamas, pp. 4101–4106.
14 Graver, J.G., R. Bachmayer, N.E. Leonard, and D.M. Fratantoni (2003). Underwater glider model parameter identification. In Proceedings of the 13th International Symposium on Unmanned Untethered Submersible Technology, Durham, NH, USA.
15 Graver, J.G. (2005). Underwater gliders: dynamics, control and design. Ph.D. thesis, Princeton University.
16 Lanchester, F.W. (1908). *Aerodonetics*. A. Constable and Co., London.
17 von Mises, R. (1959). *Theory of Flight*. Dover, New York.
18 Hoerner, S.F. (1965). *Fluid Dynamic Drag*. Published by the author, Midland Park, NJ.
19 Lamb, H. (1932). *Hydrodynamics*, 6th edition. Dover, New York.
20 Roskam, J. (1971). *Methods for Estimating Stability and Control Derivatives of Conventional Subsonic Airplanes*. Published by the author, Lawrence, KS.
21 Berman, S.M. (2003). Comparison of the lift, drag and pitch moment coefficients of a Slocum glider wind tunnel model with computational results by Vehicle Control Technologies, Inc. Technical Report, Princeton University Mechanical and Aerospace Engineering Department.
22 Jenkins, S.A., D.E. Humphreys, J. Sherman, J. Osse, C. Jones, N.E. Leonard, R. Bachmayer, J.G. Graver, T. Clem, P. Carroll, P. Davis, J. Berry, P. Worley, and J. Wasyl (2003). Underwater glider system study. Technical Report, Office of Naval Research.
23 Leonard, N.E. and E. Fiorelli (2001). Virtual leaders, artificial potentials and coordinated control of groups. In Proceedings of the 40th IEEE Conference on Decision and Control, pp. 2968–2973.
24 Ögren, P., E. Fiorelli, and N.E. Leonard (2004). Cooperative control of mobile sensor networks: adaptive gradient climbing in a distributed environment. *IEEE Transactions on Automatic Control*, 29(8), pp. 1292–1302.
25 Fiorelli, E., P. Bhatta, N.E. Leonard, and I. Shulman (2003). Adaptive sampling using feedback control of an autonomous underwater glider fleet. In 13th International Symposium on Unmanned Untethered Submersible Technology, Durham, NH, USA.
26 Fiorelli, E., N.E. Leonard, P. Bhatta, D. Paley, R. Bachmayer, and D. Fratantoni (2004). Multi-AUV control and adaptive sampling in Monterey Bay. In Proceedings of the 2004 IEEE AUV Workshop.
27 Bretherton, F.P., R.E. Davis, and C.B. Fandry (1976). A technique for objective analysis and design of oceanographic experiments applied to MODE-73. *Deep-Sea Research*, 23, pp. 559–582.
28 Gandin, L.S. (1965). *Objective Analysis of Meteorological Fields*. Israel Program for Scientific Translations.

Index

A type AUV 251–3
acoustic positioning systems 190
actuators, torpedo-shaped AUVs 164
adaptation 72
adaptive ocean sampling 424–9
air/sea interface studies 405
Al-dente 298–300
    real-time mode 299
    replay mode 299
algorithmic singularities 229
anti-submarine warfare 313
anti-terrorist surveillance 402
AOSN II sea trials 424, 426–30
approximation error 97–8
archaeological survey 402, 404
Argo II 293
Aries AUV 360–1
ARMAX models 109
artificial potential functions 59–65, 425
    Gaussian 60, 64
attack role 313, 318–19, 325–6
attainable command set 91–2, 100–1
autonomous manipulation 217
autonomous marine vehicle 353
    vehicle control 358–75
    see also autonomous surface marine vehicle
    see also autonomous underwater vehicle
Autonomous Ocean Sampling Network 424
autonomous surface marine vehicle 317–18, 329–30
    Caravela 357–9
    Delfim 355–6
    see also MESSIN
    see also surface vehicles
autonomous underwater vehicle 1
    advanced autonomy type 251–3

Aries 360–1
biomimetic 255–74
bottom reference type 250–3
coordinated path following 373–4
cruising type 250–3
diversity 250–2
flight vehicles 161
Hammerhead 127–56
hovering vehicles 161
Infante 27–8, 30–1, 190, 354–5, 360–3
intervention 217–236
Marius 354
r2D4 239–53
REMUS 45, 58
road map 252–3
robust control 161–9
Sirene 356–7
underwater gliders 407–31
URIS 76–84
autonomous wave-piercing vehicles: *see* wave-piercing autonomous vehicles
autopilots 2, 31, 38
    Hammerhead AUV 148–56
    MESSIN 348
    robust 162–9, 181–3
    Subzero III 166–9, 181–3
    torpedo-shaped AUVs 162–9
avoidance guidance: *see* obstacle avoidance
avoidance heading 55–7

B type AUV 250–3
backstepping techniques 36, 374
bandpass filtering 200
bathymetric data 293, 300–1, 305–6
bathymetric seabed profiling 281–2, 379–80
bathymetric survey 402, 404
battle-space control 402, 405

434 Index

BAUV: *see* biomimetic-autonomous underwater vehicle
BCF locomotion 255–7
beach survey 402
bearing weighting function 54–5
behaviour activation level 74
behaviour control 68–76, 83–4, 242–3
  URIS 80–3
behaviour coordination 71–5
behaviour-based architectures 68–76, 83–4
  URIS 80–3
behaviour-based robotics 69–72
benthic stations 356
BFFAUV 256
biomimetic-autonomous underwater vehicle 255–74
  body-spline equation 267–8
  braking 270–4
  centre of mass 260
  dynamic modelling 257–65
  feedback control 271
  global controller 269–70
  guidance 266–7
  hovering 273–4
  local controller 267–9
  turning 270, 272
blade elements 261, 263–5
bladed thrusters 189
body/caudal fin 263–4, 266–7
  global controller 269–70
  local controller 267–9
body-spline equation 267–8
braking 270–4
buoyancy 18, 409
buoyancy control 409
buoyancy engine 408, 417–18
buoys, mobile 398, 405
Butterworth filters 201

C type AUV 250–3
cable dynamics 170–3
cable effects 170–7
cable route survey 404
cable tracking and inspection 127, 139, 147, 404
Carangiform swimmers 255–6
Caravela autonomous research vessel 357–9
catamaran-type craft 330–1, 355
caudal fin 263–4, 266–7
check helm 3
closed-loop control scheme 222

collision avoidance 230–1
  MESSIN 335
  USV 318
Combat Support Boat 321–5
communication link 232–3
competitive behaviour coordination 71, 73
composite control 169–70, 179–81
composite waterfall imaging 289–91
computer vision 79–80
constrained control subset 90–1
constrained motion testing 283–91
control
  autonomous marine robots 358–75
  BAUV 267–71
  Hammerhead AUV 127
  intervention AUVs 218–32
  MESSIN 343–5
  underwater gliders 414–15
  UUVs 31–40, 189–90
  see also behaviour control
  see also composite control
  see also gain scheduling control
  see also robust control
  see also supervisory control
  see also switching control
control action 74
control allocation 87–92
  hybrid approach algorithm 96–8
  thruster 98–103
control architectures 67–9, 72–6, 88
cooperative behaviour coordination 71, 73
cooperative motion control 360–1, 424–6
coordinated motion control 361, 373–5
  gliders 424–30
coordinated path following 360–2, 373–5
  multiple vehicles 374–5
coordination 71–5
  competition 71, 73
  cooperation 71, 73
CORAL mission control system 376–9
correlation techniques 198
correlation-based window tracking 283
covariance-matching 129–30
cross-currents 44, 49–50
cross-track error 47–51
current measuring 330, 348–9
currents: *see* ocean currents

damped least-squares method 224, 227
damping terms 25–31
  linear 26–7
data telemetry 335
data visualisation 293–300

Index 435

dead reckoning 139–40
deep ocean 293–4
Delfim ASC 355–6, 362
   coordinated path following 360–1
   missions at sea 379
   path following 371
deliberative architectures 72
depth control
   Hammerhead 150–1, 155–6
   Infante AUV 366
   Sirene AUV 367
   Subzero III 168–9
   UUVs 32–7
discrete event systems 377–8
distributed computer architecture 378–9
Dolphin 389–93
   derivatives 390
Dorado 394–5
drifter tracking 427, 429–30
drinking water 330
drones 312–14, 323–4, 327
   semi-submersible 393
DVLNAV 295, 302–3
dynamic imaging 288–91
dynamic modelling, BAUV 257–65
dynamics 14–16
   rigid body 258–62
   underwater glider 408–11

electrohydraulic arms 218
electromechanical arm 218
electronic warfare 402
embeddedness 70
emergence 70–1
emergency countermeasures 243
engines 316–17
equations of motion
   AUVs 362–3
   BAUV 261–2
   ocean currents 19–20
   Subzero III 163–4
   underwater glider 410–13
   UUVs 16–24
estimation-based switching control 106–7, 109–10, 112–13, 122
   ROV 118–20
Euler-Lamb equation 263
evolutionary techniques 72
exploration 82, 239, 293
extended Kalman filter 129, 137
extended recursive least squares estimation algorithm 112

FALCON
   thruster control allocation 98–103
feathering motion 263, 268
feedback control 179–181
   BAUV 271
FENRIR USV 325–6
fire ships 311
fish behaviour studies 402, 404–5
fish robot 256
fish stock assessment 390, 402, 404
fish swimming movements 255–6
fixed-point iteration method 95–8, 101
flight vehicles 161–2
floats 408
force multipliers 327
fourth-order track model 338–42
fuel systems 316–17
fuzzy extended Kalman filter 137
fuzzy inference system 130
fuzzy Kalman filter 129–30
fuzzy logic observer 130–1
fuzzy membership functions 131–2

gain scheduling control 193, 364–5, 369–70
general inverse 87–8
genetic algorithms 131
   multi-objective 131
geological fault 379–80
GeoZui3D 295–300, 303
   Jason 2 306–7
glider formations 427–8
gliders: *see* underwater gliders
global positioning system 128–9, 317, 335–7
GPS/INS navigation
   Hammerhead AUV 136–45
guidance
   BAUV 266–7
   Hammerhead AUV 147–8
   line of sight 46–8, 51
   MESSIN 334
   Romeo ROV 193, 195–6
   USV 317–18
   UUV 190
guidance functions 44
guidance laws 43–52
   Hammerhead 147–8
gyrocompass 3
gyropilot 3

H_∞ design 161–2, 181, 183, 364–5
Hammerhead AUV 127–56
  autopilot design 148–56
  data and image acquisition system 280
  FKF-MOGA algorithms 131–6
  guidance 147–8
  laser-assisted vision sensor 279–81, 283–91
  modelling 145–7
  navigation system 129–45
  surface mission 137–9
  surface-depth mission 139–45
harbour protection 313
heading control
  Hammerhead AUV 153–5
  Infante AUV 366
  Sirene AUV 367
  Subzero III 167–8
  UUVs 37–40
heading sensor 51
heave 14
hierarchical architecture 193
  deliberative 67
hierarchical hybrid coordination nodes 74–5
hierarchical steering system 333–5
hierarchically supervised switching control 107–8, 111–14, 122
  ROV 120–1
high-frequency LBL 188
horizontal plane control 359, 363–6
hovering 273–4
hulls 314–16
human-computer interface 294, 308
hybrid architectures 68–9, 72–6
hybrid control theory 367
hybrid guidance strategy 147–8
hybrid structured lighting 277
hydrodynamic added mass 25
hydrodynamics 263–5
  underwater glider 409–11, 419–25
hydrographic survey 388–91, 402, 404
hydrography 402
hydrothermal chimneys 252
hydrothermal ore deposits 252
hydrothermal plumes 247–8, 250
hydrothermal vents 240, 379

identification
  MESSIN 337, 340
  model based 29–31
  underwater glider 419–25
  UUVs 24–31
image depth 199

image template 198, 200
Imagenex 881A pencil beam sonar 298–9, 302
inertia matrix 17
inertial navigation system 128–9, 245
Infante AUV 27–8, 30–1, 190, 354–5
  control 366
  coordinated path following 360–1
  dynamic models 362–3
  missions at sea 379–80
innovation adaptive estimation 129
integrated programmable controllers 345
intervention AUV 217–36
  control system 218–32
inverse kinematics 223, 227–30

Jason 1 ROV 188, 296
Jason 2 ROV 293, 295, 305–8
jetski 315–16
JHU ROV 301–5
joint limits avoidance 231–2
joint offsets calibration 234
joint potential function 231–2

Kalman filtering techniques 128–9
  fuzzy logic 129–31
kinematic control 218, 220–3
kinematic equations 14–16
  lateral 23–4
  longitudinal 21–3
kinetics 16

Labriform locomotion 255–6
Lagrangian drifter 427, 429
Lagrangian function 261
Lagrangian sensor array 429
laser altimeter 283–6
laser bathymetry sensor 281–2
laser line scan methods 278
laser scan 302–5
laser scan bathymetry 305–6
laser stripe generator 281
laser stripe illumination systems 139, 278
laser stripes 281–2, 288–9
laser triangulation 191, 197–9, 202, 281–2
laser-assisted vision sensor 277, 279–81
  constrained motion testing 283–91
launch and recovery
  USV 319
  WASV 397, 401–2
lead-lag motion 263, 268
least squares optimisation 28

legal issues 11
line of sight 43–4
line of sight guidance 46–8, 51
  cross-track error controller 49–50
line of sight tracking 149–50
linear matrix inequalities 365
linear quadratic Gaussian 149–51
local variances 200
long baseline positioning system 356
loop transfer recovery 149–51
LQG/LTR controller 149–51
Lyapunov function 415–16
Lyapunov method 29, 32, 36–7, 193, 372, 374
Lyapunov stability theory 375

manipulability 224, 231
manipulation, autonomous 218
manipulation variable 223
manipulators 217–20
  autonomous 217–18
  communication 232
  control system 218–32
  MARIS 7080 218, 221, 230
  tests 233–5
  user interface 232–3
manoeuvrability 331, 347
manoeuvring tests 346–8
mapping 82
Mariana Trough 244–5
marine biology 429
marine chemistry 429
Maris 7080 218, 221, 230
Marius AUV 354
mathematical models 337–42
  underwater gliders 408–11
  see also modelling
Matlab/Simulink 345
measure of manipulability 224–30
Measuring Dolphin 330
  see also MESSIN
MESSIN 330–49
  control 343–5
  dynamic yaw stability 331
  electrical development 332–3
  energy management 333
  guidance 342
  hydromechanical conception 330–2
  manoeuvrability 331, 347
  manoeuvring tests 346–8
  mathematical models 337–42
  positioning 336–7
  route planning 342–3

simulation 344–6
steering 333–5
meteorological survey 402
MIMIR USV 319–21
mine avoidance, real-time 392
mine countermeasure systems 312, 318, 321, 324, 402, 404
mine sweeping 43, 312, 318, 391–6
  shallow water 318, 321
mission control 342–4, 353, 375
  CORAL system 376–9
mission design 376–7
  multi-vehicle mission 377
mission map 376
mission planning 242–3, 249
  USV 317
missions at sea 379–80
mobile buoys 398, 405
model based identification 29–31
model predictive control 150–6
  genetic algorithm based 151–2
modelling
  lateral 20–1, 23–4
  longitudinal 20–3
  MESSIN 337–42
  UUVs 14–24
modularity 70
mosaicking techniques 78, 80, 82–4, 191
mother-ship 398, 403
motion control, ROVs 187–211
motion estimation
  Romeo ROV 202–11
  vision-based 196–201
MPF locomotion 255–6, 268

naval operations
  SWIMS USV 321–5
  USV game changing potential 326–7
  USVs 311–13, 319, 321–6
  wave-piercing vehicles 393–5
navigation, guidance and control systems 127
  Hammerhead AUV 128
navigation systems
  DVLNAV 295
  Hammerhead AUV 129–45, 279
  inertial 128–9, 245
  MESSIN 336–7
  optical 283
  seabed-relative 277–9
  three-dimensional 139–45
  USV 317
  vision-based 284

Index

neural networks 76
Nomato models 37–9
nonlinear backstepping controller 31–2
nonlinear control 31–2, 371–2
nonlinear damping 27–31
nonlinear modelling 13
non-synchronous scanning systems 278
NPS AUV II 20–1

obstacle avoidance 52–60, 81
   multiple point 58–60, 62–3
   planned avoidance deviation 52–4
   potential function 61–5
   reactive avoidance 54
   tests 56–7
ocean currents 19, 187, 330, 348–9
   cross-currents 44, 49–50
ocean exploration 293–4
oceanographic data collection 407–8, 429
oceanographic survey 127, 402, 404–5
oceanographic vessels 358
optical flow 191
optical laser triangulation-correlation sensor 197–9, 202
optical navigation 283
optical sensing devices 191
over-actuated open-frame UVs 87–8
   thruster control allocation 98–103
Owl programmes 313

parallelism 70
parameter identification 25
   see also identification
path following 360, 369–73
   coordinated 360–2, 373–5
pectoral fins 256, 258–9, 261, 263–7
   global controller 269–70
   local controller 267–9
Petri nets 377–8
phugoid-mode model 415–16
PILOTFISH 256
pitch 14
pitch and roll disturbance rejection 201
pitch control 32–4
planned avoidance deviation 52–4
PLATYPUS 256
pole placement algorithm 40
port security 405
pose control 359, 366–7
positioning systems 336–7
   see also global positioning system
posture error 266, 269

power spectral density 204
Predator unmanned aircraft systems 313
prediction error method 340
profiling floats 408
propeller-based actuators 189
propeller-hull interactions 189
Protector USV 313, 318
pseudoinverse 87, 92–6, 101
pure pursuit guidance 147–8

Q-learning 76

r2D4 239–253
   behaviour control 242–4
   emergency countermeasures 243–4
   hardware 241–2
   Rota Underwater Volcano 244–9
radio data transmission system 335
range gated viewing system 278
range weighting function 55–6
reactive avoidance 54
real-time human interface 294–300
real-time spatio-temporal data display 294–300
   Al-dente 298–300
   Jason 2 ROV 305–8
   JHU ROV 301–6
   navigation data 297–8
   scientific sensor data 297–8
redundancy resolution 223
reflectivity seabed profiling 281–2
region-based tracker 283, 286–8
reinforcement learning 72, 75–6
remote control 335
   robotics 293–4
Remote Minehunting Operational Prototype 393–4
remotely operated vehicles 2
   Jason 1 188, 296
   Jason 2 293, 295, 305–8
   JHU 301–5
   models 114–16, 189
   motion control 187–211
   switching control 116–121
   vision-based motion estimation 196–201
   see also Romeo ROV
REMUS AUV model 45, 58
Rescue Dolphin 329
rescue vehicles 329
resolved motion rate control 223–4
revisiting strategies 277
rigid inflatable boats 313, 315–16, 389
rigid-body dynamics, BAUV 258–62

rigid-body parameters 25
robotics, exploration 293–308
robust control 161–2
    torpedo-shaped AUVs 162–9
    verification 181–3
ROGUE 412
roll 14
Romeo ROV 189–90, 192–3
    guidance 195–6
    motion control 202–11
    velocity control 194–5
    vision-based motion estimation 196–201
R-One project 240
R-One Robot 240–2
Rota Underwater Volcano 239–41, 244–9
    r2D4 dive 245–9
route planning 342–3
ROV support platform 396
R-Two project 240–1

sample extraction 330
sampling 251–3, 424–9
SASS: *see* survey autonomous semi-submersible vehicles
SAUVIM 217–18, 233–4, 236
schnorkelling submarine 389–90
scientific data collection 354–5
sea currents: *see* ocean currents
sea trials
    gliders 426–30
    MESSIN 346
seabed-relative navigation 277, 280
Seaglider 408
Seakeeper 395–6
second-order course model 338, 341
sediment research 330
seismic survey platform 396
semi-autonomous operation 233, 313–14, 316, 327
semi-submersible vehicles 390, 393, 397–402
sensing technologies 190–1
sensor data visualisation 293–4
sensor fusion 134–6, 144
shallow water 318, 321, 329–30
shallow-draught vehicle 330
shipwreck sites 293, 296, 300–1
SIMMESS 346
singularity avoidance 223, 225–30
singularity-robust inverse 227, 230–1
Sirene underwater shuttle 356–7, 362
    control 367
situatedness 70

skew-symmetric matrix 13
sliding mode control 40, 45–6, 50–1, 189–90
sliding surface 36
Slocum glider 408, 417–25
    coordinated control 426–30
    identification 419–25
small water-plane area single hull vehicle 390
smuggling surveillance 402
Society for Underwater Technology 11
software architecture 78–9
sonar 298–300
    Doppler 245
sonar bathymetry 300–1, 305
SONQL algorithm 76–7
Spartan USV programme 313, 319
spatial data visualisation 293–4
spatio-temporal human interface 294–300
speed control
    Subzero III 166–7
    UUVs 32–5
Spray 408
stability analysis 112–14, 412–16
statics 14
station-keeping 187, 191, 210
stealth 316
steering 333–5
steering engine 3
steering model 45–6
stereovision systems 191
submarines 1, 20–1, 167
subsumption architecture 68
Subzero III 162
    autopilots 166–9, 181–3
    dynamics 163–5
    tether compensation 169–81
supervisory control 105–9
    multiple models 106–9
    stability analysis 112–14
surface vehicles 312–19, 388–9
    autonomy 317–18, 329–30
    auxiliary structures 316
    collision avoidance 318
    engines 316–17
    FENRIR 325–6
    fuel systems 316–17
    game changing potential 326–7
    guidance 317–18
    hulls 314–16
    launch and recovery 319
    MESSIN 329–49
    MIMIR 319–21

surface vehicles (*continued*)
  mission planning 317
  navigation 317
  payload systems 318–19
  propulsion systems 316–17
  SWIMS 321–5
  system partitions 313–14
  see also autonomous surface marine vehicle
surge 14
surveillance 313, 318, 389, 402, 405
survey autonomous semi-submersible vehicles 390, 397–402
  applications 402, 404–5
  command and control 400–1
  daughter AUVs 403–5
  daughter ROVs 403–4
  launch and recovery 401–2
  SASS-6M 390, 398–401
surveying 127, 329–30, 336–7, 348, 388–91, 402, 404
  shallow water 329–30
SWATH 331, 349
sway 14
SWIMS USV 321–5
switching control 105–6
  estimation-based 106–7, 109–10
  hierarchically supervised 107–8, 111–12
  stability analysis 112–14
synchronous scanning 278
system identification 146–7, 189

target localisation 236
target tracking 80–3
task priority 227–30
task reconstruction method 225–7, 230
tele-operation 232–3, 236
teleprogramming 232–3
template detection and tracking 200
temporal data visualisation 293–4
tether compensation 169–81
  cable dynamics 170–3
tethered flight vehicle 169–81
  cable dynamics 170–3
  simulations 179–81
  tether effects 170–7
three-dimensional display 296–7
three-dimensional human interface 294–300
thruster control allocations 98–103
thruster control matrix 98
token tracking 200–1
torpedo-shaped AUVs, robust control 162–9
towfish 391–3, 395–6

track deviation transfer function 339–40
track following 44–52
track following potential 62–3
track path transitions 52
trajectory tracking 43–4, 359–60, 368–9
Tri-Dog 1 239, 251–2
trimming trajectory 368–9
Troika 312
turning 270, 272
Turtle 1

ultrasonic motion tracker 236
underactuated UUV 32
underwater gliders 407–31
  control 414–15
  coordinated control 424–30
  mathematical model 408–11
  Seaglider 408
  Slocum 408, 417–25
  Spray 408
  stability 412–16
  steady gliding 420–1, 423–5
underwater observation 239
underwater robotics 293
underwater structured lighting 277
underwater vehicles
  behaviour control 68–76
  data visualisation 293
  identification 24–31
  modelling 14–24
  nonlinear control 31–40
  over-actuated open-frame 87–8
  switching-based supervisory control 105–9
  see also autonomous underwater vehicle
underwater vision systems 277–8
underwater volcano 239–49, 379
unexploded ordnance hunting 127
URIS AUV 76–84
  software architecture 78–9
  target tracking 80–3
  thruster control allocation 98–9, 101–3
user interface 232–3

vehicle primitive 377–8
velocity control, Romeo ROV 194–5
versor lemma equations 222
vertical plane control 359, 363–6
virtual bodies 425
virtual bodies and artificial potentials methodology 424–6
virtual control space 102–3
virtual leaders 425–7

vision sensor 279–80
vision-based motion estimation 196–201
    Romeo ROV 202–11
vision-based navigation 283–4
volcanos, underwater 240, 379
    see also Rota Underwater Volcano

water trials
    robust autopilots 181–3
    see also sea trials
waterfall images 289–91
wave-piercing autonomous vehicles 387–406
    daughter vehicles 403–5
    Dolphin 389–92
    Dorado 394–5
    mobile buoys 398, 405
    RMOP 393–4
    SASS 390, 397–403
    Seakeeper 395–6
    surface 390, 396–402, 405
    underwater 389–96
weapon launch platform 405

yaw 14

δ-implementation 365–6, 369